THE LAW OF INFORMATION CONFLICT

The Law of Information Conflict
National Security Law in Cyberspace

Thomas C. Wingfield

Aegis Research Corporation

International Standard Book Number (ISBN): 0-9670326-1-X

Library of Congress Catalog Card Number: 99-95501

This book may be ordered from:
Aegis Research Corporation
ATTN: Book Sales and Permissions
7799 Leesburg Pike, Suite 1100 North
Falls Church, Virginia 22043
(703) 847-6070
http://www.aegisresearch.com

Dedication

To Maryann Frost Wingfield, my mother, Samuel Griffin Wingfield, III, my late father, and Walter Gary Sharp, Sr., my mentor—between the three of you, I cannot think of anything you failed to teach me.

—TCW

Note From the Author

The opinions and conclusions expressed herein are those of the author and do not necessarily reflect the views of any governmental agency or private enterprise. This research is intended to provide as complete and useful an exploration as possible of the application of national security law in cyberspace. This text is not intended, however, to substitute for the advice of legal counsel. Legal advice should only be sought from an attorney who can apply all relevant legal principles with a complete working knowledge of all of the relevant facts, purposes, and desires. Accordingly, the publisher and author make no warranty or representation, express or implied, as to the completeness, correctness, or utility of the information in this publication—and they assume no liability of any kind whatsoever resulting from actions based upon the contents of this book.

THE LAW OF INFORMATION CONFLICT
National Security Law in Cyberspace

TABLE OF CONTENTS

PART III: *JUS IN BELLO*

PART IV: CONCLUSION

List of Illustrations

His art is eccentricity, his aim
How not to hit the mark he seems to aim at.

His passion how to avoid the obvious,
His technique how to vary the avoidance.

The others throw to be comprehended. He
Throws to be a moment misunderstood.

Yet not too much. Not errant, arrant, wild,
But every seeming aberration willed.

Not to, yet still, still to communicate
Making the batter understand too late.[1]

—Robert Francis, "Pitcher"

---

[1] Robert Francis, "Pitcher," *reprinted in* SPRINTS AND DISTANCES: SPORTS IN POETRY AND THE POETRY IN SPORT (1965) at 4.

PREFACE

Information and information systems permeate everyone's daily life. Individuals, private enterprises, and governments at all levels rely upon systems that involve the acquisition, transmission, storage, or analysis of information. In a few short decades, the global networking of computers will very likely be viewed as the one invention that had the greatest impact on human civilization. The synergy created by this global networking has and will continue to revolutionize how governments operate, interact with each other, and govern their citizens. Advances in computer technology and networking will revolutionize future warfare as well as reshape the rule of law and traditional notions of sovereignty.

Whether the concern is with a state's economic well-being and the protection of its citizens' intellectual property in the international community, or the use of information that threatens a state's territorial integrity or political independence, protecting information and information systems must become a priority for states. Part of the protection will involve the extension of creation of legal doctrines. The applicable combination of domestic, foreign, and international law is determined by whether the threat is domestic or sponsored by a foreign state, what means the actors employ, and their effects in the physical world. Accordingly, a state cannot determine the relevant law or tailor an appropriate response until it first determines whether the threat is simply a crime committed by a non-state actor or an unlawful use of force by a state under international law. When a state affects another state's information or information systems, it is engaged in information conflict.

The law of information conflict—drawn from the peacetime regime of public international law and the law of armed conflict—is that discipline of international law that governs a state's rights and responsibilities when it conducts operations that affect another state's information or information systems. This text is intended to serve as a modest beginning to the task of providing a detailed analysis of the law of information conflict. It is designed to serve as a framework of analysis for the legal regime that governs the law of information conflict, and to be a useful desk reference for lawyers, policy makers, warfighters, and other professionals concerned with a state's use and misuse of another state's information and information systems.

Although a more thorough introduction to the substance of this book is presented in its first chapter, it is worthwhile to address the principle questions to be answered in evaluating an information operation. First, is this a use of force under international law? Second, if it is, how does the law of armed conflict apply? Answering the first question is the role of the *jus ad bellum*, the law of conflict management, which governs the transition from war to peace and sets the standards for when that crucial line has been crossed. The second question is governed by the *jus in bello*, the law of armed conflict, which regulates the actions of belligerents in a conflict and the neutrals around them. This book is structured to reflect these two key questions. Part I is introductory, Part II covers the *jus ad bellum*, Part III addresses the *jus in bello*, and Part IV draws conclusions from the material surveyed.

The analytical framework suggested by this text, however, is only the beginning of a lawyer's task. The legal analysis of any planned activity is factually driven. Lawyers must also fully understand and account for the technology and capabilities involved in order to conduct a competent, much less a comprehensive, legal analysis. Accordingly, policy makers and warfighters must involve their legal advisers in all stages of acquisition, training, planning, and operation. To best support national objectives and remain within the bounds of the law, information operations must be conducted as a team effort—with lawyers in a supporting, mission enabling role.

THOMAS C. WINGFIELD

Falls Church, Virginia
May 2nd, 2000

About the Author and Publisher

THOMAS C. WINGFIELD is Principal National Security Policy Analyst at Aegis Research Corporation. A national security attorney, he is the author of numerous journal articles on international and national security law. He holds a Master of Laws (International and Comparative) and a Doctor of Laws degree from Georgetown University Law Center, and a Bachelor of Arts in History (with a minor in Russian language) from Georgia State University. He is currently a candidate for a Doctor of Juridical Science degree, studying under Professor John Norton Moore, at the University of Virginia School of Law. Mr. Wingfield has chaired the American Bar Association's Committee on International Criminal Law, and, following ten years of active duty as a naval intelligence officer, holds a commission as a Lieutenant Commander in the Naval Reserve.

AEGIS RESEARCH CORPORATION is a leading provider of world-class expertise in information operations (policy, strategy, and analysis), information security, threat analysis, threat characterization and countermeasures, mission planning and simulation, security implementation support, comprehensive enterprise strategy development, enterprise protection planning, and professional technical services for the federal national security community, state and local governments, and US corporations.

ACKNOWLEDGEMENTS

To the extent that this book is helpful, the author owes a special debt of gratitude, albeit anonymously, to those lawyers, military commanders, operators, government officials, and scholars we have had the pleasure of working with over the years who have collectively helped shape our views on national security law, deterrence, and the use of force in cyberspace. To the extent that this book fails to fulfill your expectations, I accept full responsibility.

Of particular note, I would like to thank Rae Huffstutler, Don Hardison, and Virginia Mullin, colleagues at Aegis Research Corporation, and Scott D. Billigmeier, of MindSim Corporation, for their insightful review and superb suggestions from an operational and technical standpoint, which helped tremendously to shape the usefulness and readability of the final product.

The work undertaken by Gary Clemmons, editor; Margaret Peterlin, legal editor; Timothy P. Sullivan, treaty editor; and Richard Moore, technical editor, was as thoughtful as it was thorough. Their substantive and technical input, over many months, was the making of the project.

Professor Michael Schmitt's work on use of force analysis in the world of information operations was the basis for forward-looking section near the end of chapter seven. At the time of this writing, his writings define the state of the art on the topic, and are likely to remain so for some time to come. The author assumes full responsibility for any applications or extensions of Professor Schmitt's work which are not fully in accord with his original theoretical work.

The author wishes to acknowledge the absolutely fundamental work done in the preparation of this book by Professor Walter Gary Sharp, Sr., who coined the title, *The Law of Information Conflict: National Security Law in Cyberspace*, and who conceived the basic structure of the text.

Chapters one through six, and chapters ten and eleven, draw substantially from his earlier book, *Cyberspace and the Use of Force,* also published by Aegis Research Corporation. I am also indebted to him for portions of chapters seven and nine, and appendix I. In addition, his insightful use-of-force graphics appear in this book as he designed them. Although the press of business prevented Professor Sharp from remaining on this project through its completion, his imprimatur is indelibly fixed on this book. To the extent that the remaining sections of the book, especially its overall conclusions, diverge from his work, the responsibility is solely the author's.

Finally and most importantly, a very special thanks goes to William H. Geiger, President and CEO of Aegis Research Corporation, who commissioned the work of this book as a small part of his dedication to, and respect for, the rule of law in national security affairs.

Appreciation is also gratefully acknowledged to the following organizations that provided electronic versions of two documents in this book:

The Fletcher School of Law and Diplomacy of Tufts University for an electronic copy of the CHARTER OF THE UNITED NATIONS found in Appendix C.

The United States Army Judge Advocate General's School for an electronic copy of Enclosure A to the STANDING RULES OF ENGAGEMENT FOR US FORCES found in Appendix E.

The Law of Information Conflict

PART I. INTRODUCTION

"Be sure you are right, then go ahead."[2]
—David Crockett,
Motto during War of 1812

"Success is the sole earthly judge of right and wrong."[3]
—Adolf Hitler, Mein Kampf

◆

CHAPTER I: APPLYING INTERNATIONAL LAW IN CYBERSPACE†

The American military tradition favors Davey Crockett's model for behavior in conflict over Adolf Hitler's, with military commanders seeking and receiving legal advice on all aspects of proposed operations, to be sure they are right before going ahead. Although military and intelligence operations in cyberspace employ means and methods undreamt of even a few years ago, these information operations are governed by a comprehensive legal regime. This regime includes the domestic law of the acting state, the domestic law of the target state, and the relevant international law. This last category has three distinct regimes, depending on the phase of any given conflict: the law of peace (jus in pace), *the law of conflict management* (jus ad bellum), *governing the transition from peace to war, and the law of war* (jus in bello). *Before evaluating the efficacy of any policy choice, decision-makers must have sound, fact-based legal advice that is connected to clearly*

---

† This chapter is drawn substantially from Chapter I of WALTER GARY SHARP, SR., CYBERSPACE AND THE USE OF FORCE (1999).

[2] David Crockett, *reprinted in* THE POCKET BOOK OF QUOTATIONS 319 (Henry Davidoff ed., 1952).

[3] Adolf Hitler, *reprinted in id.*

articulated principles of law. This book is intended to frame this initial discussion, focusing on the corpus of international law that applies to information operations, and following its transition from peace to crisis to war.

Historically, when new technology provides one state a potential military advantage over another, the disadvantaged will quickly seek to outlaw or regulate those military applications that may be used as a new means or method of warfare [*i.e.*, a use of force between states]. When the military potential of hot air balloons frightened the international community, for example, an attempt was made to outlaw aerial bombardment. At the First Hague Peace Conference,[4] the Russians proposed a prohibition on "the discharge of any kind of projectile or explosive from balloons or by similar means."[5] This proposal resulted in an international norm signed and ratified by 25 states that prohibited "for a term of five years, the launching of projectiles and explosives from balloons, or by other new methods of a similar nature."[6]

This norm was reaffirmed during the Second Hague Peace Conference in 1907, when 27 states signed a similar declaration that was also intended to be limited in time.[7] Although the initial reaction was to

---

[4] The First Hague Peace Conference was held in 1899. It was the first major international effort to regulate the means and methods of warfare since codification of international law first began in 1856. DOCUMENTS ON THE LAWS OF WAR 2-3 (Adam Roberts & Richard Guelff eds., 2d ed. 1989) [hereinafter ROBERTS & GUELFF].

[5] THE LAWS OF ARMED CONFLICTS: A COLLECTION OF CONVENTIONS, RESOLUTIONS AND OTHER DOCUMENTS, 201, Dietrich Schindler and Jiři Toman, eds. [hereinafter SCHINDLER & TOMAN, ARMED CONFLICTS].

[6] Declaration (IV, 1) to Prohibit for the Term of Five Years the Launching of Projectiles and Explosives from Balloons, and other new Methods of a Similar Nature, Jul. 29, 1899, *reprinted in* SCHINDLER & TOMAN, ARMED CONFLICTS, *supra* note 5, at 201, 202-05.

[7] Declaration (XIV) Prohibiting the Discharge of Projectiles and Explosives from Balloons, Oct. 18, 1907, *reprinted in* SCHINDLER & TOMAN, ARMED CONFLICTS, *supra* note

outlaw the activity, the perceived advantages of aerial bombardment caused a more reserved attitude within the international community as aerial navigation advanced.[8] The attempts in 1907 to prohibit aerial bombardment permanently resulted only in the insertion of the language "by whatever means" to Article 25 of the Hague Regulations on land warfare which prohibits the attack or bombardment of undefended towns and villages.[9]

As silly as this historical example may now seem in hindsight after a century during which aircraft have played such a critical role in warfare, it showcases two approaches of trying to restrain the feared but undemonstrated military potential of advancing technology. At first, the international community attempted to establish an outright prohibition on aerial bombardment. Although some would describe this form of prohibition as arms control, it was actually an attempt to make aerial bombardment an unlawful method of warfare under the law of armed conflict. When an outright prohibition failed to gain consensus, the international community resorted to regulating the military potential by restricting the technology's use under the law of armed conflict.[10]

5, at 201, 202-06. This declaration was subsequently ratified by 20 states, but was only intended to remain in force until the Third Hague Peace Conference. There was never a third conference, and the declaration is still formally in force. Both the United States and Great Britain ratified this declaration. *Id.* at 201, 206.

[8] *See* SCHINDLER & TOMAN, ARMED CONFLICTS, *supra* note 5, at 201.

[9] *Id.*

[10] For a more contemporary example, consider the initial outcry of the international community to outlaw horrific environmental damage like that which occurred during the Persian Gulf War. When the debate settled, the international community accepted that an absolute prohibition on environmental damage during armed conflict was unworkable, that such destruction was already prohibited by the laws of armed conflict, and that additional proscriptions were not necessary. *See, e.g.*, Walter Gary Sharp, Sr., *The Effective Deterrence of Environmental Damage During Armed Conflict: A Case Analysis of the Persian Gulf War*, 137 MIL. L. REV. 1 (1992).

Despite the initial reaction of the international community to prohibit outright aerial bombardment, the end result of the First and Second Hague Peace Conferences was simply an application of existing international law to new technology. Beginning with the presumption that the activity in question was a use of force, both of these Hague Conferences attempted to utilize the law of armed conflict as a way to regulate or even prohibit it. In 1923, a commission of international jurists completed the Hague Rules of Air Warfare.[11] Although these rules were never adopted in any legally binding form, they were an application of the customary norms and general principles of the law of armed conflict to the use of aircraft in war.[12]

Only when the military applications of new technology are truly abhorrent and unacceptable under all circumstances, such as chemical and biological warfare, has the international community reached consensus and prohibited those applications as means or methods of warfare.[13] When only some of the consequences of new technologies, however, fall beyond the defined parameters of acceptable activity, as was the case with aerial bombardment, the new means and methods of warfare are simply

---

[11] Hague Rules of Air Warfare, Feb. 1923, *reprinted in* SCHINDLER & TOMAN, ARMED CONFLICTS, *supra* note 5, at 207.

[12] *Id.*; For a detailed and excellent discussion of the application of international law to the use of aircraft during armed conflict that is widely used within the Department of Defense as the definitive reference on the law of air warfare, *see* W. Hays Parks, *Air War and the Law of War,* 32 A.F. L. REV. 1, 1-225 (1990).

[13] *See, e.g.*, Convention on the Prohibition of the Development, Production, Stockpiling and Use of Chemical Weapons and on Their Destruction, Jan. 13, 1993, 32 I.L.M. 800; S. Treaty Doc. No. 21, 103d Cong., 1st Sess. (1993); *excerpted in* NATIONAL SECURITY LAW DOCUMENTS 348 (John Norton Moore *et al.* eds., 1995) [hereinafter Chemical Weapons Convention]; Geneva Protocol for the Prohibition of the Use in War of Asphyxiating, Poisonous or Other Gases, and of Bacteriological Methods of Warfare, June 17, 1925, 26 UST 571 [hereinafter 1925 Geneva Protocol, Bacteriological Methods of Warfare], *reprinted in* NATIONAL SECURITY LAW DOCUMENTS 328 (John Norton Moore *et al.* eds., 1995).

regulated by existing law, and when necessary, treaty law and evolving state practice.

The prospect of hostile state activities and military operations in cyberspace has elicited a pattern of concern in the international community reminiscent of the balloon warfare experience. When the legal community first considered the legal regime that governed state activities and military operations in cyberspace, some US government attorneys stated rather boldly that the application of modern information systems technology to military purposes was so new that *no* law applied. It now seems almost universally accepted that a considerable body of international law applies to the use of force by states in cyberspace.

In parallel to the initial reaction of the international community to the military potential of hot air balloons, the epigraph to this book on page one reflects the international community's concern over the potential for damage and disruption caused by military operations against and through information systems linked by the internet. It is excerpted from a United Nations General Assembly resolution that resulted from an initiative of the Russian Federation. This resolution calls upon Member States to promote multilateral "consideration of existing and potential threats in the field of information security" and invites them to inform the Secretary-General of their views on several detailed issues relating to information security.[14] Although couched in terms of criminal and terrorist threats, *i.e.*, the use of force by non-state actors, it was initiated in the First Committee, which concerns itself with disarmament.[15] The

---

[14] *Developments in telecommunications and information in the context of international security*, G.A. Res. 53/70, U.N. GAOR, 53rd Sess., U.N. Doc. A/RES/53/70 (1998). General Assembly resolutions are not binding on Member States. *See* U.N. Charter arts. 10-11, 14.

[15] This point was also implied in the United States *Explanation of Vote* after it joined in the adoption of G.A.Res. 53/70 by consensus. The United States explained:

> The General Assembly's adoption of the resolution in plenary will launch the international community on a complex enterprise encompassing many interrelated factors which delegates to the First Committee do not

unavoidable momentum of the international community in the First Committee will likely yield an attempt to prohibit certain military applications of the internet, *i.e.*, the use of force by states in cyberspace. The likely result, however, will not be new international norms that substantially outlaw state activities in cyberspace, but rather a debate, and perhaps a partial consensus, on the application of international law to the use of force by states in cyberspace.

For the purpose of analyzing what constitutes a use of force or an armed attack, there are three relevant categories of international law. First, the *peacetime regime of international law* governs the conduct of states during peacetime, and it remains in place during armed conflict to the extent it is not inconsistent with a state of hostilities. While treaties requiring certain levels of trade between two states would be terminated when those two states become engaged in armed conflict, for example, peacetime environmental treaties would continue to impact a naval commander's decision to sink an oil tanker running a blockade during armed conflict. Second, the *law of conflict management* is a part of the peacetime regime of international law that defines and governs the use of force between states during peacetime. It is also a part of the peacetime regime that remains in place during armed conflict. Before armed conflict, the law of conflict

---

ordinarily address. For example, the topic includes technical aspects that relate to global communications—as well as non-technical issues associated with economic cooperation and trade, intellectual property rights, law enforcement, anti-terrorist cooperation, and other issues that are considered in the Second or Sixth Committee.

The General Assembly distributes most of its agenda items to one of seven main committees: the *First Committee*, which considers disarmament and related security issues; the *Special Political Committee*, which discusses political questions not dealt with by the First Committee; the *Second Committee*, which considers economic and financial matters; the *Third Committee*, which considers social, humanitarian, and cultural questions; the *Fourth Committee*, which is responsible for decolonization issues; the *Fifth Committee*, which considers administrative and budgetary matters; and, the *Sixth Committee*, which considers legal issues. U.N. DEPARTMENT OF PUBLIC INFORMATION, EVERYONE'S UNITED NATIONS: A HANDBOOK ON THE WORK OF THE UNITED NATIONS 15-16 (1986), U.N. Sales No. E.85.I.16.

management governs the use of force between states, and during armed conflict the law of conflict management principally imposes an obligation on states to terminate the hostilities as soon as possible consistent with their own self-defense requirements. Under the law of conflict management, a use of force that reaches a certain scope, duration, and intensity will establish an armed conflict as a matter of law. Third, the *law of armed conflict* governs the actual conduct of hostilities. This body of law specifically authorizes the use of a wide-range of force during armed conflict that is otherwise unlawful during peacetime.

Although a considerable body of international law applies to the use of force by states in cyberspace, its application is not always obvious, leaving unanswered the many questions surrounding precisely how international law relates to the use of force in cyberspace. For example, what constitutes a use of force in cyberspace? When are states allowed to use force against each other in cyberspace? How does international law prohibit or govern the use of force and other state activities in cyberspace? Does international law impose any restrictions on state activities in cyberspace that are not considered a use of force? What constitutes an armed attack in cyberspace? When does an armed conflict exist between two or more states? Can state activities in cyberspace cause an armed conflict? When does a state have the right to use force in self-defense? Is computer espionage lawful? Are computer network attacks lawful? Do computer network attacks constitute a use of force? When can a state use force to defend itself against computer network attacks? These, and other questions, will be answered in Chapters 6 and 7 of this book.

Beyond these threshold questions, this book will address several subsequent issues. Of prime importance is the broad issue of how the peacetime regime of international law, other than the law of conflict management, may restrict state activities in cyberspace. Depending upon the circumstances, there may be treaties, and international regulations that have the force of law, that may restrict or prohibit certain state activities in cyberspace. These treaties include, for example, those that regulate the use of satellites, telecommunications, outer space, and the law of the sea.

In addition to these issues, this book will address the application of the law of armed conflict to the use of force by states in cyberspace once hostilities have commenced. At a point along the spectrum of interstate activities called the line of belligerency, a use of force by a state establishes an international armed conflict as a matter of law and the law of armed conflict applies.[16] Even though the law of conflict management continues to apply during armed conflict, the law of armed conflict specifically authorizes a state to use all necessary and proportional force not otherwise prohibited by the law of armed conflict that is required for the partial or complete submission of the enemy with a minimum expenditure of time, life, and physical resources.

There are a number of important issues, however, that will *not* be addressed in this book as it is intended to analyze only use of force issues between states under the international law of conflict management. This book, for example, will not discuss the use of force by a state against recreational hackers, terrorists, organized criminals, and other non-state actors. The use of force by a state against these non-state actors is a law enforcement issue that must, at least at present, be principally addressed through cooperative bilateral and multilateral extradition and mutual legal assistance treaties.[17] These treaties are a part of the peacetime regime of

---

[16] *See* Figure 10, *infra*, for a complete graphical representation of the concept.

[17] Despite that a non-state actor can cause *identical* damage to a state's information infrastructure as can a state actor, a hostile, transnational activity in cyberspace committed by a non-state actor remains a law enforcement issue. The issue of state and non-state sponsorship, however, may be very *factually* complicated by a number of circumstances such as the activities of state-owned commercial enterprises and surrogate-actors, as well as the anonymity afforded by technology. Nevertheless, the legal analysis remains rather straightforward. Determining when state-owned commercial enterprises, for example, are acting as a commercial enterprise or at the direction of a state is a determination surrounding *facts* such as who controls the enterprise, who directed the activity, and the nature of the activity. It is not an issue of law.

Consequently, from a legal perspective, *all* hostile, transnational activities in cyberspace are either non-state-sponsored and thus a crime addressed by national and peacetime treaty law, or they are state-sponsored and thus a use of force governed by the law of conflict management and the law of armed conflict. Notwithstanding what the international law, but they are not a part of the law of conflict management.

Furthermore, this book will not address the effect of United States domestic or of foreign domestic law on state activities in cyberspace. Any number of state activities in cyberspace that are lawful under international law may nevertheless be highly regulated or unlawful under the national law of one or more states. For example, if the military forces of one state are conducting activities in cyberspace that originate in their own state and target another state, but are transmitted through the telecommunications infrastructures of five other states, the national laws of seven states may apply in addition to relevant international law. In the United States, activities that involve intelligence collection, espionage, the nonconsensual penetration of computers to determine the identity of an attacker, and the counterattack of hackers through servers and systems not belonging to the hackers, are highly regulated and under some circumstances, may be unlawful. The internet is like a global party line, and it can be very challenging under the domestic law of the United States

---

destructive effect may be, hostile, transnational activities in cyberspace committed by recreational trespassers, terrorists, criminals, and other non-state actors do not constitute a use of force within the meaning of the law of conflict management. Even a hostile, destructive act by a military service member or other government employee is a law enforcement issue *if* the act was conducted in his or her private capacity and not at the direction of a state. Similarly, the ineffectiveness of law enforcement arrangements among *cooperative* states to address the problem of hostile, transnational activity in cyberspace by non-state actors does not change a crime to a use of force or the basic nature of the legal issues.

The complete refusal or unwillingness of a state, however, to cooperate in the suppression or prevention of an acknowledged non-state-sponsored hostile, transnational activity in cyberspace that originates in its sovereign territory constitutes state-sponsorship of a use of force *ipso facto*—thereby invoking the law of conflict management which may authorize a use of force in self-defense against such a state or the non-state actors in that state. In the absence of any state-sponsorship of terrorist or criminal activities, a use of force by a state against those non-state actors in the sovereign territory of another state without that state's consent may rise to the level to be an unlawful use of force against that territorial state.

for the law enforcement and intelligence communities to find the authority to monitor or intercept transmissions when US citizens are online.

Finally, this book will not address the special circumstances of state activities that have been authorized by the United Nations Security Council under its Chapter VII coercive authority. Article 39 of the Charter of the United Nations empowers the Security Council to authorize the use of force to maintain international peace and security at a threshold below international law, even lower than when states are authorized to use force in self-defense. A blockade, for example, that would otherwise be unlawful under international law when committed by a state would be lawful if authorized by the Security Council under its Chapter VII authority. Similarly, most state activities in another state are unlawful without the territorial state's consent, however, state activities conducted during a United Nations peace operation in a state without the territorial state's consent is lawful if authorized by the Security Council under its Chapter VII authority. Furthermore, Article 103 of the Charter preempts most international obligations of states if they conflict with their obligations under the Charter. Consequently, a state activity that otherwise may be deemed an unlawful use of force under international law may be lawful if authorized by the Security Council under its coercive Chapter VII authority.

The primary focus of this book, therefore, is to examine the two legal questions of:

(1) which interstate activities in cyberspace constitute a threat or use of force under international law, and

(2) when such a threat or use of force does constitute an armed attack under international law, how does the law of armed conflict apply to the lawful exercise of the inherent right of self-defense in cyberspace?

These questions are foundational to the law of information conflict (LOIC), which is the composite of the peacetime regime of international law, the law of conflict management, and the law of armed conflict that regulates the conduct of all state activities in cyberspace. It embodies the application of the entire peacetime regime as well as the entirety of the laws of conflict management and armed conflict to state activities in cyberspace.

There are a number of other foundational questions, however, that should be addressed prior to the discussion of these two issues. What is cyberspace, for example? What is it about cyberspace that makes states vulnerable targets? Why do states want to use force in cyberspace? What is the law of conflict management? When do states have the right to use force? What responsibility does the Security Council have to use force? What is the line of belligerency? What is the law of armed conflict, and how does it help define what constitutes a use of force and an armed attack under the law of conflict management? These, and other questions, will be answered in Chapters 2 through 5.

Chapter 2 provides a factual foundation. It defines cyberspace. It explains why cyberspace makes a state vulnerable; it provides examples of how those vulnerabilities can be exploited; and, it illustrates how cyberspace is simply another medium or environment through which states can either use force or support the use of force. Finally, it concludes that states are very likely to develop military applications of cyberspace.

The next chapter provides the legal foundation necessary to understand what constitutes a use of force or an armed attack in cyberspace under international law. After a brief historical review of the evolution of the law of conflict management, Chapter 3 provides a detailed discussion of the contemporary norms under the Charter of the United Nations that regulates the use of force between states during peacetime. It details when states are authorized to use force under international law against another state as well as the responsibility of the United Nations Security Council to authorize the use of force against states to maintain international peace and security. Chapter 3 also provides a brief discussion of the principles of necessity, proportionality, unnecessary collateral damage or injury to civilians, and anticipatory self-defense, sufficient for a contextual understanding of use of the force analysis. These principles are addressed in greater detail in Part II of the book, which specifically addresses their applicability to information operations.

Any meaningful legal foundation of the use of force between states must discuss the interrelationship of the law of conflict management with the law of armed conflict. Although the law of armed conflict regulates only the use of force *during* hostilities, Chapter 4 examines in depth the one provision within the law of armed conflict that is widely accepted as the threshold test for when armed conflict exists between states. The application of this provision by states is critical in determining what constitutes a use of force and an armed attack under the law of conflict management. The terms hostilities, armed conflict, and war are used throughout these discussions. While these terms are frequently used interchangeably, the term "hostilities" only generally refers to a state of enmity between states[18] and it has no specific legal meaning under international law. It is used to generically refer to either armed conflict or war. As explained in more detail in Chapter 4, the terms armed conflict and war actually refer to the same state of hostilities between states, but war is determined as a matter of law and armed conflict is determined as a matter of fact. Chapter 5 completes the legal foundation provided in this section, melding Chapters 3 and 4 together to form the basic principles that support the analysis that follows.

Chapters 6 and 7 constitute the core use of force analysis of this book. They detail what constitutes a use of force and an armed attack under international law. Chapter 6 analyzes and then relates the existing international norms that govern the use of force between states to the activities of states in cyberspace. It dissects the obligations of states to settle their disputes by peaceful means and analyzes when political, economic, and non-military force can constitute an unlawful use of force and an armed attack. This chapter also explains when a threat to use

---

[18] BLACK'S LAW DICTIONARY 665 (5th ed. 1979) [hereinafter BLACK'S LAW DICTIONARY].

force is unlawful and when violations of international law or the indirect use of force can constitute an unlawful use of force. The topic of threats to use force is challenging because it moves beyond an analysis of the legal issues based upon objective facts into the realm of the intent of the acting state and the perceptions of a potential victim state, *i.e.*, a discussion of hostile intent. This chapter then details when states can lawfully use force to protect their nationals abroad. Chapter 6 concludes with a detailed discussion of what constitutes an armed attack under international law and a graph that summarizes Chapters 3 through 6.

Part III addresses the second broad issue of this book, the application of the law of armed conflict to activities in cyberspace. After providing a historical review of the development of the law of armed conflict, Chapter 7 examines the four basic principles of the law of armed conflict: discrimination, necessity, proportionality, and chivalry, which are sketched briefly below.

The principle of *discrimination* requires that, prior to any use of force, the attacker distinguish between combatants and noncombatants, between military objectives and civilian objects. Noncombatants include civilians and "protected persons," such as medics and chaplains. While it is unlawful to attack civilian objects as such, it is not unlawful to collaterally damage civilian objects located so close to lawful military objectives that no reasonable precautions could have prevented the damage. In cyberspace, such distinctions will require the collection and analysis of information to produce the intelligence required to make such determinations. The military commander must consider foreseeable secondary effects in his target selection and mission planning.

The principle of *necessity* permits the use of all force required for mission accomplishment and force protection, but not superfluous force or unnecessary suffering. Superfluous force is that which is inflicted solely for the sake of causing damage. Unnecessary suffering has two components: quantitatively, it is simply the complement of military necessity (necessary force); qualitatively, it includes those means and methods of war that are by nature inhumane (chemical weapons, biological weapons, exploding bullets, transparent bullets which cannot be detected in the body by x-ray equipment).

The principle of *proportionality* requires the military commander to balance the collateral damage (against civilians and their property) of a planned attack against the concrete and direct military advantage expected to be gained. Proportionality is *not* the requirement to respond with only as much force as was used in the precipitating attack, and is *not* the requirement to employ the same means and methods as in the provoking attack. Whatever other policy preferences may be expressed later in the national security decision-making process, the basic legal requirement is to ensure the balance of military advantage gained outweighs the damage done to noncombatants and their property.

The principle of *chivalry* permits ruses of war, but forbids perfidy. Perfidy is unlawfully deceiving an opponent that he is entitled to enjoy, or required to accord, protections of law, which are, in fact, inapplicable in the situation. Attacking an enemy facility from a vehicle displaying the Red Cross or the U.N. flag would be examples of perfidiously deceiving the enemy into according protections under law (not attacking the vehicle) which are not applicable. Broadcasting a false notice of cease-fire or calling for a *parlementaire* before attacking the unwary opponent would be an example of perfidiously deceiving one's opponent that he is entitled to a protection of law, which in fact, he is not. Ruses of war, on the other hand, are legal. They are wartime deceptions that convince an enemy to launch or withhold an attack based on tactical reasons, rather than perceived legal reasons. Numerous examples will be presented, assisting the reader in analogizing to various information operations.

Chapter 8 examines twenty-nine treaties on the law of armed conflict that bear, to a greater or lesser extent, on information operations. Treaty law reflects the four basic principles of discrimination, necessity, proportionality, and chivalry, described above. Treaties and other international agreements codify and implement customary international law. They occasionally lead international law by affirmatively proposing new norms and standards, at first applicable to only those nations which sign the agreements. Such new standards may become binding on other nations over time, if those other nations adopt the new norms, and do so out of a sense of legal obligation. Treaties applicable to one age or type of warfare may nonetheless be useful in charting the course of the law of information conflict, to the extent that analogies may be drawn from them and applied to operations in cyberspace. This chapter is, to a great extent, intended to spur the reader's imagination, by raising potential issues and questions.

Part III concludes with Chapter 9, a survey of the effect of hostilities on the peacetime legal regime, a portion of which survives hostilities between the belligerents, and informs the legal analysis of proposed information operations. The principle areas are divided geographically into the law of the sea, air law, space law, and foreign domestic (or host nation) law. The law of the sea addresses the metaphor of "innocent passage" in cyberspace, unauthorized broadcasting, and liability for damage done in wartime. Air law covers the metaphor of defining airspace (as applied to defining cyberspace), the legal restrictions on interfering with aircraft communications or navigation, as well as physical attacks on aircraft. Space law is useful in the distinction of air and space, the use of space platform for peaceful and non-peaceful purposes, and the liability of those who cause damage against or through the use of space platforms. Finally, foreign domestic law is most applicable in the area of stationing arrangements and the limitations they place on the employment of weapons and forces forward deployed.

Part IV concludes the book. Chapter 10 addresses Computer Network Espionage (CNE) and Computer Network Attack (CNA), which are located on two places along the spectrum of conflict, and have distinctly different legal characteristics. Computer Network Espionage, like any form of pure espionage, is lawful under international law, but is usually not lawful under the domestic law of the target state. CNE usually involves little or no force, and involves only as much intrusion as is necessary to collect the required information from the adversary's systems. Computer Network Attack, on the other hand, involves some kind of destruction with consequences in the physical world. CNA should be analyzed as any other type of use of force, and, depending on the scope, duration, and intensity of the force employed, may rise to the level of an armed attack. Finally, Chapter 11 summarizes the author's findings, clarifying and restating them in a convenient, condensed format.

Having outlined the structure of the book, it is time to turn to the first question facing national policymakers, military operators, and the attorneys advising them: at what point do information operations rise to the level of a threat or use of force, implicating a move across the spectrum of conflict from peace toward armed conflict?

CHAPTER 2: MILITARY APPLICATIONS OF CYBERSPACE†

"I think, Sir, those aeroplanes are going to spoil war. When they come over I can only tell my men to cover their heads with hay and make a noise like a mushroom."[19]

—*General Grierson to King George V, following army maneuvers in 1912*

The enormous efficiencies offered by the high-speed automated data processing of computers and their ability to transmit and receive data near-instantaneously from anywhere in the world via terrestrial and orbital links have catapulted our society into the information age. The efficiencies, however, are not without a price. The increased centralization of many processes, which had been less efficiently but more securely dispersed, has created attractive targets for malefactors wishing to disable a victim state. The "critical infrastructures" of a society are now vulnerable to attacks that transcend oceans and continental distances, bypass traditional military units, and confound 20th century strategic thought. New information weapons require a re-examination of the legal principles that will shape their lawful use.

Cyberspace is not a physical place—it defies measurement in any physical dimension or time-space continuum. It is a shorthand term that refers to the environment created by the confluence of cooperative networks of computers, information systems, and telecommunication infrastructures commonly referred to as the internet and the World Wide Web. Information is the valued commodity of cyberspace, but nothing actually exists in cyberspace. When something is said to be in cyberspace, it is in fact physically resident in a computer or an information system, or it is in transit within a telecommunications infrastructure.

---

† This chapter is drawn substantially from Chapter 2 of WALTER GARY SHARP, SR., CYBERSPACE AND THE USE OF FORCE (1999).

19 General Grierson, *quoted by* Major-General J.A. Thompson, *in* THE FUTURE OF BRITISH SEA POWER 163 (Geoffrey Till, ed., 1984).

Modern telecommunication systems evolved from the first electric telegraph in 1837[20] to the first transatlantic cable in 1866,[21] the first articulating telephone in 1876,[22] the first wireless telegraph in 1895,[23] the first commercial public radiotelephone service in 1946,[24] and the first telecommunications satellite, which was launched by the United States in 1958.[25] The electronic computer was not invented until 1943, and the ENIAC, the first electronic vacuum tube computer, was built in 1946.[26] In the 1950s IBM imagined a world market for computers and began marketing business computers.[27] Finally, in 1975, the world's first personal computer was announced.[28] Computers have evolved from electronic monoliths that were 80 feet long, weighed 30 tons, and had 17,000 tubes, to today's desktop computers that can store a million times more information than the first computer, are 50,000 times faster,[29] and can weigh as little as 2.2 pounds.

---

[20] THE NEW YORK PUBLIC LIBRARY DESK REFERENCE 129 (1998). [hereinafter DESK REFERENCE].

[21] *Id.* at 131.

[22] *Id.*

[23] *Id.* at 133.

[24] Steven Levy, *The Computer*, NEWSWEEK, Winter 1997-98 Special Issue, at 28, 30.

[25] DESK REFERENCE, *supra* note 20, at 138.

[26] *Id.* at 136-37.

[27] Levy, *The Computer* at 28, 29, *supra* note 24.

[28] *Id.*

[29] DESK REFERENCE, *supra* note 20, at 109, 146.

The internet was originally a network of computers linked by telecommunication infrastructures managed by the US Department of Defense for military and government purposes in the 1970s.[30] Universities and other research facilities soon created their own internal computer networks.[31] As these intranets began merging with one another to form a cooperatively run global collection of computer networks with a common addressing scheme, the internet was born.[32] The World Wide Web refers to a collection of files and databases linked by hypertext that was created in 1989 as the primary platform of the internet that translates diverse computer protocols into standard format.[33]

Those who rely upon the benefits of the internet and the World Wide Web also create vulnerabilities for themselves. The Honorable John J. Hamre, former US Deputy Secretary of Defense, had focused on the World Wide Web as "the heart of the Defense Reform Initiative and as key to the reengineering and streamlining of our business practices."[34] He also recognized, however, that at the same time the

> Web can also provide our adversaries with a potent instrument to obtain, correlate and evaluate an unprecedented volume of aggregated information regarding DoD capabilities, infrastructure, personnel and operational procedures. Such information, especially when combined with information from

---

[30] *Id.* at 108.

[31] *Id.*

[32] *Id.* at 108, 112.

[33] *Id.* at 108, 115. Hypertext is a method of connecting World Wide Web sites through text-based links rather than the menu-oriented systems. *Id.* at 112.

[34] The Honorable John J. Hamre, Deputy Secretary of Defense Memorandum, *Information Vulnerability and the World Wide Web* 1 (Sep. 24, 1998).

> other sources, increases the vulnerability of DoD systems and may endanger DoD personnel and their families.[35]

For example, the memorandum just quoted is publicly available at the Department of Defense (DoD) web site,[36] as well as detailed information on the DoD total force anthrax immunization program,[37] and Operation Desert Fox.[38]

To further demonstrate the information available on the World Wide Web, consider that the following overhead imagery—limning in remarkable detail the Maryland neighborhood of the author, and, more ominously, the District of Columbia neighborhood of the President of the United States—are easily available on the internet and can be downloaded in a matter of minutes.[39] Access to this type of public information can provide the beginning of a poor-state's reconnaissance intelligence agency.

---

[35] The Honorable John J. Hamre, Deputy Secretary of Defense Memorandum, *Information Vulnerability and the World Wide Web* 1 (Sep. 24, 1998).

[36] *See* http://www.defenselink.mil/other_info/depsecweb.pdf.

[37] *See* http://www.defenselink.mil/specials/Anthrax/.

[38] *See* http://www.defenselink.mil/specials/desert_fox/. The mission of Operation Desert Fox was to strike military and security targets in Iraq that contribute to Iraq's ability to produce, store, maintain and deliver weapons of mass destruction. *Id.*

[39] *See* http://www.terraserver.com. Microsoft® TerraServer utilizes georectified digitized aerial photographs from the United States Geological Survey and high-resolution satellite images that are products of a joint Russian/American venture to market declassified satellite photographs from sophisticated Russian mapping satellites.

Figure 1. Morningside Photo

In addition to the vulnerabilities posed by publicly available information, states and businesses must also be concerned about those who unlawfully penetrate, or hack into, intranets and computer systems closed to the public. With their open architecture, the internet and the World Wide Web are ideally suited for asymmetrical warfare and corporate espionage. They can be used by states and non-state actors to *anonymously* pry into a state's public, sensitive, and classified computers; to collect a wide range of government and business information; to

Figure 2. White House Photo

manipulate data; to deceive decision makers; to influence public opinion; and, to even cause physical destruction from remote locations abroad. An attacking state that understands computers and information systems technology does not need a large land force, a navy to move its forces to hostile shores, or an air force to maintain air superiority over the battlefield to inflict damage on another state. With a few thousand dollars invested in a personal computer, a dedicated and persistent hostile threat can gain access to almost any internet-linked information infrastructure of any state in the world. The technology of cyberspace adds new meaning to over-the-horizon warfare. Execution of an organized, large-scale attack can anonymously begin with distant key commands on a computer keyboard.

During the military exercise *Eligible Receiver* in June 1997, for example, the National Security Agency demonstrated that a hostile enemy state could disrupt computer operations at major military commands, cause large-scale blackouts, and interrupt emergency phone service in Washington, D.C. and several other cities in the United States.[40] Then, in February 1998, two California teenagers with an Israeli mentor broke into sensitive DoD systems and eluded US law enforcement authorities for nearly one month.[41] Although they did not compromise national security or penetrate any classified systems, their attacks were very disturbing because they highlighted sensitive vulnerabilities during the planning for military airstrikes in Iraq.[42]

The Honorable Jamie S. Gorelick, while serving as the Deputy Attorney General of the United States, provided a few other excellent

---

[40] Bradley Graham, *US Studies New Threat: Cyber Attack*, WASH. POST, May 24, 1998, at A1.

[41] *Id.*

[42] *Id.*

real-world examples that demonstrate the vulnerabilities of governments and businesses:

> In 1994, nine people, including an MCI employee, were indicted for a scheme involving a $50-million telephone calling card fraud. Using a sniffer program (which monitors network traffic), they captured and used more than 150,000 calling card numbers. The scheme had been directed by hackers in Germany who then made international calls to attack US computer networks.
>
> A computer hacker broke into files at a bank and a credit union, and then used the information to apply for credit cards in the victim's name. The criminal then used these cards to go on a buying spree. The victim's ability to obtain credit was ruined and had to be painstakingly reestablished.
>
> Con artists have used electronic bulletin board systems to hype recently purchased penny stocks, driving up the price and giving the con artists a profit.
>
> Between June and October in 1994, approximately 40 wire transfers were attempted from Citibank's cash management system through the use of a computer and phone lines from St. Petersburg, Russia, by compromising the password and user identification code system. Citibank was successful in blocking most of the transfers or recovering the funds from recipient banks, limiting its losses. But the potential loss was enormous.
>
> In 1989, the "Legion of Doom" in Atlanta, Georgia, remotely accessed the administrative computers of Bell South and wiretapped calls and altered phone services. It could have shut down the phone network for the Southeastern United States.
>
> From 1993 to 1995, a man in California gained control of the computers running local telephone switches, and discovered information concerning US government wiretaps conducted pursuant to the Foreign Intelligence Surveillance Act (FISA). He also uncovered a criminal wiretap and warned the target.
>
> In 1992, a computer intruder was arrested for tampering with the Emergency 911 systems in Virginia, Maryland, and New

> Jersey in order to introduce a virus and bring down the systems.
>
> Also in 1992, a fired employee of an emergency alert network sabotaged the firm's computer system by hacking into the company's computers, causing them to crash for about 10 hours. During that time, there was an emergency at an oil refinery. The disabled system was therefore unable to alert thousands of nearby residents to a noxious release from the refinery.
>
> Finally, a sniffer was introduced into computers of NASA's Goddard Space Flight Center, permitting someone to download a large volume of complex calibration telemetry calculations transmitted from satellites. The sniffer remained undetected for an unprecedented length of time.[43]

Although these particular vulnerabilities were discovered in the United States, all states share the vulnerabilities of the internet. To date, it *appears* that states have only been attacked by recreational and criminal hackers. The Indonesian government, however, is being blamed for a highly organized attack during January 1999 on non-government computers in Ireland, which brought down the East Timor virtual country domain and internet service to over 3,000 customers.[44] This is the first *reported* hostile use of cyberspace by a state against computers within another state.[45]

In addition to their exposure to corporate espionage, commercial businesses and computer-dependent commercial infrastructures are likely the most lucrative targets during armed conflict between states in

---

[43] The Honorable Jamie S. Gorelick, then Deputy Attorney General of the United States, at the US Air Force Academy, Colorado Springs, Colorado (Feb. 29, 1996). Available at http://www.lawyernet.com/members/jimfesq/wca/1996/28/ deep.html.

[44] Email news article from Chris Nuttall, internet correspondent, *First case of state-sponsored IW?* (Jan. 27, 1999) (on file with author).

[45] *Id.*

cyberspace where the opportunity exists to bring a stock exchange, and perhaps the economy of an entire state, to its knees.[46] In Executive Order 13010, President Clinton declared that certain "national [commercial] infrastructures are so vital that their incapacity or destruction [by either physical or cyber attack] would have a debilitating impact on the defense or economic security of the United States." The Executive Order detailed eight categories of critical infrastructures:

- telecommunications;
- electrical power systems;
- gas and oil storage;
- transportation;
- banking and finance;
- water supply systems;
- emergency services (including medical, police, fire, and rescue); and,
- continuity of government.

Because so many of these critical infrastructures are owned and operated by the private sector, it is vital that the public and private sector work together to protect them. A hostile state could correctly assume that a state's businesses and critical infrastructures are its soft underbelly that is easier to hit than hardened military targets.

Joint doctrine on the subject is promulgated in two publications, JOINT DOCTRINE FOR INFORMATION OPERATIONS and JOINT DOCTRINE FOR COMMAND AND CONTROL WARFARE (C2W).[47] The former addresses the fundamentals of information operations, offensive

---

[46] David Hughes, *Cyber Raid, Wall Street: This Is No Drill*, AVIATION WEEK & SPACE TECHNOLOGY 98 (Dec. 22/29, 1997).

[47] JOINT DOCTRINE FOR INFORMATION OPERATIONS, Joint Pub 3-13, 9 October 1998 [hereinafter JOINT PUB 3-13] and JOINT DOCTRINE FOR COMMAND AND CONTROL WARFARE (C2W), Joint Pub 3-13.1, 7 February 1996 [hereinafter JOINT PUB 3-13.1].

information operations, and defensive information operations. As for the fundamentals:

> Information operations (IO) involve actions taken to affect adversary information and information systems while defending one's own information and information systems. They apply across all phases of an operation, the range of military operations, and at every level of war. They are a critical factor in the joint force commander's (JFC's) capability to achieve and sustain the level of information superiority required for decisive joint operations.
>
> IO capitalize on the growing sophistication, connectivity, and reliance on information technology. IO target information or information systems in order to affect the information-based process, whether human or automated. Such information-dependent processes range from National Command Authorities-level decision making to the automated control of key commercial infrastructures such as telecommunications and electric power.
>
> Many different capabilities and activities must be integrated to achieve a coherent IO strategy. Intelligence and communications support are critical to conducting offensive and defensive IO. The thoughtful design and correct operation of information systems are fundamental to the overall conduct of IO. Additionally, to achieve success, IO must be integrated with other operations (air, land, sea, space, and special) and contribute to national military objectives.
>
> Intelligence support is critical to the planning, execution, and assessment of IO. The joint staff intelligence representative(s) assigned to support the IO cell should be the liaison for intelligence support for all IO planning. Intelligence must be timely, accurate, usable, complete, relevant, objective, and sufficiently detailed to support an array of Department of Defense (DoD) IO requirements, including research, development, and acquisition and operational support. Intelligence preparation of the battlespace is vital to

successful IO. Support from non-DoD and non-US sources also may be required.[48]

Beyond these fundamental considerations, the specifics of offensive and defensive information operations are addressed:

> Offensive IO involve the integrated use of assigned and supporting capabilities and activities, mutually supported by intelligence, to affect adversary decision makers and achieve or promote specific objectives. These assigned and supporting capabilities and activities include, but are not limited to operations security (OPSEC), military deception, psychological operations (PSYOPS), electronic warfare (EW), physical attack/destruction, and special information operations (SIO), and may include computer network attack.
>
> Offensive IO may be conducted in a variety of situations and circumstances across the range of military operations and may have their greatest impact in peace and the initial stages of a crisis. Beyond the threshold of crisis, offensive IO can be a critical force enabler for the JFC. Offensive IO may be conducted at all levels of war—strategic, operational, and tactical—throughout the battlespace.
>
> Defensive IO integrate and coordinate policies and procedures, operations, personnel, and technology to protect and defend information and information systems. Defensive IO are conducted through information assurance, OPSEC, physical security, counterdeception, counterpropaganda, counterintelligence, EW, and SIO. Defensive IO ensure timely, accurate, and relevant information access while denying adversaries the opportunity to exploit friendly information and information systems for their own purposes. Offensive IO also can support defensive IO.

---

[48] JOINT PUB 3-13, *supra* note 47, at vii-viii.

> Defensive IO ensure the necessary protection and defense of information and information systems upon which joint forces depend to conduct operations and achieve objectives.
>
> Four interrelated processes support defensive IO: information environment protection, attack detection, capability restoration, and attack response. Because they are so interrelated, full integration of the offensive and defensive components of IO is essential. JFCs and their subordinate commanders should plan, exercise, and employ available IO capabilities and activities to support integrated defensive IO.[49]

The doctrine for command and control warfare fits within this larger framework, and provides further definition to the weapons available:

> The elements of C2W are as follows: OPSEC denies critical information necessary for the adversary commander to estimate the military situation accurately; psychological operations are vital to a broad range of US political, military, economic, and informational activities, including support of C2 during C2-attack and C2-protect operations; military deception focuses on causing the adversary commander to estimate incorrectly the situation in the operational area with respect to friendly force dispositions, capabilities, vulnerabilities, and intentions; electronic warfare includes electronic warfare support, electronic attack, and electronic protection; and physical destruction in support of C2W refers to the use of "hard kill" weapons or other means, such as sabotage or covert actions, against designated targets as an element of an integrated C2W effort.[50]

---

[49] JOINT PUB 3-13, *supra* note 47, at viii-ix.

[50] JOINT PUB 3-13.1, *supra* note 47, at vi.

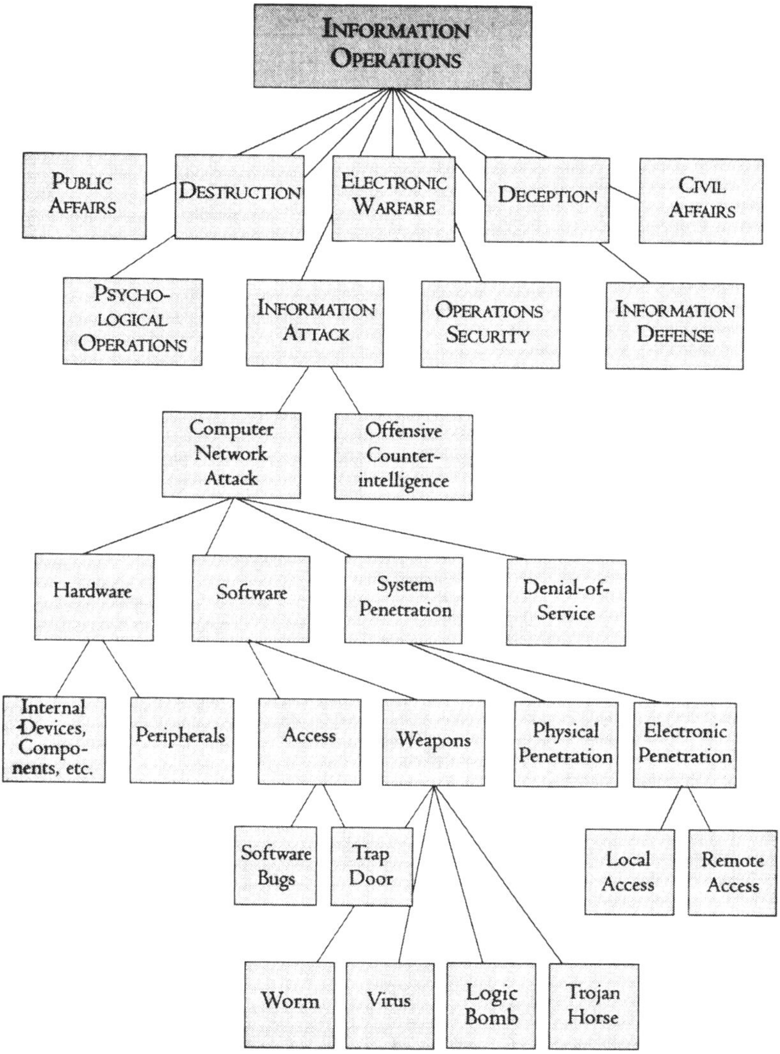

Figure 3. Information Operations Taxonomy

This doctrinal organization is reflected in Figure 3, which shows the complexity of IO means and methods. All of these complex subdisciplines of IO, however, present the same analytical challenge to the operator and her legal advisor: first, a determination must be made whether the proposed action is a use of force under international law, and, if the law of armed conflict is found to apply, then the operation must be thoroughly reviewed in light of the four basic principles of the law of armed conflict and any applicable treaties. In short, while each fact pattern of each information operation is unique, the systematic process for analyzing any given operation is, in itself, fairly straightforward.

Technology and techniques associated with information systems on the internet are evolving so rapidly that it is not possible to know the exact nature and magnitude of the threat posed by military applications in cyberspace. The examples provided in this chapter that depict the attempts and successes of non-state actors to acquire, disrupt, manipulate, or destroy data are illustrative, however, of the vulnerabilities which can be exploited by states for military or other hostile purposes. The National Security Agency demonstrated during a military exercise that the United States—and thus, probably all countries—are vulnerable to such actions, cyberspace being simply another medium or environment through which states can either use force or support the use of force. Suffice it to say that the disruption, manipulation, and destruction of data via the internet promises to be sufficiently crippling that information weapons are sure to be developed and utilized in the near future. Some, although not all, of these military applications in cyberspace will rise to the level of a use of force that constitutes an armed attack against a state—and these will be discussed in Chapter 6 of this book.

PART II. *JUS AD BELLUM*[51]

◆

CHAPTER 3: THE UNITED NATIONS AND THE CONTEMPORARY *JUS AD BELLUM*[†]

Every society throughout history has put some limitation on the resort to force, and the current international legal regime began to take shape in the mid-nineteenth century. Over the next hundred years, lofty humanitarian theory and brutal aggressive practice shaped the customary legal standards for the conduct of war, the jus in bello. *As for* jus ad bellum, *the law governing the transition from peace to war, what had been entirely the prerogative of the sovereign came under the aegis of international law with the drafting and ratification of the UN Charter at the end of the Second World War. This new legal regime set thresholds for threats to the peace (article 39), threats and uses of force, (article 2(4)), and armed attacks (article 51). While the precise location of these thresholds can be determined only by analyzing them in the context of specific factual situations, it is possible to articulate an intellectual framework describing the relationship of these thresholds, and their general applicability to any given fact pattern. Although the body of law which applies once the last of these thresholds has been breached is a distinct subject analyzed later in this book, a preliminary look at the fundamental doctrines of the* jus in bello *is useful in determining their role during the transition from peace to war.*

[†] This chapter is drawn substantially from Chapter 3 of WALTER GARY SHARP, SR., CYBERSPACE AND THE USE OF FORCE (1999).

[51] Chapters 3 and 4 of Part II are based upon previously published work that can be found at: Walter Gary Sharp, Sr., *Revoking an Aggressor's License to Kill Military Forces Serving the United Nations: Making Deterrence Personal*, 22 Md. J. Int'l L. & Trade 1 (1998). The Maryland Journal of International Law and Trade granted the standing right of republication on December 23, 1997.

Evolution of the *Jus ad Bellum*

War has existed during the entire history of human society.[52] Indeed, a study in 1968 calculated "that there had been only 268 years free of war in the previous 3,421."[53] From the organized food-gathering and wife-seeking raids of primitive man, to the Persian Wars of the fifth century B.C. and the Second World War of the twentieth century,[54] war was accepted as a legitimate form of violence.[55] Although another study in 1998 reflects a continued downward trend for armed conflicts since 1989, it estimates that there were still 25 major armed conflicts ongoing in 1997.[56] While only the conflict between India and Pakistan was interstate, other states supported at least one party in these intrastate conflicts.[57]

Motives for war were wide-ranging, and included the acquisition of territory, the domination of another people, the access to scarce resources, glory, prestige, and revenge.[58] Other motivations included the

---

[52] RICHARD A. PRESTON & SYDNEY F. WISE, MEN IN ARMS: A HISTORY OF WARFARE AND ITS INTERRELATIONSHIPS WITH WESTERN SOCIETY 1 (4th ed. 1979). [hereinafter PRESTON & WISE, MEN IN ARMS].

[53] DONALD KAGAN, ON THE ORIGINS OF WAR AND THE PRESERVATION OF PEACE 4 (1995).

[54] PRESTON & WISE, MEN IN ARMS, *supra* note 52, at 8, 19, 295.

[55] *See id.* at 5-14.

[56] STOCKHOLM INTERNATIONAL PEACE RESEARCH INSTITUTE, SIPRI YEARBOOK 1998: ARMAMENTS, DISARMAMENT AND INTERNATIONAL SECURITY chap. 1 (1998). [A summary of this study can be found at http://www.SIPRI.org]

[57] *Id.*

[58] *See* PRESTON & WISE, MEN IN ARMS, *supra* note 52, at 5-6.

desire to pillage, rape, and murder because war was simply a culture and a way of life for some peoples, such as the Cossacks.[59]
In many respects, war simply began as organized armed robbery by plundering hordes among dissimilar peoples and tribes.[60]

The law of conflict management, *jus ad bellum,* is a set of rules that govern the resort to armed conflict and determine whether the conflict is lawful or unlawful in its inception.[61] A study of the law of conflict management is central to understanding what actions are considered a use of force under international law because war is the legal consequence of a use of force by one state against another.[62] For four millennia, the *jus ad bellum* generally allowed states to engage in war at any time; however, virtually every recorded civilization developed some rules governing the initiation of war.[63] The Egyptians and the Sumerians developed rudimentary *jus ad bellum* during the second millennium B.C.[64] As a general rule, the Hittites of the fourteenth century B.C. formally exchanged letters and demands before commencing hostilities.[65]

---

[59] *See* JOHN KEEGAN, A HISTORY OF WARFARE 7-9 (1993).[hereinafter KEEGAN, WARFARE].

[60] LOTHAR KOTZSCH, THE CONCEPT OF WAR IN CONTEMPORARY HISTORY AND INTERNATIONAL LAW 25 (1956). [hereinafter KOTZSCH, CONCEPT OF WAR].

[61] ROBERTS & GUELFF, *supra* note 4, at 2-3.

[62] *See* The Prize Cases, 67 US (2 Black) 635 (1862), *excerpted in* NATIONAL SECURITY LAW 78, 78-80 (Stephen Dycus *et al.* eds.) (1997). In this 1862 case, the United States Supreme Court defined war under the law of nations as "[t]hat state in which a nation prosecutes its right by force."

[63] I THE LAW OF WAR: A DOCUMENTARY HISTORY 3 (Leon Friedman ed., 1972) [hereinafter FRIEDMAN].

[64] *Id.*

[65] *Id.*

During the approximate period 335 B.C. to A.D. 1800, war was examined on a moral, philosophical level that approved of war if the cause were just.[66] This philosophy of war as an instrument of justice was replaced in the early 1800s by the acceptance of war as a political instrument of national policy.[67] The Prussian philosopher and military theorist General Karl von Clausewitz is often quoted as describing war as the continuation of national policy by other means.[68] Two fundamental tenets of Clausewitz's philosophy are the legitimacy of war as a normal phase in the relations among states and the absolute sovereignty of states.[69]

The first attempt to place juristic restrictions on this freedom to resort to war began at the Hague Peace Conferences of 1899 and 1907.[70]

---

[66] NATIONAL SECURITY LAW 51-57 (John Norton Moore *et al.* eds., 1990).

[67] *Id.* at 57. Chapter 3 of this text provides a more detailed discussion of the development of the law of conflict management divided into six approximate time periods: (1) just war, 335 B.C. to A.D. 1800; (2) war as fact, 1800-1918; (3) early League of Nations, 1919-1925; (4) Kellogg-Briand Pact and late League, 1928-1945; (5) early United Nations Charter, 1945-1958; and (6) contemporary Charter, 1959-present.

[68] *See, e.g.*, KEEGAN, WARFARE, *supra* note 59, at 3. (While noting that variations of the concept that war is the continuation of policy by other means is frequently quoted, Keegan challenges the simplicity of the translation. The original German, Keegan explains, expresses a more subtle and complex idea that "war is the continuation 'of political intercourse' ... 'with the intermixing of other means'..."); NATIONAL SECURITY LAW 57 (John Norton Moore *et al.* eds., 1990) ("War is only a part of political intercourse...War is nothing but a continuation of political intercourse, with a mixture of other means."); US MARINE CORPS, FMFM 1, WARFIGHTING 19 (1989) ("War does not exist for its own sake. It is an extension of policy with military force."). General Karl von Clausewitz (1780-1831) served as the director of the War College in Berlin from 1818 to 1830 where he began his work on his three-volume masterpiece ON WAR, which was published posthumously in 1832. *Karl von Clausewitz*, MICROSOFT ENCARTA '95 (CD-ROM Multimedia Encyclopedia, 1994).

[69] CLAUSEWITZ ON WAR 63-64 (Anatol Rapoport ed., 1968).

[70] THE CHARTER OF THE UNITED NATIONS: A COMMENTARY 109-111 Bruno Simma ed., 1994). [hereinafter THE CHARTER OF THE UNITED NATIONS: A COMMENTARY].

Article I of the 1907 Hague Convention III, for example, required parties to present a "previous and explicit warning, in the form either of a declaration of war, giving reasons, or of an ultimatum with conditional declaration of war."[71] This was nothing more, however, than a formal recognition that the resort to war was legal.[72] Similarly, the series of nineteen Bryan Treaties concluded by the United States and a number of other states between 1913 and 1916 imposed an obligation on contracting parties to submit all of their disputes to a conciliation commission, and not to begin hostilities prior to the commission's report.[73]

After World War I, the League of Nations also attempted to restrict the right of states to resort to war.[74] As in the Bryan Treaties, members of the League of Nations were required to submit their disputes to judicial settlement, arbitration, or to the Council of the League, and were prohibited to begin war within a period of three months from the judicial decision, arbitral award, or the Council's report.[75] The only prohibition on the resort to war was against any state that complied with the judicial decision, arbitral award, or a unanimous decision of the Council.[76]

---

[71] Hague Convention III Relative to the Opening of Hostilities, Oct. 18, 1907, art. I, 36 Stat. 2259, I Bevans 619, *reprinted in* SCHINDLER & TOMAN, ARMED CONFLICTS, *supra* note 5, at 57.

[72] THE CHARTER OF THE UNITED NATIONS: A COMMENTARY, *supra* note 70, at 109.

[73] *Id.* at 109-10.

[74] *Id.* at 110.

[75] *Id.*

[76] *Id.*

The League of Nations was unsuccessful in its attempt to prohibit or, in practice, restrict the resort to war.[77] The international community attempted to correct the shortcomings of the League of Nations by adopting the 1924 Geneva Protocol for the Pacific Settlement of International Disputes.[78] This Protocol prohibited the resort to war except in self-defense or in the case of collective enforcement measures, but it never became binding law.[79]

The decisive turning point in the history of the *jus ad bellum* was the adoption of the Kellogg-Briand Pact on August 27, 1928.[80] Article I of this Pact provides that "[t]he High Contracting Parties solemnly declare in the names of their respective peoples that they condemn recourse to war for the solution of international controversies and renounce it as an instrument of national policy in their relations with one another."[81] Nearly all existing states of the time became parties to the Kellogg-Briand Pact, and it soon became customary international law.[82] Despite its significance in principle, the Pact had its weaknesses.[83] Most notably, it merely prohibited war and not the use of force, allowing states to engage in

---

[77] *Id.*

[78] *Id.*

[79] *Id.*

[80] *Id.*

[81] General Treaty on the Renunciation of War as an Instrument of National Policy, Aug. 27, 1928, art. I, 46 Stat. 2343, 94 L.N.T.S. 57, *quoted in* THE CHARTER OF THE UNITED NATIONS: A COMMENTARY, *supra* note 70, at 110.

[82] *Id.* at 110-11.

[83] *See id.*

devastating hostilities while claiming that no offenses under the Pact were being committed.[84]

Contemporary *Jus ad Bellum*

Article 2(4) of the Charter of the United Nations[85] was the *coup de grâce* for the theory and practice of the *jus ad bellum* that recognized the right of states to resort to war. This article requires all Member States to "refrain in their international relations from the threat or use of force against the territorial integrity or political independence of any state . . . ."[86] The Charter clearly outlaws the aggressive use of force[87] while recognizing a state's inherent right of individual and collective self-defense in Article 51 and the Security Council's obligation under Article 39 to maintain or restore international peace and security. Articles 2(4), 39, and 51 of the Charter now redefine and codify the contemporary *jus ad bellum* in its

---

[84] *Id.* at 111.

[85] U.N. CHARTER. Excerpts of the Charter are provided as Appendix C.

[86] U.N. CHARTER art. 2(4).

[87] Generally, aggressive refers to beginning a dispute by being the first to either threaten or use force. *See* BLACK'S LAW DICTIONARY, *supra* note 18, at 60. However, an aggressive use of force, *i.e.*, aggression, is not defined by the Charter, and despite 54 years of efforts, the international community has failed to formulate a generally acceptable definition of aggression. *See generally, e.g.*, ANN VAN WYNEN THOMAS AND A.J. THOMAS, JR., THE CONCEPT OF AGGRESSION IN INTERNATIONAL LAW (1972) [hereinafter THOMAS & THOMAS, CONCEPT OF AGGRESSION]. Accordingly, an aggressive use of force can best be identified by studying state practice. For the purposes of this book, an 'aggressive use of force' is a shorthand term used to refer to any use of force within the meaning of Article 2(4) that is not justified by a state's right of self-defense or authorized by the Security Council under its coercive Chapter VII powers. While this definition is somewhat circular in its reasoning, it is useful shorthand to facilitate a discussion of the basic principles of international law necessary for the more detailed discussion in parts III and IV of this book that define heuristically what state activities constitute an aggressive use of force.

entirety.[88] If a state activity is a use of force within the meaning of Article 2(4), it is unlawful unless it is an exercise of that state's inherent right of self-defense, or unless it is authorized by the Security Council under its coercive Chapter VII authority.

With the creation of the Charter's norm prohibiting the right of states to resort to the aggressive use of force, the historical concept of the *jus ad bellum* evolved to one of the *jus contra bellum, i.e.*, the law against the aggressive use of force.[89] The underlying principle of the *jus contra bellum* is now penal law[90] that prohibits all aggressive use of force while retaining a state's right of individual and collective self-defense. The aggressive use of force by a state is now a crime against peace that has been outlawed by the international community.[91] Accordingly, international law imputes individual criminal responsibility for those who operate at a policy-decision level of an aggressor state[92] for a crime against peace; *i.e.*, the

---

[88] *See* THE CHARTER OF THE UNITED NATIONS: A COMMENTARY, *supra* note 70, at 111.

[89] KOTZSCH, THE CONCEPT OF WAR, *supra* note 60, at 83.

[90] *Id.*

[91] *E.g.*, THE CHARTER OF THE UNITED NATIONS: A COMMENTARY, *supra* note 70, at 111; NATIONAL SECURITY LAW 85, 369-70 (John Norton Moore *et al.* eds., 1990); IAN BROWNLIE, INTERNATIONAL LAW AND THE USE OF FORCE BY STATES 112, 154 (1963) [hereinafter BROWNLIE, INTERNATIONAL LAW].

[92] *E.g.*, NATIONAL SECURITY LAW 369-70 (John Norton Moore *et al.* eds., 1990), which quotes the opinion of the International Military Tribunal in *US v. Leeb (The High Command Case)* at 370:

> If and as long as a member of the armed forces does not participate in the preparation, planning, initiating, or waging of aggressive war on a policy level, his war activities do not fall under the definition of crimes against peace.

"planning, preparation, initiation or waging of a war of aggression, or a war in violation of international treaties, agreements or assurances, or participation in a common plan or conspiracy for the accomplishment of any of the foregoing..."[93]

Articles 2(4), 39, and 51 must be read together to determine the scope and content of the Charter's prohibition on the aggressive use of force, the responsibility of the Security Council to enforce this prohibition, and the right of all states to use force in self-defense. Article 2(4) of the Charter prohibits the threat or use of force by any state against the territorial integrity or political independence of another state except in individual or collective self-defense as authorized by international law and recognized by Article 51 of the Charter.[94] Articles 2(4) and 51 provide:

> *Article 2*
> The Organization and its Members, in pursuit of the Purposes stated in Article 1, shall act in accordance with the following Principles: ...
> (4) All Members shall refrain in their international relations from the threat or use of force against the territorial integrity or political independence of any state, or in any other manner inconsistent with the Purposes of the United Nations.
>
> *Article 51*
> Nothing in the present Charter shall impair the inherent right of individual or collective self-defence if an armed attack

---

See also Comments of the ILC on the Seven Nuremberg Principles, 1950, § 117, U.N. GAOR 5th Sess., supplement 12, U.N. Doc. A/1316 (1950) *reprinted in* SYDNEY D. BAILEY, PROHIBITIONS AND RESTRAINTS IN WAR 163, 168 (1972).

[93] *See* Charter of the International Military Tribunal, Aug. 8, 1945, art. 6, 59 Stat. 1545, 82 U.N.T.S. 280, *reprinted in* SCHINDLER & TOMAN, ARMED CONFLICTS, *supra* note 5, at 911, 914.

[94] *See* NATIONAL SECURITY LAW 85 (John Norton Moore *et al.* eds., 1990).

> occurs against a Member of the United Nations, until the Security Council has taken the measures necessary to maintain international peace and security. Measures taken by Members in the exercise of this right of self-defense shall be immediately reported to the Security Council and shall not in any way affect the authority and responsibility of the Security Council under the present Charter to take at any time such action as it deems necessary in order to maintain or restore international peace and security.[95]

Article 2(4) is so important to the maintenance of international peace and security, it has been described by one scholar as a "critical legal obligation" outlawing aggressive war that permits no derogation and as "the single most important fundament of world order . . . ."[96] Article 51 is equally important, because it recognizes the distinction between an unlawful, aggressive use of force and a lawful, defensive use of force that is the inherent right of all states.[97] Both Articles 2(4) and 51 are also customary international law.[98]

Self-Defense and the Use of Force by States

It is a fundamental principle of the contemporary law of conflict management that states may only use force lawfully in individual or collective self-defense. States recognize this principle, and have invoked their right of self-defense as justification for almost every use of force

---

[95] U.N. CHARTER arts. 2(4), 51.

[96] JOHN NORTON MOORE, CRISIS IN THE GULF: ENFORCING THE RULE OF LAW 22 (1992). [hereinafter MOORE, CRISIS IN THE GULF].

[97] *Id.* at 151.

[98] THE CHARTER OF THE UNITED NATIONS: A COMMENTARY, *supra* note 70, at 126-27, 666-67.

since the Charter has been in effect.[99] Since the text of Article 51 recognizes this right of self-defense at the time an armed attack occurs, the pivotal focal point of any self-defense analysis is on the meaning of an armed attack.[100] It is generally accepted, however, that the Article 51 right of self-defense is coextensive with a state's right under customary international law.[101] The remainder of this section will detail the basic rules of self-defense necessary for a full understanding of what self-defense options are available as a matter of international law should an armed attack occur. Chapter 6, *infra*, will analyze in detail what state activities constitute a use of force that rises to the level of an armed attack.

Necessity and Proportionality

Customary international law requires that all uses of force be necessary and proportional,[102] and it prohibits the use of force for retaliatory or punitive actions.[103] For this reason, the punitive aspect of many military operations (the US raid on Libya in 1986, Israeli responses to terrorist attacks) is often downplayed, and the missions are characterized as self-defense in the form of deterring future aggression in a low-intensity, ongoing conflict. The principles of necessity and proportionality apply under both the law of conflict management and the law of armed conflict. The law of conflict management requires that a state's use of force be necessary for either individual or collective self-

---

[99] *See id.* at 663.

[100] *See id.*

[101] MOORE, CRISIS IN THE GULF, *supra* note 96, at 151. *But see* THE CHARTER OF THE UNITED NATIONS: A COMMENTARY, *supra* note 70, at 666, which concludes that Article 51 excludes any right of self-defense "other than that in response to an armed attack."

[102] MOORE, CRISIS IN THE GULF, *supra* note 96, at 156.

[103] THE CHARTER OF THE UNITED NATIONS: A COMMENTARY, *supra* note 70, at 677.

defense.[104] For example, the requirement of necessity for the international community to use force in collective self-defense was clearly met when Iraq invaded and brutally occupied Kuwait in 1990.[105] If Iraq had simply entered Kuwait, destroyed a number of its oil fields, and then quickly left, the principle of necessity would very likely not have justified the use of force in the absence of a continuing threat.

Similarly, the law of conflict management requires that a state's use of force be proportional in intensity and magnitude to what is reasonably necessary to promptly secure the permissible objectives of self-defense.[106] It is "emphatically *not* a requirement that the defense limit itself to weapons systems employed by the aggressor or to employ force only at levels employed by the aggressor or to employ force only at levels sufficient to be ineffectual."[107] Indeed, during a debate on the Viet Nam War, one noted scholar concluded that if a state party to an armed conflict has greater military resources potentially available to it, an insufficient use of force by that state to end the war quickly would violate general norms of international law.[108]

A necessary use of force is defined by the law of armed conflict as only

> that degree and kind of force, not otherwise prohibited by the law of armed conflict, required for the partial or complete

---

[104] *See* MOORE, CRISIS IN THE GULF, *supra* note 96, at 156-57.

[105] *Id.* at 157.

[106] *Id.* at 158.

[107] *Id.*

[108] *Id.* at 181 n.38.

> submission of the enemy with a minimum expenditure of time, life, and physical resources ... .[109]

Similarly, the law of armed conflict defines a proportional use of force as that level of force required to destroy a military objective,[110] but which does not cause unnecessary collateral destruction of civilian property or unnecessary human suffering of civilians.[111]

The principle of proportionality is frequently misunderstood as limiting the use of force that can be used to destroy a military objective to the strength or firepower of that objective—or in some other way limiting the use of force between combatants. It does not, however, require any such parity of force. Proportionality is a limitation on the use of force against a military objective only to the extent that such a use of force may cause unnecessary collateral destruction of civilian property or unnecessary human suffering of civilians. The principle of proportionality is a balancing of the need to attack a military objective with the collateral damage and human suffering that will be caused. Proportionality categorically imposes *no* limitations on the use of force between

---

[109] US DEP'T OF THE NAVY, NWP 1-14M, THE COMMANDER'S HANDBOOK ON THE LAW OF NAVAL OPERATIONS § 5.2 (1995) [hereinafter COMMANDER'S HANDBOOK].

[110] Lawful military objectives include combatants and

> those objects which, by their nature, location, purpose, or use, effectively contribute to the enemy's war-fighting or war-sustaining capability and whose total or partial destruction, capture, or neutralization would constitute a definite military advantage to the attacker under the circumstances at the time of the attack.

COMMANDER'S HANDBOOK, *supra* note 109, at § 8.1.1.

[111] *Id.* at § 8.1.

combatants in the absence of any potential effect on civilians or civilian property.[112]

Collateral Damage and Injury to Civilians

Proportionality limits the use of force that can be used to destroy a military objective to that which does not cause unnecessary collateral destruction of civilian property or unnecessary human suffering of civilians, but it does not prohibit *any* damage to civilian property or injury to civilians. If civilian property and civilians support a war effort, they are subject to attack, and they are subject to incidental damage during an attack on a lawful military objective. These two corollaries to the principles of necessity and proportionality are very important to highlight in the context of the impact of state activities on civilian property and civilians.

While civilian property and civilians may not be the *object* of an attack as such, states may use force against civilian property and activities that support or sustain an enemy state's warfighting capability during armed conflict.[113] States may use force during armed conflict, for example, against economic targets such as enemy lines of communication, rail yards, bridges, rolling stock, barges, industrial installations producing

---

[112] Indeed, one of the four strategic concepts of the national military strategy of the United States is to use decisive force to overwhelm an adversary. CHAIRMAN OF THE JOINT CHIEFS OF STAFF, NATIONAL MILITARY STRATEGY OF THE UNITED STATES OF AMERICA 3 (1997). Even more notably, in December, 1990, US Secretary of Defense Dick Cheney threatened Saddam Hussein that the US response to an Iraqi use of weapons of mass destruction would be "absolutely overwhelming and ... devastating." US DEP'T OF DEF., CONDUCT OF THE PERSIAN GULF WAR: FINAL REPORT TO CONGRESS PURSUANT TO TITLE V OF THE PERSIAN GULF CONFLICT SUPPLEMENTAL AUTHORIZATION AND PERSONNEL BENEFITS ACT OF 1991 (PUBLIC LAW 102-25), app. Q, at 2 (Apr., 1992) [hereinafter CONDUCT OF THE PERSIAN GULF WAR: FINAL REPORT].

[113] COMMANDER'S HANDBOOK, *supra* note 109, at § 8.1.

warfighting products, and power generation plants.[114] In today's modern society, much of a state's civilian infrastructure is used for military purposes, and is thus subject to lawful attack during armed conflict if there is a military advantage to be gained by such an attack.[115]

In its final report to Congress on the conduct of the Persian Gulf War, the US Department of Defense noted that

> [a] bridge or highway vital to daily commuter and business traffic can be equally crucial to military traffic, or support for a nation's war effort. Railroads, airports, seaports, and the interstate highway system in the United States have been funded by the Congress in part because of US national security concerns, for example; each proved invaluable to the movement of US military units to various ports for deployment to Southwest Asia (SWA) for Operations Desert Shield and Desert Storm. Destruction of a bridge, airport, or port facility, or interdiction of a highway can be equally important in impeding an enemy's war effort.
>
> The same is true with regard to major utilities; for example, microwave towers for everyday, peacetime civilian communications can constitute a vital part of a military command and control (C2) system, while electric power grids can be used simultaneously for military and civilian purposes.[116]

A state has always been dependent upon its private industry to support its war efforts, and to the extent private industry supported a war effort, private industry has always been a lawful target under the law of armed

---

[114] *Id.* at § 8.1.1.

[115] CONDUCT OF THE PERSIAN GULF WAR: FINAL REPORT, *supra* note 112, at 10, 11.

[116] *Id.*

conflict. Technology, however, is increasing this interdependence of a state's commercial sector with its government functions, making more and more of the civilian infrastructure subject to lawful attack during armed conflict. Except for shipping, private industry may be well protected within a state's borders. Unfortunately, technology such as computers and the internet is exposing the civilian infrastructure—making it more accessible and vulnerable to attack.

Furthermore, it is not unlawful to cause *incidental* injury to civilians, or collateral damage to civilian property, during an attack on a legitimate military objective.[117] The balancing of proportionality does require, however, that such incidental injury or collateral damage not be excessive in light of the advantage anticipated by the attack.[118] The law of armed conflict requires a military commander to take all reasonable precautions, based upon all the facts known or reasonably available at the time, to keep civilian casualties and damage to the minimum consistent with mission accomplishment and the security of his or her personnel.[119]

Anticipatory Self-Defense

There is no consensus over the point in time at which self-defense against an armed attack may be taken.[120] Some scholars conclude that the customary international law right of anticipatory self-defense is incorporated by Article 51, while others conclude that an anticipatory

---

[117] *Id.* at § 8.1.2.

[118] *Id.*

[119] *Id.*

[120] THE CHARTER OF THE UNITED NATIONS: A COMMENTARY, *supra* note 70, at 675. For an excellent survey and discussion of the issues surrounding anticipatory self-defense, see the compilation of five articles at 19 CASE W. RES. J. INT'L L. (1987).

right is contrary to Article 51.[121] Opposing schools of thought, however, usually converge when applied to a specific fact situation. Anticipatory self-defense authorizes the use of force in self-defense at a point in time *before* an attack occurs. As defined by then Secretary of State Daniel Webster in the *Caroline* case, this point in time occurs when "the necessity of that self-defence is instant, overwhelming and leaving no choice of means, and no moment for deliberation."[122] One opponent of anticipatory self-defense concludes that Article 51 requires that an armed attack occurs—but he also concludes that self-defense is "permissible only after the armed attack has already been launched."[123] The fact that this opponent acknowledges that force may be used in self-defense before the attack occurs but after it is launched acknowledges the existence of the right of anticipatory self-defense by his very argument. The real debate is not over whether the right of anticipatory self-defense exists, but to what extent it exists under the conditions of the *Caroline* case.

Articles 2(4) and 51: Closing the Gap

The thresholds of Articles 2(4) and 51 are different in scope; *i.e.*, the Article 2(4) concept of a threat or use of force is much broader than the more restrictive Article 51 threshold of armed attack.[124] This correct observation, however, causes one author to arrive at the "stunning conclusion" that an *unlawful* use of force that does not meet the threshold of an armed attack and does not give the victim state the right to respond with a use of force.[125] If this conclusion simply means that the right to

---

[121] THE CHARTER OF THE UNITED NATIONS: A COMMENTARY, *supra* note 70, at 675-76.

[122] *Id.* at 675.

[123] *Id.* at 676.

[124] *See id.* at 663.

[125] *Id.* at 663-64.

self-defense is invoked somewhere along the spectrum of state activities between threats of force and the actual use of force, then it is a reasonable interpretation of Articles 2(4) and 51. After all, the low threshold of Article 2(4) includes mere verbal threats, and international law requires that all uses of force, including those in self-defense, to be necessary and proportional. Accordingly, such an interpretation does not prohibit a use of force in self-defense in response to a minor use of force that occurs over a very short period of time, but the right of self-defense under customary international law may not justify an armed response.

In sharp contrast, if this "stunning conclusion" means that states do not always have the right, albeit limited to narrowly circumscribed situations, to use necessary and proportional force in self-defense, it is highly unacceptable—and in practice, it has not been accepted by any consensus of states. Notwithstanding the purposes and principles of the Charter international law does not require a state to sit idly by while another state uses unlawful force against it. To do so would undermine international peace and security by allowing states to unlawfully use force with impunity and to gain from their unlawful actions.

It is not clear textually just how large a gap may exist between the two thresholds. In practice, the analysis within this gap focuses on what is a lawful, defensive response to an unlawful use of force. For example, if one state unlawfully uses force by using a single missile to destroy a telecommunications tower just across its border that was broadcasting unwanted pop videos into its territory, the victim state could use force in self-defense by shooting down the incoming missile. Once the missile strike occurs, however, international law would prohibit a punitive use of force in retaliation as such for the strike. An appropriate response by the territorial state after such an *isolated* missile strike would be to seek civil redress, although a proportional strike on the launch facility would also be appropriate if the attack were viewed in the context of *continuing* hostilities.

Similarly, consider a situation where one state admittedly uses armed military force to rescue its nationals whose lives are in danger, and does so over the objection of the territorial state. Chapter 6, *infra*, concludes that this is a use of force within the meaning of Article 2(4) that is a lawful act of self-defense. The rescue operation, however, is very short in duration—perhaps less than an hour. Once the rescue operation is completed, the situation exists where a use of force has occurred, but the territorial state has no right to respond with force in self-defense because the rescue operation is now complete.[126] An appropriate response by the territorial state *after* the rescue operation would be to publicly denounce the rescue operation.

In contrast, consider a pattern of missile attacks by a state that originate from the same launch site. In addition to having the right to shoot down incoming missiles in self-defense, the victim state also has the right of anticipatory self-defense to preemptively destroy the launch site to prevent future attacks.

Those who embrace the opposite conclusion that "any state affected by another state's unlawful use of force not reaching the threshold of an 'armed attack,' *is bound, if not exactly to endure the violation, then at least to respond only by means falling short of the use or threat of force, which are thus often totally ineffective*"[127] are frequently forced to temper their harsh norm with liberal interpretations of the definitions of use of force and armed attack. For example, one scholar concludes it is within a state's *lawful* discretion to deny transit rights to a land-locked state.[128] In extreme cases, however, the same scholar concludes that such a denial may lead "to an impairment equating to a military invasion (*e.g.*, by cutting off all communications routes) [and that] an 'armed attack' can be taken to exist."[129] This conclusion is problematic, since such a denial is an exercise

---

[126] The justification for a territorial state to use armed force *during* the rescue operation is discussed in Chapter 6 under the section concerning the protection of nationals abroad.

[127] THE CHARTER OF THE UNITED NATIONS: A COMMENTARY, *supra* note 70, at 663-64.

[128] *Id.* at 670.

[129] *Id.*

of a state's lawful right to exclude others from its territory. Similarly, the same scholar concludes that the use of armed force by a state against commercial vessels and aircraft of another state is not an armed attack, unless it is on the whole of the fleets.[130]

Frequently, but not always, the end result of the application of the harsh-norms approach is an analysis of what is a necessary and proportional response in self-defense. The same scholar who makes the exceptions discussed above nevertheless concludes that "forcible rescue operations for nationals endangered in another country are no longer lawful under the UN Charter."[131] In the final analysis, both sides interpret armed force very broadly to include non-military physical force and indirect force that includes the passive acquiescence of the use of a state's territory by a belligerent.[132]

Finally, consider the conclusion of another scholar who interprets a use of force to be the same as an armed attack. He concludes that a use of force must reach a certain "gravity" and that minor frontier incidents are not *per se* uses of force that rise to the Article 2(4) threshold because "the minor nature of an attack is prima facie evidence of absence of intention to attack, of honest mistake, or simply the limited objectives of an attack."[133] His explanation also returns the analysis to the issue of what is a necessary and proportional response in self-defense under the circumstances.

The principle that a state never loses its right of self-defense may be inconsistent with the views of some, but it is consistent with state

---

[130] *Id.* at 671.

[131] *Id.* at 672.

[132] *Id.* at 111-13.

[133] *See* BROWNLIE, INTERNATIONAL LAW, *supra* note 91, at 366.

practice and the following *dicta* found in the *Nicaragua* judgment of the International Court of Justice:

> Since the Court is here dealing with a dispute in which a wrongful use of force is alleged, it has primarily to consider whether a State has a right to respond to intervention with intervention going so far as to justify a use of force in reaction to measures which do not constitute an armed attack but may nevertheless involve a use of force. ... However, since the Court is bound to confine its decision to those points of law which are essential to the settlement of the dispute before it, it is not for the Court here to determine what direct reactions are lawfully open to a State which considers itself the victim of another State's acts of intervention, possibly involving the use of force. ... It might however be suggested that, in such a situation, the United States might have been permitted to intervene in Nicaragua in the exercise of some right analogous to the right of collective self-defence, one which might be resorted to in a case of intervention short of armed attack.[134]

Accordingly, a state never loses its right to use force in self-defense in response to a use of force within the meaning of Article 2(4), however, the right of self-defense under customary international law may not always justify an armed response. This is a very important point of international law to remember throughout any discussion of a state's right of self-defense—although a state never loses its right of to use force in self-defense, the principles of necessity and proportionality alone may *not* justify the use of force in self-defense.

---

[134] Military and Paramilitary Activities in and against Nicaragua (Nicar. v. US), 1986 I.C.J. 14, 110 (June 27), para. 210 (emphasis added). For an excellent statement of the facts which led up to this I.C.J. case and a comprehensive discussion of the international legal issues implicated by this conflict, *see* JOHN NORTON MOORE, THE SECRET WAR IN CENTRAL AMERICA (1987).

Security Council Authority to Use Force

As an exercise of the international community's inherent right of collective self-defense, Article 39 of the Charter imposes an obligation on the Security Council to maintain international peace and security. Article 39 provides that

> The Security Council shall determine the existence of any threat to the peace, breach of the peace, or act of aggression and shall make recommendations, or decide what measures shall be taken in accordance with Articles 41 and 42, to maintain or restore international peace and security.[135]

Decisions taken by the Security Council under Article 39 are binding on all Member States.[136] Every threat or use of force proscribed by Article 2(4) is, *per se*, a threat to international peace and security within the meaning of Article 39.[137] Accordingly, the Security Council has the coercive authority to authorize the use of force in response to any violation of Article 2(4).[138]

The authority of the Security Council to use force, however, extends beyond violations of Article 2(4). Indeed, the Article 39 threshold extends considerably below the Article 2(4) threshold,[139] giving the Security Council the power to authorize states to use force under

---

[135] U.N. CHARTER art. 39.

[136] *Id.* at art. 25; THE CHARTER OF THE UNITED NATIONS: A COMMENTARY, *supra* note 70, at 407-18.

[137] THE CHARTER OF THE UNITED NATIONS: A COMMENTARY, *supra* note 70, at 119.

[138] *Id.*

[139] *Id.* at 119.

circumstances where states do not independently have the right to use force in self-defense. For example, scholars have concluded that threats to the peace within the meaning of Article 39 include extreme intrastate violence or human rights violations, the failure of a state to surrender terrorists in accordance with the order of the Security Council, an illegal racist regime, cross-frontier expulsion of refugees, diversion of a river by an up-stream state, and serious violations of international law that may provoke an armed response.[140]

Article 41 authorizes the Security Council to shape an international community response to a threat to international peace and security that falls short of deploying military forces under Article 42. This article provides that

> The Security Council may decide what measures not involving the use of armed force are to be employed to give effect to its decisions, and it may call upon the Members of the United Nations to apply such measures. These may include complete or partial interruption of economic relations and of rail, sea, air, postal, telegraphic, radio, and other means of communication, and the severance of diplomatic relations.[141]

It is very important to emphasize two aspects of Article 41 that are frequently overlooked or misunderstood. First, Article 41 is an explicit recognition that the Security Council has the authority to require Member States to cease any or all forms of economic relations, communications, or diplomatic relations with a state in order to coerce that state to conform to internationally accepted standards of behavior. Its authority is directed toward Member States, not the state that has created a threat to international peace and security. Second, Article 41 does *not* provide that the enumerated examples may not constitute a use of force within the

---

[140] *Id.* at 113, 611-12.

[141] U.N. CHARTER art. 41.

meaning of Article 2(4). It simply provides that they are 'measures not involving the use of armed force' that states can be required to employ in contrast to measures involving the use of armed force in Article 42 that the Security Council does not have the authority to require Member States to deploy. As discussed in Chapter 6, *infra*, political and economic coercion may be a use of force within the meaning of Article 2(4) which may constitute an 'armed attack' within the meaning of Article 51, thus invoking a state's right of self-defense.

In contrast, Article 42 authorizes the Security Council to conduct or authorize belligerent military operations against an aggressor state when such actions are necessary to maintain or restore international peace and security. This article provides that

> Should the Security Council consider that measures provided for in Article 41 would be inadequate or have proved to be inadequate, it may take such action by air, sea, or land forces as may be necessary to maintain or restore international peace and security. Such action may include demonstrations, blockade, and other operations by air, sea, or land forces of Members of the United Nations.[142]

In contrast to Article 41, which explicitly authorizes the Security Council to *require* Member States to enforce economic and political sanctions against an aggressor state, Article 42 only permits the Security Council to *authorize* Member States to use armed military force against an aggressor state.[143]

---

[142] U.N. CHARTER art. 42.

[143] THE CHARTER OF THE UNITED NATIONS: A COMMENTARY, *supra* note 70, at 633.

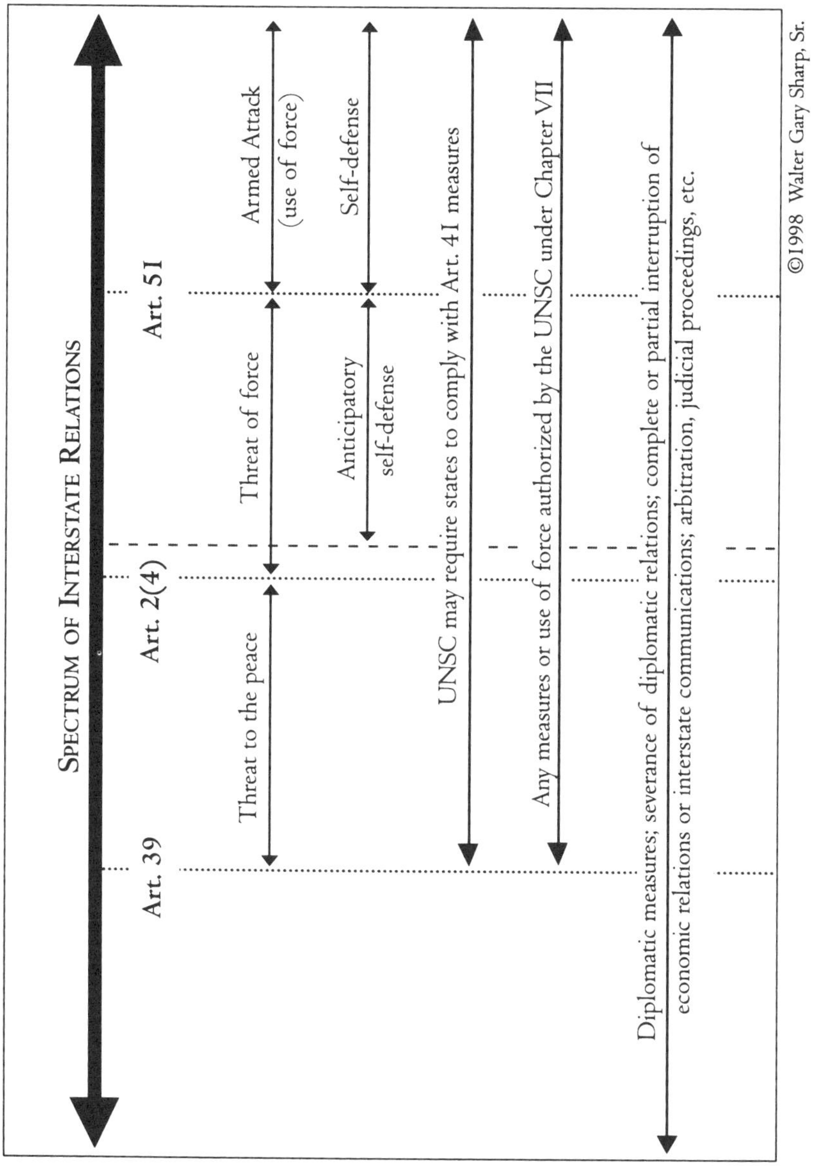

Figure 4. The Charter *Jus ad bellum* Paradigm

Conclusion

There is no mechanical rule in the law of conflict management that clearly defines what a use of force is under all circumstances. What constitutes a prohibited 'threat or use of force' is a question of fact that must be subjectively analyzed in each case in the context of all relevant law and circumstances. Such a question of fact defies rote, categorical definition. For example, a state's activity that is lawful against another state of equal size may nevertheless be significant enough against a smaller state to cross Article 2(4)'s "political independence" threshold. Accordingly, a state's activities in cyberspace, like all of its activities anywhere else, must be analyzed within the framework of the law of conflict management and state practice.

Figure 4 graphically depicts the interrelationships of Articles 2(4), 39, 41, 42, and 51. This figure will serve as the foundation upon which model will be built in later chapters to analyze use of force issues. It highlights along the spectrum of interstate relations the varying thresholds of Articles 2(4), 39, and 51. It also portrays graphically the corresponding rights of states to use force in self-defense and the authorities of the Security Council to take actions to maintain international peace and security.

This chapter was intended to provide a historical context and an understanding of the basic principles of the law of conflict management. A detailed analysis of how the law of conflict management and state practice defines a use of force and an armed attack is provided in Chapter 6, *infra*. During all subsequent discussions of the law of conflict management, it must be remembered that although a state never loses its right to use force in self-defense in response to a use of force within the meaning of Article 2(4), the right of self-defense under customary international law may not always justify an armed response.

CHAPTER 4: THE COMMON ARTICLE 2 THRESHOLD†

The four Geneva Conventions contain several common articles. Article 2 in each describes the circumstances under which the Conventions apply—that is, the circumstances of armed conflict. The three circumstances are a declared war, a partial or total occupation, or the existence of de facto *hostilities. As the first two situations are relatively straightforward, most debate centers on the nature of* de facto *hostilities. In his Commentary to the Conventions, Jean Pictet formulated his famous "scope/duration/intensity" analysis, stating that when each of these three criteria rise above a certain threshold, then a state of armed conflict does exist. This is the international standard for judging whether any given threat or use of force, or combination of them, rise to the level of armed hostilities and implicate the* jus in bello.

Evolution and Significance of the Threshold

The law of war, *jus in bello,* also commonly referred to as the law of armed conflict, governs the actual conduct of hostilities and has developed as customary international law through the practice of almost all societies over thousands of years—from the era of the Greeks and Romans to the Middle Ages and the twentieth century.[144] The practice of codifying the law of armed conflict in binding international agreements began in the nineteenth century with the 1856 Paris Declaration on Maritime War.[145] Codification accelerated at the turn of the 20th century,[146] and the law of armed conflict has generally developed in two regimes: the Hague

† This chapter is drawn substantially from Chapter 4 of WALTER GARY SHARP, SR., CYBERSPACE AND THE USE OF FORCE (1999).

[144] ROBERTS & GUELFF, *supra* note 4, at 1-2; I FRIEDMAN, *supra* note 63, at 3.

[145] ROBERTS & GUELFF, *supra* note 4, at 2-4.

[146] ROBERTS & GUELFF *supra* note 4, at 3.

regulations that govern the means and methods of warfare, and the Geneva conventions that govern the protection of victims of war.[147]

Most notable of the regime of Geneva conventions are the four Geneva Conventions of 1949 developed at the initiative of the

---

[147] *See generally id.* Nevertheless, despite the development of these extensive regimes and the threat of prosecution, barbaric cruelty remains a characteristic of twentieth century warfare. One study commissioned in 1913 by the Carnegie Endowment for International Peace recorded the atrocities among the Turks, Serbs, and Greeks during the Balkan war that ended in the early twentieth century:

> These soldiers all state that everywhere they burned the Bulgarian villages. Two boast of the massacre of prisoners of war. One remarks that all the girls they met with were violated. Most of the letters dwell on the slaughter of noncombatants, including women and children. "Here we are burning the villages and killing the Bulgarians, both women and children. We picked out their eyes (five Bulgarian prisoners) while they were still alive." ... From Kukush to the Bulgarian frontier the Greek Army devastated the villages, violated the women and slaughtered the noncombatant men.

United States v. List, 11 TRIALS OF WAR CRIMINALS BEFORE THE NUERNBERG MILITARY TRIBUNALS 759, 1218-19 (1950). Even during the war in the Balkans that began in 1992, as many as 100,000 women were taken hostage and systematically raped in an effort to defile and impregnate them so they would not be accepted back into their community. Patricia Forestier, *Psychiatric Genocide! How the Barbarities of 'Ethnic Cleansing' Were Spawned by Psychiatry*, FREEDOM, May 1993, at 6, 6-11, 34-35. Other atrocities included soldiers who burned families alive in their homes, crushed the heads of young children, and raped pregnant mothers in front of their families. Rod Nordland, *'Let's Kill the Muslims!'*, NEWSWEEK, Nov. 8, 1993, at 48, 48-49.

All of these and many other atrocities are clearly prohibited by the *jus in bello*. *Their continued existence is not a failure of the rule of law, but a failure of the international community to enforce the law.* Despite the serious war crimes that were committed during a number of international armed conflicts such as Korea, Vietnam, Palestine, Pakistan-Bangladesh-India, Cyprus, Lebanon, and the Persian Gulf, the international community remains reluctant to take war crimes prosecutions seriously. The creation and work of the International Criminal Tribunal for the former Yugoslavia and the recent developments on an International Criminal Court are hopeful steps.

International Committee of the Red Cross (ICRC).[148] These four conventions apply during international armed conflict, and deal with the following four categories of victims of war, respectively: wounded and sick in armed forces in the field; wounded, sick and ship-wrecked in armed forces at sea; prisoners of war; and, civilians.[149] The four Geneva Conventions of 1949 are adhered to by more states than any other agreement on the laws of armed conflict and are declaratory of customary international law.[150]

These conventions are linked by certain general principles and by common articles that are found throughout each of the four conventions.[151] Common Article 2 governs the application of the four Geneva Conventions of 1949 and is widely accepted as the threshold test

---

148 ROBERTS & GUELFF *supra* note 4, at 169. Collectively, these four conventions are referred to as the four Geneva Conventions of 1949. Individual citations are as follows: Geneva Convention for the Amelioration of the Condition of the Wounded and Sick in Armed Forces in the Field, Aug. 12, 1949, 6 U.S.T. 3114, 75 U.N.T.S. 31 [hereinafter 1949 Geneva Convention No. I], *reprinted in id.* at 171; Geneva Convention for the Amelioration of the Condition of the Wounded, Sick, and Shipwrecked Members of Armed Forces at Sea, Aug. 12, 1949, 6 U.S.T. 3217, 75 U.N.T.S. 85 [hereinafter 1949 Geneva Convention No. II], *reprinted in* ROBERTS & GUELFF *supra* note 4, at 194; Geneva Convention Relative to the Treatment of Prisoners of War, Aug. 12, 1949, 6 U.S.T. 3316, 75 U.N.T.S. 135 [hereinafter 1949 Geneva Convention No. III], *reprinted in* ROBERTS & GUELFF, *supra* note 4, at 216; Geneva Convention Relative to the Protection of Civilian Persons in Time of War, Aug. 12, 1949, 6 U.S.T. 3516, 75 U.N.T.S. 287 [hereinafter 1949 Geneva Convention No. IV], *reprinted in* ROBERTS & GUELFF, *supra* note 4, at 272.

149 ROBERTS & GUELFF, *supra* note 4, at 169.

150 *Id.* at 169-170.

151 *Id.* at 169.

for when an international armed conflict exists, and consequently, the application of the law of armed conflict in its entirety.[152]

Common Article 2 is thus the transition point between peace and armed conflict that determines when the law of armed conflict applies. This transition point, which determines when an armed conflict begins under the Common Article 2 threshold, corresponds to the Charter's Article 51 threshold of an armed attack. Consequently, an understanding of the Common Article 2 threshold helps to define what is a use of force and an armed attack under the law of conflict management.

Defining an Armed Attack

Common Article 2 invokes the provisions of the Conventions upon one of three factual conditions. Accordingly, an international armed conflict exists upon the declaration of war, the occurrence of "any other armed conflict" between two or more contracting parties even if the state of war is not recognized by one of them, and in all cases of partial or total occupation even if met with no armed resistance.[153] The existence of international armed conflict and the corresponding application of the law of armed conflict in cases of declared war[154] or occupation[155] is normally self-evident.

---

[152] *See id.* at 169-170; Commentary on the Geneva Convention Relative to the Protection of Civilian Persons in Time of War 17-21 (Jean S. Pictet ed., 1958) [hereinafter 1949 Geneva Convention No. IV, Commentary].

[153] *See* 1949 Geneva Convention No. IV, *supra* note 148, *reprinted in* ROBERTS & GUELFF, *supra* note 4, at 272; 1949 Geneva Convention No. IV Commentary, *supra* note 152, at 20.

[154] Although a declaration of war is not required to establish the existence of international armed conflict, such a declaration does define a legal state of armed hostilities between states even in the absence of the use of force. US DEP'T OF THE ARMY, FIELD MANUAL 27-10, THE LAW OF LAND WARFARE, § 8 (1956) [hereinafter FM 27-10].

Although the terms 'war' and 'armed conflict' are frequently used interchangeably and refer to a state of hostilities that invokes the law of armed conflict, war refers to a state of *de jure* hostilities invoked by a formal declaration of one party[156] that creates an international armed conflict as a matter of law.[157] In contrast, 'any other armed conflict" refers to a state of *de facto* hostilities invoked by the use of force by one party without any formal declaration of war.[158] Under exceptional circumstances, such as the invasion of Kuwait by three Iraqi Republican Guard Forces Command divisions on August 2, 1990,[159] *de facto* hostilities may also be self-evident. Without such exceptional circumstances, however, determining when "any other armed conflict" exists is a factual, subjective determination that centers on the use of force between two states.

The commentary of the 1949 Geneva Convention No. IV published by the ICRC describes the Common Article 2 threshold for *de facto* hostilities as follows:

> *Any difference* arising between two States and leading to the intervention of members of the armed forces is an armed

---

[155] Military occupation is a question of fact that "presupposes a hostile invasion, resisted or unresisted, as a result of which the invader has rendered the invaded government incapable of publicly exercising its authority, and that the invader has successfully substituted its own authority for that of the legitimate government in the territory invaded." *Id.* at § 355.

[156] ROBERTS & GUELFF, *supra* note 4, at 1-2.

[157] FM 27-10, *supra* note 154, at § 8. For an excellent historical discussion of the evolution and meaning of *de jure* and *de facto* wars, *see* KOTZSCH, CONCEPT OF WAR, *supra* note 60, at 36-65.

[158] ROBERTS & GUELFF, *supra* note 4, at 1-2.

[159] For a detailed account of the Iraqi invasion of Kuwait and the response of the international community, *see* CONDUCT OF THE PERSIAN GULF WAR: FINAL REPORT, *supra* note 112.

> conflict within the meaning of Article 2, even if one of the Parties denies the existence of a state of war. It makes *no difference how long the conflict lasts,* or *how much slaughter takes place.*[160]

Although not very clearly defined, this explanation depicts a low Common Article 2 threshold for the application of the Conventions designed to afford maximum protection to nonbelligerents and belligerents by ensuring the Conventions apply to as many hostile interventions between the members of the armed forces of two states as possible.[161]

This ICRC Commentary of Common Article 2 does offer, however, three criteria that add structure to a factual analysis of the existence of *de facto* hostilities. The emphasized terms '[a]ny difference', 'no difference how long the conflict lasts', and 'no difference ... how much slaughter takes place' provide that the *scope, duration,* and *intensity* of a use of force between the members of the armed forces of two states are the central factors in determining the existence of *de facto* hostilities. Based upon this ICRC Commentary of the Common Article 2 threshold, *de facto* hostilities exist, and consequently, the *jus in bello* applies, when almost <u>any</u> use of force—regardless of its scope, duration, <u>or</u> intensity—occurs between the members of the armed forces of two states.

Although this ICRC description is intended to create a very low threshold for the application of the law of armed conflict, the test remains situational and must be applied in the context in which the use of force between members of the armed forces of two states is actually used. Clearly, for example, the authorized use of force by the military police of

---

[160] 1949 Geneva Convention No. IV Commentary, *supra* note 152, at 20. (emphasis added).

[161] *See id.* at 17-21; HOWARD S. LEVIE, 59 US NAVAL WAR COLLEGE INTERNATIONAL LAW STUDIES: PRISONERS OF WAR IN INTERNATIONAL ARMED CONFLICT 14-26 (1977).

one state on one of its military installations to apprehend a visiting member of the armed forces of another state who is intoxicated and trespassing does not create *de facto* hostilities between the two states.

The United States currently utilizes a conjunctive scope, duration, and intensity analysis in its determination of belligerent status. This approach has evolved over the last fifteen years from the disjunctive threshold of the ICRC commentary on Common Article 2. On December 3, 1983, two unarmed US Navy planes flying in support of the U.N. Multinational Force in Lebanon were fired upon by Syrian anti-aircraft guns and surface-to-air missiles.[162] When the United States responded the next day with airstrikes against the Syrian positions from which anti-aircraft fire had come, US Navy Lieutenant Robert O. Goodman, Jr., was shot down and held by Syria.[163] The issue of his status as a prisoner of war under the 1949 Geneva Convention No. III was raised, and the US Department of State issued the following press guidance:

> *Question*—Is the captured airman a "prisoner of war" under the Third Geneva Convention?
>
> *Answer*—Yes. The Third Geneva Convention accords "prisoner-of-war" status to members of the armed forces who are captured during "armed conflict" between two or more parties to the Convention. "Armed conflict" includes any situation in which there is hostile action between the armed forces of two parties, regardless of the duration, intensity or scope of the fighting and irrespective of whether a state of war exists between the two parties. The Third Convention also provides for the prompt return of prisoners when "active hostilities" have ceased. As we have made clear, the incident that occurred between our forces and those of Syria has

---

[162] 3 CUMULATIVE DIGEST OF US PRACTICE IN INTERNATIONAL LAW: 1981-88, at 3456 (Marian Nash (Leich) ed., 1995) [hereinafter CUMULATIVE DIGEST].

[163] *Id.* at 3456.

> terminated, and we are hopeful that Syria will promptly return our airman and cooperate with the Government of Lebanon in resolving the problems of that country.[164]

Syria released Lieutenant Goodman on January 3, 1984.[165]

In contrast, a situation, occurring ten years earlier, similar to Lieutenant Goodman's resulted in a determination that a downed pilot was not a prisoner of war, but an unlawful detainee. On October 3, 1993, US Army Rangers operating in support of the expanded United Nations Operations in Somalia raided a house in Mogadishu in an attempt to arrest General Mohamed Farah Aidid for previous attacks on United Nations personnel conducting humanitarian relief operations.[166] When the raid was over, nineteen United Nations personnel were killed, and one US Army helicopter pilot, Chief Warrant Officer 3 Michael Durant, was shot down and held by Somali clansmen.[167] The issue of his status as a prisoner of war under the 1949 Geneva Convention No. III was raised, and the United States determined that he was an unlawful detainee because the United Nations operations did not rise to the level of a *de facto* international armed conflict that would trigger the *jus in bello*.[168] Warrant Officer Durant was released on October 14, 1993.[169]

---

[164] *Id.* at 3456-57. (emphasis added).

[165] *Id.* at 3457.

[166] U.N. DEP'T OF PUBLIC INFO., THE UNITED NATIONS AND SOMALIA: 1992-1996, at 12, 50-54, U.N. Doc. DPI/1677, U.N. Sales No. E. 96.I.8 (1996) [hereinafter THE UNITED NATIONS AND SOMALIA]. This text is Vol. VIII of the United Nations Blue Book Series.

[167] *Id.* at 54-55.

[168] *See* Steven J. Lepper, *The Legal Status of Military Personnel in United Nations Peace Operations: One Delegate's Analysis*, 18 Hous. J. Int'l L. 359, 362-64 (1996).

[169] THE UNITED NATIONS AND SOMALIA, *supra* note 166 at 331.

During the ratification process of the Chemical Weapons Convention, the United States adopted its current position, which raises the threshold described by the ICRC commentary. The Chemical Weapons Convention provides that "[e]ach State Party undertakes not to use riot control agents as a method of warfare."[170] On April 24, 1997,[171] the US Senate conditioned its consent to the Chemical Weapons Convention on the requirement that:

> (26) RIOT CONTROL AGENTS.
> (A) PERMITTED USES. Prior to the deposit of the United States instrument of ratification, the President shall certify to Congress that the United States is not restricted by the Convention in its use of riot control agents, including the use against combatants who are parties to a conflict, in any of the following cases:
> (i) UNITED STATES NOT A PARTY. The conduct of peacetime military operations within an area of ongoing armed conflict when the United States is not a party to the conflict (such as recent use of the United States Armed Forces in Somalia, Bosnia, and Rwanda).
> (ii) CONSENSUAL PEACEKEEPING. Consensual peacekeeping operations when the use of force is authorized by the receiving state, including operations pursuant to Chapter VI of the United Nations Charter.
> (iii) CHAPTER VII PEACEKEEPING. Peacekeeping operations when force is authorized by the Security Council under Chapter VII of the United Nations Charter.[172]

---

[170] Chemical Weapons Convention, *supra* note 13.

[171] Helen Dewar, *Senate Approves Chemical Arms Pact After Clinton Pledge*, WASH. POST, Apr. 25, 1997, at 1. The treaty was approved by the Senate by a vote of 74 to 26. *Id.*

[172] S. Exec. Res. 75, 105th Cong., CONG. REC. S3378 (daily ed. Apr. 17, 1997). The Senate gave its advice and consent to the Chemical Weapons Convention subject to a total of 28 conditions. *Id.* at S3373-79.

In response, the President's certification of April 25, 1997, provided that:

> In accordance with Condition (26) on Riot Control Agents, I have certified that the United States is not restricted by the Convention in its use of riot control agents in various peacetime and peacekeeping operations. These are situations in which the United States is not engaged in a use of force of a scope, duration and intensity that would trigger the laws of war with respect to US forces.[173]

Based upon this "shared understanding" between the Senate and the President,[174] the United States specifically adopted a scope, duration, and intensity analysis to determine when *de facto* hostilities exist.

The United States' position is an example of the evolution of state practice that reflects the contemporary difficulty of a factual determination for *de facto* hostilities that arises because military forces are

---

[173] President's Certifications and Report to the Congress in Connection with the US Senate Resolution of Ratification of the Chemical Weapons Convention (Apr. 25, 1997), *available in White House Virtual Library* (last modified May 1, 1997) <http://library.whitehouse.gov>. The United States deposited its instrument of ratification on April 25, 1997, and the Chemical Weapons Convention entered into force on April 29, 1997, with eighty-eight of 165 signatory states having ratified. *Signatory States to the Chemical Weapons Convention* (last modified Apr. 29, 1997) <http://www.opcw.nl/memsta/namelist.htm> (emphasis and italics added).

[174] As a result of the reinterpretation controversy of the *Treaty on the Limitation of Anti-Ballistic Missile Systems* in the early seventies, a controversial constitutional principle was asserted by the Senate, and subsequently agreed upon by the Executive Branch, that "the meaning [of a treaty] is to be determined in light of what the Senate understands the treaty to mean when it gives its advice and consent." NATIONAL SECURITY LAW 194-196 (Stephen Dycus *et al.* eds.) (1997). Although a considerable debate continues as to the breadth of the binding effect of this principle, *see id.* at 195-98, the Executive Branch acknowledged and was very aware of the binding effect of a "shared understanding" that would be created by its communications with the Senate during the ratification process of the Chemical Weapons Convention.

authorized to use force in self-defense[175] and to accomplish their mission during the conduct of peacetime military operations[176] that may occur in an area of ongoing hostilities.[177] These peacetime military operations include law enforcement, peacekeeping, humanitarian and disaster relief, counter-terrorist, hostage rescue, and noncombatant evacuation operations.[178] A rote application of the Common Article 2 threshold allows a limited and intermittent use of force to invoke a state of *de facto* hostilities and unintentionally causes military personnel to become belligerents and lawful targets even though they are conducting a peacetime military operation. State practice has demonstrated this rote application of international law to be politically unacceptable to the international community.[179]

As a result of this unintended consequence when the Common Article 2 threshold is applied to peacetime military operations, particularly those under the authority of the United Nations, the international community has rejected the disjunctive threshold of the

---

[175] *See* U.N. CHARTER art. 51.

[176] Walter Gary Sharp, Sr., *Protecting the Avatars of International Peace and Security,* 7 DUKE J. COMP. & INT'L L. 93, 105 (1996) [hereinafter *Protecting the Avatars*]; Umesh Palwankar, *Applicability of International Humanitarian Law to United Nations Peace-keeping Forces,* 294 INT'L REV. OF THE RED CROSS 227 (May-June 1993) *excerpted in* WALTER GARY SHARP, SR., UNITED NATIONS PEACE OPERATIONS 145, 146 (1995).

[177] Although cast as a contemporary issue, the international community was also concerned in the early 1960s over the possibility that the United Nations peace-keeping forces in the Congo may have become involved in hostilities that invoked the *jus in bello.* FINN SEYERSTED, UNITED NATIONS FORCES IN THE LAW OF PEACE AND WAR 60, 180 (1966).

[178] *See* JOINT PUB 3-07, JOINT DOCTRINE FOR MILITARY OPERATIONS OTHER THAN WAR, III-1 (1995).

[179] For a more detailed examination of this state practice, see Sharp, Sr., *Protecting the Avatars, supra* note 176, at 93.

ICRC Commentary as too restrictive. For example, one scholar concludes that an armed attack occurs, and thus an armed conflict exists, "when force is used on a relatively large scale and with substantial effect."[180] Another scholar concludes that a use of force must reach a certain "gravity" as a matter of law, and that minor frontier incidents are not *per se* a use of force that rises to the Article 2(4) threshold because "the minor nature of an attack is prima facie evidence of absence of intention to attack, of honest mistake, or simply the limited objectives of an attack."[181]

The international community now subjectively analyzes the use of force by an independent state, and whether such force constitutes an armed attack, in the context of its scope, duration, and intensity. Accordingly, *de facto* hostilities exist between two states when the scope, duration, and intensity of force between them reaches the level of an armed attack within the meaning of Article 51 of the Charter. The conjunctive threshold does not add much more clarity to what precise level or quantum of force is required to trigger the Common Article 2 threshold. In practice, however, the application of the conjunctive test recognizes a more reasoned balancing of all of the circumstances surrounding the use of force.

Conclusion

Figure 5 graphically summarizes these principles of existing international law. It depicts the line of belligerency as that point on the use of force spectrum at which international armed conflicts begin and the law of armed conflict applies. This point is triggered by the declaration of war, the occurrence of *de facto* hostilities, and all cases of partial or total occupation. *De facto* hostilities exist when the use of force between two states reaches the scope, duration, and intensity that state practice defines

---

[180] THE CHARTER OF THE UNITED NATIONS: A COMMENTARY, *supra* note 70, at 669.

[181] *See* BROWNLIE, INTERNATIONAL LAW, *supra* note 91, at 366.

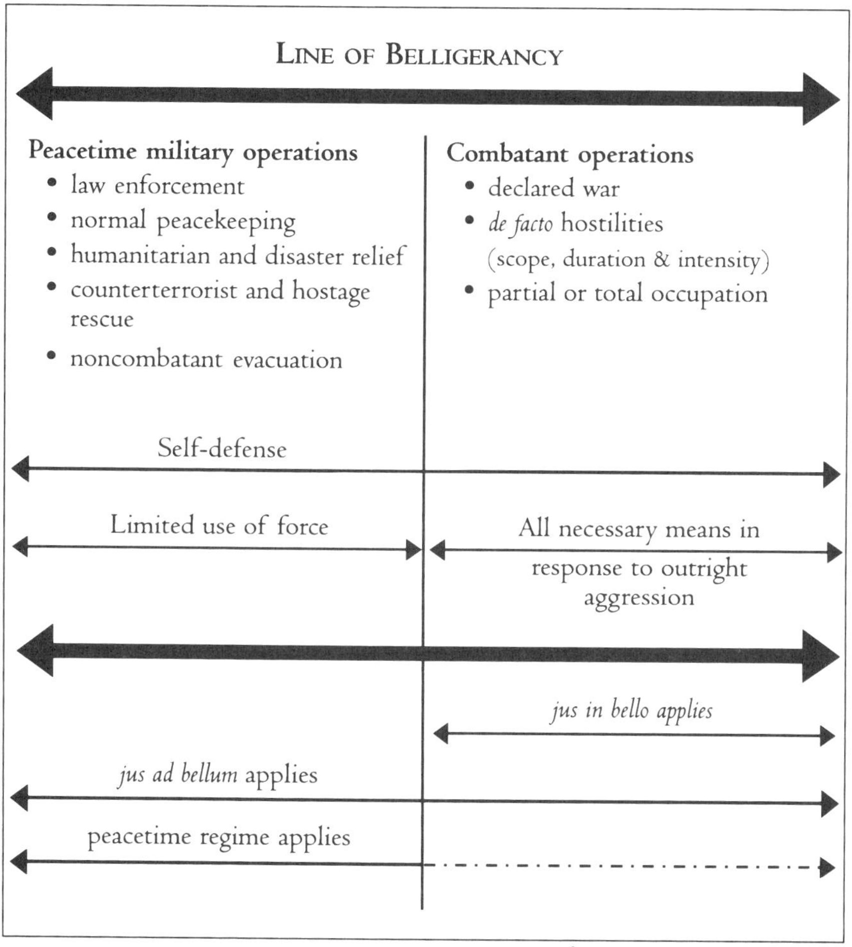

Figure 5. The Application of International Law

as an armed attack. The law of conflict management applies during peacetime and armed conflict, and the obligations of the parties to an armed conflict under the peacetime regime of international law remain in place to the extent they are not inconsistent with a state of hostilities. Accordingly, the line that represents the application of the peacetime regime during armed conflict is dotted.

In summary, short of an actual declaration of war or a case of partial or total occupation, a state does not become a party to an international armed conflict until such time as it becomes engaged in a use of force of a scope, duration, and intensity between them that reaches the level of an armed attack within the meaning of Article 51 of the Charter. What constitutes a use of force of a scope, duration, and intensity that constitutes an armed attack and triggers the law of armed conflict is a question of fact that must be subjectively analyzed in each and every case in the context of all relevant law and circumstances.

This question of fact also defies rote, categorical definition. For example, military forces conducting a noncombatant evacuation operation do not become a party to an armed conflict when they use limited force to rescue personnel. Similarly, military forces serving under the authority of the United Nations do not become a party to an armed conflict when they use limited force to accomplish an assigned humanitarian relief or peace operation. In contrast, individual or collective military action in response to outright aggression, such as the coalition response to the Iraqi aggression that led to the Persian Gulf War, does cross the Common Article 2 threshold and trigger the application of the *jus in bello*. This chapter was intended to provide a historical context and an understanding of the basic principles surrounding the Common Article 2 threshold and the interrelationship between an armed attack and armed conflict. A detailed analysis of how this threshold specifically applies in cyberspace is provided in Chapter 6, *infra*.

CHAPTER 5: DELINEATING THE RULES[†]

Before starting an exhaustive review of the rules governing uses of force in cyberspace, it is important to understand how international lawyers determine just what those rules are. Domestic law in most countries is the product of a carefully worked out civil code; in the English-speaking world, case law, supported by statute, is determinative. At the international level, however, the law is often less well defined and frequently more poorly enforced. For these reasons, it is important to have an agreed-upon system for separating the aspirational from the authoritative.

Sources of International Law

The general rule of international law and sovereignty is that states are only bound by those rules of international law to which they themselves have agreed.[182] Even the coercive authority of the Security Council is based upon the Member States' agreement that it has the authority to direct nonconsensual, binding measures in matters of international peace and security.[183] This principle is limited only to the extent that state practice develops into customary international law, which is binding on all states.[184]

---

† This chapter is drawn substantially from Chapter 5 of WALTER GARY SHARP, SR., CYBERSPACE AND THE USE OF FORCE (1999).

182 THE CHARTER OF THE UNITED NATIONS: A COMMENTARY, *supra* note 70, at 83.

183 *See* U.N. CHARTER arts. 25, 39-42, 48-49.

184 THE CHARTER OF THE UNITED NATIONS: A COMMENTARY, *supra* note 70, at 83. Customary international law results from a general and consistent practice of states followed by them out of a sense of obligation. This creates an interesting cycle of policy decisions that create precedent, which in turn influences policy and creates customary international law over time.

Article 38(I) of the Statute of the International Court of Justice, one of the six principal organs of the United Nations, requires the Court to apply

> a. international conventions, whether general or particular, establishing rules expressly recognized by the contesting states;
> b. international custom, as evidence of a general practice accepted as law;
> c. the general principles of law recognized by civilized nations;
> d. subject to the provisions of Article 59, judicial decisions and the teachings of the most highly qualified publicists of the various nations, as subsidiary means for the determination of rules of law.[185]

This article is recognized as the most authoritative description of the sources of international law.

The first three categories of Article 38(I) are not listed in an order of legal hierarchy which mandates the application of one over the other,[186] but generally speaking, international conventions may be more specific and on point in a given interstate controversy than international custom. Similarly, international custom may be more relevant to an interstate dispute than general principles of national law. The unique nature of a dispute between two contiguous states, however, may make the general principles of the national law of the parties not only more

---

[185] STATUTE OF THE INTERNATIONAL COURT OF JUSTICE, art. 38(I) (emphasis added). The Statute was adopted in 1945 as an annex to the Charter. *See* U.N. CHARTER art. 92. All Member States are parties to the Statute of the ICJ by operation of art. 93 of the Charter. Only states may be parties in cases before the International Court of Justice. STATUTE OF THE ICJ , art. 34. Article 59 of the STATUTE OF THE ICJ provides that the decision of the ICJ has "no binding force except between the parties and in respect of that particular case," *i.e.*, the principle of *stare decisis* does not apply to decisions of the ICJ.

[186] THE CHARTER OF THE UNITED NATIONS: A COMMENTARY, *supra* note 70, at 992.

relevant, but controlling. The final category of Article 38(1) considers the judicial decisions and the teachings of the most highly qualified publicists of various nations as a subsidiary means for the determination of the rules of international law. These decisions and teachings "are considered to be indicative of the interpretation and application" of international law.[187]

Reliance Upon State Practice

No international convention, to include the Charter of the United Nations, defines "threat or use of force," "armed force," or "armed attack,"[188] and despite decades of earnest work, the international community has failed to define aggression with any clarity in any other document.[189] Existing international law that governs the use of force must therefore be derived from a heuristic analysis of how the Charter's use of force paradigm has been interpreted through state practice. Based upon this study of state practice, and how it is interpreted by highly qualified publicists, a profile of what is acknowledged by the international community as a use of force can be derived by studying the legal justifications and conclusions states make with regard to their activities.

States may often *explicitly* acknowledge what they consider a use of force, for example, when they declare the lawfulness of their activities in terms of the Charter's conflict management paradigm. For example, the United States justified the lawfulness of its August 1998 missile strikes against terrorist camps in Afghanistan and Sudan as an exercise of its right

---

[187] *Id.* at 992.

[188] *Id.* at 111-12.

[189] *See generally, e.g.*, THOMAS & THOMAS, CONCEPT OF AGGRESSION, *supra* note 87.

to self-defense as recognized by Article 51 of the Charter.[190] Chapter 3, *supra*, concluded that what constitutes a prohibited "threat or use of force" is a question of fact that must be subjectively analyzed in each and every case in the context of all relevant law and circumstances. A questioned state activity that is considered a use of force within the meaning of Article 2 is unlawful under international law unless it is

- an exercise of that state's right of individual or collective self-defense as authorized by international law and recognized by Article 51 of the Charter, or

- authorized by the Security Council acting under its Chapter VII coercive authority,

regardless of whether the activity occurred partially or completely in the physical world or cyberspace. Chapter 3 also concluded that a state never loses its right to use force in self-defense in response to a use of force within the meaning of Article 2(4), however, the right of self-defense under customary international law may not justify an armed response.

At other times, states may *implicitly* acknowledge what they consider a use of force when they declare the lawfulness of their activities by asserting conclusions, such as whether their activities have resulted in an international armed conflict or whether the law of armed conflict applies. From these conclusions, we can determine what states consider a use of force because a questioned state activity that has been generally accepted by the international community as having created the legal status of an international armed conflict or as having invoked the law of armed conflict, is for that reason considered a use of force prohibited by Article 2(4) of the Charter. The most obvious example that clearly supports this proposition is the 1990 Iraqi invasion of Kuwait. Not only did the

---

[190] Sam Skolnik, *The Law Behind the Bombs: Experts Debate legality of US Airstrikes Against Terrorist*, LEGAL TIMES, Aug. 24 and 31, 1998. Sudanese and Pakistani leaders denounced the missile strikes as violations of international law. *Id.*

international community immediately declare the invasion to be an unlawful use of force, it also concluded that Iraq's use of force invoked the legal status of an international armed conflict and the law of armed conflict. Chapter 4, *supra*, concluded that what constitutes a use of force of a scope, duration, and intensity that triggers the law of armed conflict is a question of fact that must be subjectively analyzed in each and every case in the context of all relevant law and circumstances.

The precise converse of this last proposition is not true under all circumstances for two reasons. First, those state activities that do not invoke the law of armed conflict obviously include noncoercive activities such as those inherent in normal diplomatic relations. Second, since international law evolves through state practice, which is motivated principally by political considerations and national interests,[191] the converse proposition may simply be politically unacceptable and rejected by United Nations and state practice. Accordingly, a questioned state activity which has been generally accepted by the international community as not having created the legal status of an international armed conflict or as not having invoked the law of armed conflict, will usually, but not always, be considered below the use of force threshold prohibited by Article 2(4).

For example, when Chief Warrant Officer 3 Michael Durant was shot down in October 1993 during a raid on General Aidid's stronghold and held by Somali clansmen,[192] the United States asserted that he was an unlawful detainee and not a prisoner of war because the United Nations operations did not rise to the level of a *de facto* international armed conflict

---

[191] MYRES S. MCDOUGAL & FLORENTINO P. FELICIANO, LAW AND MINIMUM WORLD PUBLIC ORDER: THE LEGAL REGULATION OF INTERNATIONAL COERCION 39-49 (1961).

[192] THE UNITED NATIONS AND SOMALIA, *supra* note 166, at 12, 50-55.

that would trigger the *jus in bello*.[193] It was politically unacceptable for the United States and the international community to characterize these forces as combatants during an international armed conflict, notwithstanding specific Security Council Chapter VII approval to use armed military force to arrest General Aidid.[194]

Similarly, consider the case of noncombatant evacuation operations that do not invoke the *jus in bello* but nevertheless are considered a use of force. Later in this text, Chapter 6 concludes that international law considers such a rescue of nationals a use of force within the meaning of Article 2(4), but, as previously discussed in Chapter 4, international law does not consider such activity to be of the scope, duration, and intensity that invokes an international armed conflict.

In summary, a use of force within the meaning of Article 2(4) will always invoke a victim state's right of self-defense, but it may not always rise to the scope, duration, and intensity threshold of an armed attack that invokes a *de facto* international armed conflict and the *jus in bello*. This is consistent with the universally accepted view that the thresholds of Articles 2(4) and 51 are different in scope; *i.e.*, the Article 2(4) concept of

---

[193] *See* Steven J. Lepper, *The Legal Status of Military Personnel in United Nations Peace Operations: One Delegate's Analysis*, 18 Hous. J. Int'l L. 359, 362-64 (1996).

[194] Although this raid serves as a useful example, this is obviously an over simplification of the legal analysis. One very important factor to consider is the fact that it was very debatable that General Aidid and his forces represented any thing other than an armed criminal band of thieves and not the organized military of another state. It was also very important to the conclusion above that the raid was authorized by the coercive authority of the Security Council as a use of force that served to protect peacekeepers and other civilians during United Nations humanitarian operations in Somalia. For an excellent review of the United Nations operations in Somalia and a comprehensive collection of almost all of the relevant documents, *see* THE UNITED NATIONS AND SOMALIA, *supra* note 166.

a 'threat or use of force' is much broader than the Article 51 threshold of 'armed attack.'[195]

Conclusion

There are significant similarities between the Article 2(4), Article 51, and the Common Article 2 thresholds, however, it should be reemphasized at this point that there is but one conflict management paradigm, and that is found in the Charter of the United Nations. The Common Article 2 threshold of the four Geneva Conventions of 1949 is not a separate paradigm of conflict management. A study of state practice and the writings of publicists, jurists, and scholars that interprets Common Article 2 simply helps define what quantum of force is necessary to rise to the level of the Article 2(4) and Article 51 thresholds.

Since there is no accepted norm within the international community that defines what is a lawful or unlawful use of force, it is helpful as state practice is reviewed to map it out in a linear model of interstate activities by building upon Figure 6. Figure 9 continues this development of a linear model for a heuristic use of force analysis by adding the Common Article 2 analysis discussed in Chapter 4 and this chapter to Figure 6. This will provide a profile of what will likely be considered by the international community to be use of force in cyberspace.[196]

---

[195] *See* THE CHARTER OF THE UNITED NATIONS: A COMMENTARY, *supra* note 70, at 663.

[196] For an excellent overview and critique of the application of the Charter paradigm to real-world uses of force, *see* ANTHONY CLARK AREND and ROBERT J. BECK, INTERNATIONAL LAW & THE USE OF FORCE (1993). For a an even broader perspective on the sources of international law and rule-making, *see* ANTHONY CLARK AREND, LEGAL RULES AND INTERNATIONAL SOCIETY (1999).

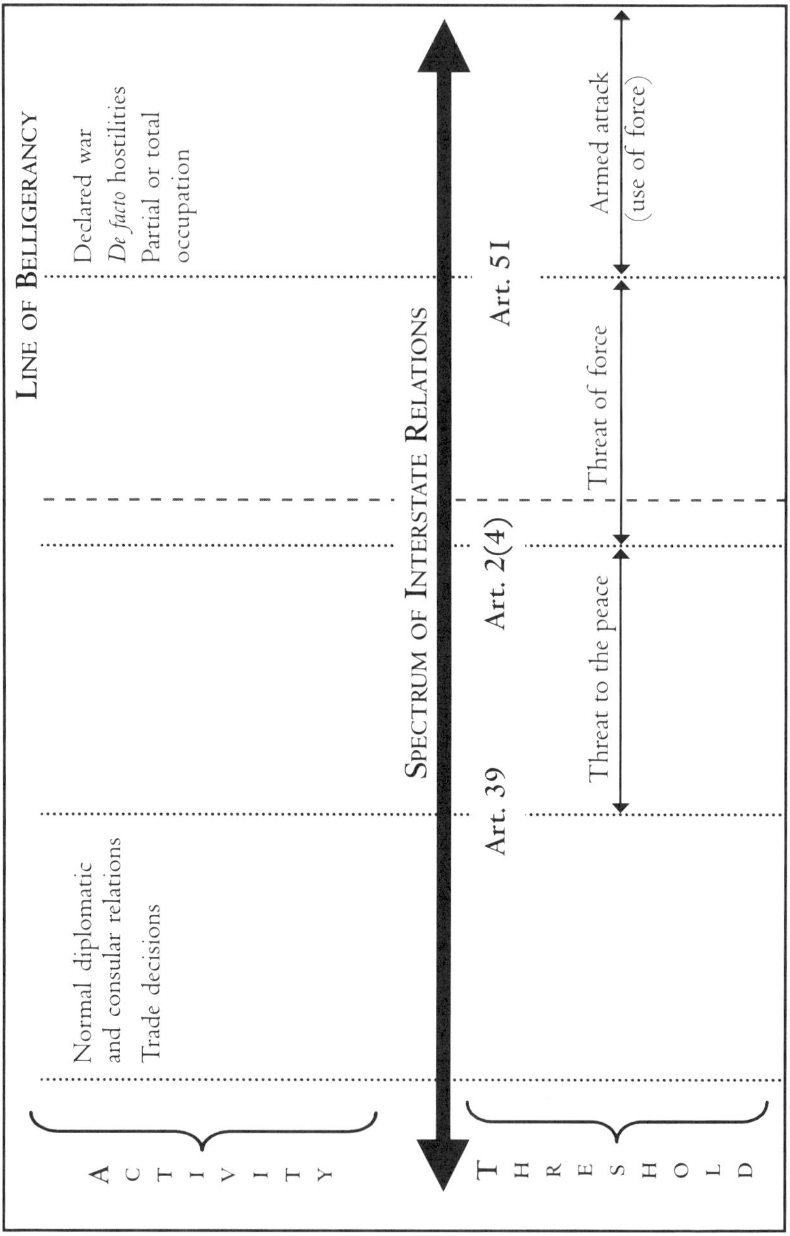

Figure 6. The Rules Defining the Use of Force

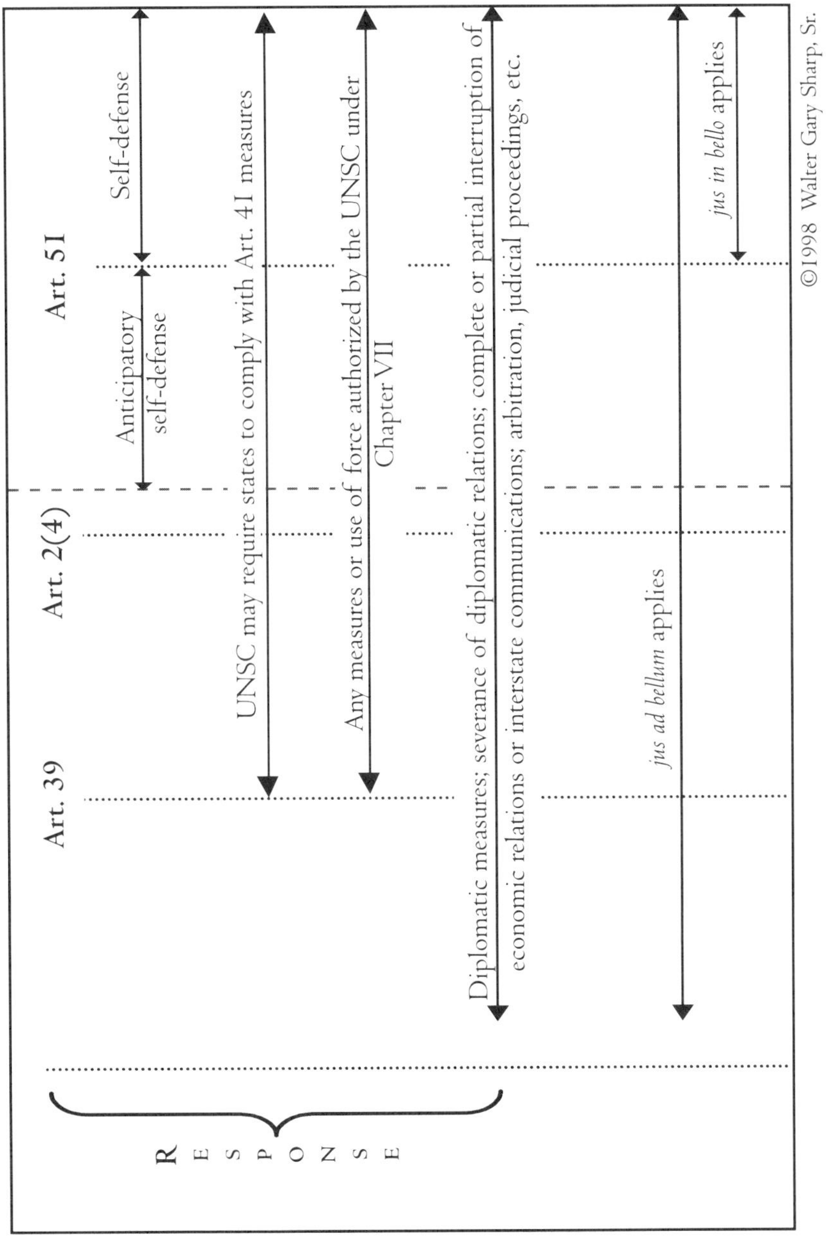

Figure 6. continued

The outer limits of this model can be plotted by starting with the polar extremes that are universally accepted examples of what does and does not constitute a use of force and an armed attack. For example, the left end should begin with full diplomatic and consular relations, because they are accepted as the desired status quo between two states and do not constitute a use of force. Similarly, states agree that a unilateral decision by a state not to trade with another is not a use of force, regardless of the potential impact of that trade decision. This end of the spectrum represents those activities that do not rise to the level of a scope, duration, and intensity that either constitutes a use of force or an armed attack.

At the other extreme, the model should end on the right side with the three factual conditions of the Common Article 2 threshold for when an international armed conflict exists. This end of the spectrum represents those activities that do clearly rise to the level of a scope, duration, and intensity that constitutes a use of force and an armed attack. What remains between these polar extremes will be profiled in the next chapter with a study of state practice.

CHAPTER 6: APPLICATION OF EXISTING RULES[†]

It is in applying the theoretical, academic principles of Articles 39, 2(4), and 51 to real-world operations that true legal standards develop. Within the meaning of Article 2(4), it is important to identify just what constitutes a threat of force, and what constitutes a use of force—both violating international law, but each allowing for a different level of response. Beyond these thresholds, the Article 51 trigger for the inherent right of self-defense is contingent on a use of force amounting to an armed attack. This chapter evaluates state practice to determine where these thresholds fall on the spectrum of conflict, and concludes with a two-page graphical representation of the author's conclusions.

The Preamble of the Charter of the United Nations provides that one of its four goals is to ensure "that armed force shall not be used, save in the common interest..."[197] To further this goal, all Member States agreed in Article 2(3) that they "shall settle their international disputes by peaceful means in such a manner that international peace and security, and justice, are not endangered."[198] The Charter provides for a number of institutional arrangements to settle disputes such as

- granting authority to the General Assembly to discuss any issue that relates to international peace and security or the welfare of the international community,
- a strong role for the Security Council in dispute settlement,

---

[†] This chapter is drawn substantially from Chapter 6 of WALTER GARY SHARP, SR., CYBERSPACE AND THE USE OF FORCE (1999).

[197] U.N. CHARTER pmbl.

[198] U.N. CHARTER art. 2(3).

- coercive authority of the Security Council when international peace and security is threatened, and

- the International Court of Justice.[199]

Article 2(3) is a legally binding *waiver* of the sovereign right of Member States[200] under international law to aggressively use force, and the states also agreed to enforce their obligation to settle their disputes peacefully with the prohibition on the aggressive use of force found in Article 2(4).

Modern international relations, however, is based on a sovereign nation-state system and is "motivated by national interest as its ultimate purpose."[201] These national interests, and the perceived national interests, of some 180 or more nation-states seem to cause interstate disputes and tensions to remain as a constant in the midst of countless diverse cultures, religions, and ideologies. An effective foreign policy takes into account these varying national interests, and "[d]iplomacy has many peaceful ways for bringing influence to bear" to resolve disputes.[202]

While the phrase "use of force" is commonly understood to include a military attack of one state by the organized military of another state, *i.e.*, an armed attack, some coercive state activities that fall short of an armed attack may also cross the thresholds of Article 2.[203] The phrase

---

199 *See* THE CHARTER OF THE UNITED NATIONS: A COMMENTARY, *supra* note 70, at 99.

200 The term "Member States" refers to states that are members of the United Nations and therefore a party to the Charter of the United Nations. The International Court of Justice explicitly held in its *Nicaragua* judgment of June 27, 1986, that the obligation under Article 2(3) is also customary international law. *Id.* at 100.

201 HENRY KISSINGER, DIPLOMACY 17 (1994).

202 EUGENE V. ROSTOW, TOWARD MANAGED PEACE 22 (1993).

203 BROWNLIE, INTERNATIONAL LAW, *supra* note 91, at 361-62.

"use of force" also applies to all agencies and agents of a state government, such as the organized military, militia, security forces, police forces, intelligence personnel, mercenaries, and other surrogate forces or volunteers.[204] Defining what constitutes a use of force in a full spectrum of potential state activity requires a thorough understanding of how Article 2 has been interpreted under international law. This chapter will explore state practice and the writings of jurists to develop an understanding of the Article 2(3) obligation as well as the Article 2(4) prohibition, and will then apply each of those discussions to state activities in cyberspace.

Obligation to Settle by Peaceful Means

A majority of writers suggest that "states are under an obligation to deploy active efforts for the settlement of their international disputes."[205] Accordingly, those actions must be conducted "in good faith and in a spirit of co-operation."[206] Article 2 does not mandate a specific result, only that the parties strive in good faith for the resolution of a dispute.[207] It is recognized that even with good will on both sides, some disputes cannot be resolved by the parties involved in the dispute.[208] Parties are expected to compromise and seek a common denominator in

---

[204] *Id.* at 361.

[205] THE CHARTER OF THE UNITED NATIONS: A COMMENTARY, *supra* note 70, at 101. A dispute arises when one state rejects a specific claim of another state; *i.e.*, this article does not impose upon states a legal obligation to live together "in full harmony and without any political antagonism." *Id.* at 102.

[206] *Id.*

[207] *Id.*

[208] *Id.*

their negotiations, but they are not required to behave in a manner that adversely affects their rights on balance.[209]

If the parties are unable to reach a resolution on their own, the General Assembly has the authority to make recommendations in the form of nonbinding resolutions.[210] Similarly, in matters that are "likely to endanger the maintenance of international peace and security," the Security Council has the authority under Chapter VI of the Charter to investigate and make nonbinding recommendations.[211] When the dispute becomes a threat to the peace or results in either a breach of the peace or an act of aggression, the Security Council has the obligation to make recommendations or decide what measures shall be taken by states to maintain or restore international peace and security.[212]

Peaceful means include seeking a solution by "negotiation, enquiry, mediation, conciliation, arbitration, judicial settlement, [and] resort to regional agencies or arrangements . . . ."[213] States have a sovereign right for other states not to intervene in matters that are essentially within their domestic jurisdiction,[214] and any activity that infringes upon that prohibition on intervention is not peaceful within the meaning of Article 2(3).[215] An exception applies to a state's obligation to settle their disputes

---

[209] *See id.* at 101-03.

[210] U.N. CHARTER arts. 14, 35.

[211] *See generally id.* at chap. VI.

[212] *Id.* at art. 39.

[213] *Id.* at art. 33.

[214] *Id.* at art. 2(1) and (7).

[215] THE CHARTER OF THE UNITED NATIONS: A COMMENTARY, *supra* note 70, at 104.

exclusively by peaceful means when the dispute has been triggered by an unlawful act of one of the parties involved.[216] In such cases, the aggrieved state's right to react is extended to permit retortion and reprisal.[217]

If Article 2(3) requires states to actively utilize peaceful means in good faith to settle their disputes, then it also prohibits states from persistently refusing "even to attempt to reach a settlement,"[218] participating in activities that further threaten international peace and security by jeopardizing the resolution of its dispute, and negotiating in bad faith. Similarly, it prohibits states from interfering in other states' disputes in such a manner that would delay settlement or otherwise threaten international peace and security. Finally, all articles must be interpreted in conjunction with the purposes of the Charter found in Article I.[219] Accordingly, any state activity that creates a dispute between two or more states would violate the Article 2(3) proscription.

---

[216] *Id.*

[217] *Id.* Retortion "connotes measures which do not infringe upon any rights of the affected state, thus constituting simple unfriendly acts." *Id.* at 104. Generally, retortion refers to when a state unlawfully subjects citizens of another state to severe and stringent regulations or harsh treatment in response to similar unlawful mistreatment of its own citizens. *See* BLACK'S LAW DICTIONARY, *supra* note 18, at 1183. A reprisal "is an act which is unlawful *per se*, unless it can be justified as a countermeasure triggered by an unlawful act and is designed to induce the offending state to return to full compliance with the law." *Id.* at 105. It should be emphasized that this right of reprisal no longer includes the right to unlawfully use force contrary to Article 2(4) of the Charter. A use of force within the meaning of Article 2(4) is unlawful unless it is justified by a state's right of self-defense or authorized by the Security Council under its coercive Chapter VII powers.

[218] THE CHARTER OF THE UNITED NATIONS: A COMMENTARY, *supra* note 70, at 101.

[219] U.N. CHARTER art. 2 provides that "[t]he Organization and its Members, in pursuit of the Purposes stated in Article I, shall act in accordance with the following Principles." *See also* THE CHARTER OF THE UNITED NATIONS: A COMMENTARY, *supra* note 70, at 50.

Since these prohibitions apply to all potential state activities, any state activity in cyberspace that

- is an act of bad faith as it pertains to that state's dispute settlement negotiations,
- otherwise threatens a peaceful settlement of that state's own dispute,
- interferes with a dispute of other states in such a manner that would delay or prevent their dispute settlement, or
- creates a dispute between other states that threatens international peace and security,

violates a Member State's obligations under Article 2(3). For example, it would violate Article 2(3) for a state to falsify one of its own computer databases and then 'allow' another state access to it via the internet in an effort to disrupt the dispute settlement negotiations. It would also violate Article 2(3) for a state to falsify one of its own computer databases and then 'allow' another state or states access to it via the internet in an effort to create a dispute between two or more states. This norm does not prohibit all forms of deception between states. During peacetime, only those forms of deception that specifically violate the obligations of Article 2(3) are unlawful under the law of conflict management. During armed conflict, deception is specifically permitted and governed by the law of armed conflict.

While these two hypothetical uses of the internet and cyberspace may violate a state's obligations under Article 2(3), they would not be a use of force under international law. If these hypothetical uses also caused, for example, physical destruction in another state, then all the factors taken together may constitute a use of force under international law. Physical destruction and other factors that may cause a state activity to become a use of force are discussed below. In sum, what is accepted as a peaceful means under Article 2(3) is not, *ab initio*, a use of force within the meaning of Article 2(4). The converse is not necessarily true, *i.e.*, a violation of a state's obligations under Article 2(3) is not *per se* a use of force.

Even if not a use of force, state activities that violate Article 2(3) may nevertheless carry with them adverse consequences. Accusations of illegality, or even impropriety, may cause a state to lose political support internationally or domestically, thereby undermining the strength of a state's negotiating position or jeopardizing a more favorable result. It may also divert attention from the actual dispute to a more collateral matter, and thereby serve the strategic advantage of an opposing state.[220]

Political and Economic Coercion

The scope of Article 2(4)'s prohibition of the use of force is subject to dispute. The prevailing view, according to one authoritative text, is that Article 2(4) does not prohibit "any possible kind of force, but is ... limited to armed force."[221] Political and economic coercion may be a lawful and "permissible means of exerting pressure on other states that violate international law."[222] Whether or not political and economic coercion is lawful depends upon an evaluation of the "means and ends of

---

[220] Among the many examples that can be given in support of this conclusion, consider the recent controversy over whether the UN Special Commission is using its good offices to conduct intelligence operations for the United States. Regardless of the truth or accuracy of these reports, the media and public focus is now on these allegations, drawing attention from much more serious violations of international law involving the manufacture and possession of chemical and biological weapons. Fueling the debate is a discussion of the purposes and principles of the United Nations that are similar to those couched in Article 2(3). For more details of this controversy, see Barton Gellman, *Annan Suspicious of UNSCOM Role*, WASH. POST, Jan. 6, 1999, at 1.

[221] THE CHARTER OF THE UNITED NATIONS: A COMMENTARY, *supra* note 70, at 112.

[222] *Id.*

the measure ... in relation to one another."[223] This view holds that political and economic coercion may be unlawful and may be a use of force, but it is simply not a use of force prohibited by Article 2(4).[224] One scholar concludes that "whilst it is correct to assume that paragraph 4 applies to force other than armed force, it is very doubtful if it applies to economic measures of a coercive nature."[225]

Most of the developing countries and the former Eastern bloc countries, however, have repeatedly claimed that Article 2(4) also prohibits political and economic coercion.[226] Perhaps the strongest argument in favor of the prevailing view is that the drafters of the Charter were only concerned with the use of military force, and that they explicitly rejected a proposal to extend the prohibitions of Article 2(4) to economic coercion.[227] Other scholars, however, have also reviewed the working documents of the drafters of the Charter, and concluded that they "do not indicate that the phrase ['use of force'] applied only to armed force..."[228]

The prevailing view also finds contextual support in the preamble, which provides that a goal of the Charter is to ensure "armed force" shall not be used except in the common interest."[229] Accordingly, Article 2(4)

---

[223] *Id.* at 113.

[224] *Id.* at 112-13.

[225] BROWNLIE, INTERNATIONAL LAW, *supra* note 91, at 362.

[226] THE CHARTER OF THE UNITED NATIONS: A COMMENTARY, *supra* note 70, at 112.

[227] *Id.*

[228] *See* BROWNLIE, INTERNATIONAL LAW, *supra* note 91, at 361-2.

[229] THE CHARTER OF THE UNITED NATIONS: A COMMENTARY, *supra* note 70, at 112.

must prohibit only the use of armed force.[230] This argument, however, is not persuasive because it focuses on only one goal and ignores the broader context of the Charter.

If the drafters required the proposal to extend the prohibitions of Article 2(4) to economic coercion they could have done so by using the term armed force. Instead, they chose to use only the phrase "threat or use of force" while at the same time using the term armed force more than once elsewhere in the Charter. More importantly, the preamble and the purposes of Article 1 describe a Charter that is concerned with far more than simply the use of armed force. The Charter is concerned with human rights, social progress, justice, standards of life, tolerance, and living together in peace as good neighbors. In light of these goals and purposes, consider the final clause of Article 2(4):

> (4) All Members shall refrain in their international relations from the threat or use of force against the territorial integrity or political independence of any state, or in any other manner inconsistent with the Purposes of the United Nations.

A more balanced, contextual interpretation of Article 2(4) concludes that political and economic sanctions may indeed be a source of threats to international peace and security and an unlawful use of force that can threaten the territorial integrity and political independence of a state in violation of Article 2(4).[231]

One scholar concludes, for example, that the prohibition in Article 2(4) may be violated when a state undertakes to use its political and economic power with such sufficient intensity it poses a genuine

---

230 *Id.*

231 *See* THE INTERNATIONAL LEGAL SYSTEM 1296-97 (Joseph Modeste Sweeney *et al.* eds., 1981).

threat to the sovereignty of the target state.[232] This is consistent with the conclusion of two other scholars that "[e]conomic aggression can be used as a tool to subjugate one nation to the will of another."[233] There is obviously a difference in scale and effect of a simple economic policy of a state, or a cartel of states, that defines national economic interests, and those organized policies that are intended to reach a magnitude that interferes with the political independence and territorial integrity of a state. Two scholars define this difference as follows:

> Although a state at international law may chart its own economic course in its relations with other states, still all states are restricted by the rule that a policy cannot be exercised for the sole purpose of causing injury to and forcing the will of another state unless such economic coercion is used in the exercise of the right of self-defense or reprisal. An illegal intervention of an economic nature becomes aggression if it jeopardizes essential rights of a state that are requisite to its security.[234]

Accordingly, a complete contextual analysis reveals that Article 2(4) is a prohibition on a spectrum of force that does not include coercive political and economic sanctions that are intended to influence another state's policy or actions, but ranges from state activities that begin with coercive political and economic sanctions that threaten the territorial integrity or political independence of another state to that of armed force. Boycotts, the severance of diplomatic relations, the interruption of communications, and economic competition or sanctions between states are lawful and not considered a threat or use of force. In contrast, political or economic aggression is a use of force within the meaning of Article 2(4).

---

[232] *Id.* at 1297.

[233] THOMAS & THOMAS, CONCEPT OF AGGRESSION, *supra* note 87.

[234] *Id.* at 90-91.

Consider the application of Article 2(4)'s prohibition in cyberspace.[235] State A is a fifth the size of State B and, except for one very small border along a lake, is geographically an enclave of State B. State A, apparently based upon insider information internal to State B, has been subsidizing its currency on the global market in such a way that it devalues the currency of State B. This is the deliberate intent of State A because its President dislikes a number of the economic policies of State B. The devaluation is not significant, but it does adversely affect the trade balance of State B.

In response, State B restricts its national companies from broadcasting any economic or financial news into State A. Such a political and economic policy, or 'sanction,' is a lawful response to State A's financial policy, however, it is unsuccessful because State A can easily get the same information from CNN.

State B then decides to further restrict the flow of information by actively jamming all CNN broadcasts that flow into State A. Assume that this expanded policy violates existing telecommunications agreements and State B's contractual arrangements with CNN, it is therefore an unlawful act under international law, but it is not a use of force within the meaning of Article 2(4). It is also not successful. State B now creates a false database and 'allows' State A to hack into it, hoping to mislead State A and undermine its ability to devalue the currency of State B. Similarly, this information operation by State B is not a use of force, and is unsuccessful.

In frustration, State B completely jams all electronic transmissions into, out of, and within State A. This has a devastating effect on State A.

---

[235] While real-world conflicts are a superb way to demonstrate these principles in practice, and are used where possible in this chapter, there are no reported information operations in cyberspace that will exemplify these principles. Accordingly, somewhat simplified hypothetical scenarios will be used to demonstrate how these principles of *jus ad bellum* may restrict state activities in cyberspace.

All telecommunications within and by State A with other states have ceased. Emergency services are significantly curtailed throughout State A, and the flow of emergency medical data between hospitals has stopped. No financial transaction can occur electronically within State A or between anyone in State A and someone in another state. The volume of imports and exports has dropped sharply 95% from the previous month. State A cannot communicate electronically with its embassies around the world, regional organizations, or the United Nations. Without a single shot being fired, State A has effectively been brought to its knees in a few short months and is at the complete mercy of State B. To survive, State A must surrender to the will of State B. State B's virtual blockade of State A is a use of force within the meaning of Article 2(4). Indeed, it is the electronic equivalent of an armed attack, and is unlawful under these circumstances because it was not justified by any right of self-defense.

Threats of Force

A threat of force "consists in an express or implied promise by a government of a resort to force conditional on nonacceptance of certain demands of that government."[236] State practice has demonstrated a "relatively high degree" of tolerance towards mere threats to use force because most were in the context of self-defense.[237] A threat to use force is unlawful if not made under conditions that otherwise make the actual use of force lawful.[238] For example, it is unlawful under Article 2(4) for a state to make "a blatant and direct threat of force, used to compel another State to yield territory or to make substantial political concessions . . . ."[239]

---

[236] BROWNLIE, INTERNATIONAL LAW, *supra* note 91, at 364.

[237] *Id.* at 118.

[238] *Id.*

[239] THE CHARTER OF THE UNITED NATIONS: A COMMENTARY, *supra* note 70, at 118.

This principle does place restrictions on state activities in cyberspace. Isolated threats are usually verbal in nature and do not implicate the use of cyberspace except as a transmission medium for the threat. To the extent, however, that unlawful verbal threats are reinforced or conveyed by state activities in cyberspace, those activities may become an integral part of the unlawful threat and thus become an unlawful use of force.

State activities that constitute a threat or cause the perception of a threat create special problems in conflict management because they constitute circumstantial evidence of impending hostile acts and hostile intent. Threats serve to give definition to the intent of the acting state and the perceptions of a potential victim state. Hostile act and hostile intent are the operational concepts that are integral to the rules of engagement (ROE) of a state that implement its inherent right of self-defense. The "Standing Rules of Engagement for US Forces" (SROE) defines hostile act and hostile intent as follows:

> e. **Hostile Act.** A hostile act is an attack or other use of force by a foreign force or terrorist unit (organization or individual) against the United States, US forces, and in certain circumstances, US citizens, their property, US commercial assets, and other designated non-US forces, foreign nationals and their property. It is also force used directly to preclude or impede the mission and/or duties of US forces, including the recovery of US personnel and vital US Government property. When a hostile act is in progress, the right exists to use proportional force, including armed force, in self-defense by all necessary means available to deter or neutralize the potential attacker or, if necessary, to destroy the threat. ...
>
> f. **Hostile Intent.** Hostile intent is the threat of imminent use of force by a foreign force or terrorist unit (organization or individual) against the United States, US forces, and in certain circumstances, US citizens, their property, US commercial assets, or other designated non-US forces, foreign nationals and their property. When hostile intent is present, the right exists to use proportional force, including armed force, in self-

> defense by all necessary means available to deter or neutralize the potential attacker or, if necessary, to destroy the threat. . . .[240]

Isolated verbal threats, and threats of the use of force that are conveyed by initial troop movements, the shaping of alliances, or initial Intermediate Range Ballistic Missile (IRBM) movements, likely do not constitute hostile intent. Threats of the use of force conveyed by the massing of troops or IRBMs on the border, the tactical use of fire control radar, and the interference with early warning (EW) or command and control (C2) systems, are very likely to be considered an indication of hostile intent. State activities that convey hostile intent constitute a threat to use force, and a state that is the object of that hostile intent has the right to use necessary and proportional force to respond in anticipatory self-defense.

In comparison, hostile acts include blockades; the destruction of early warning or command and control systems; and those uses of force against the territory, warships, military aircraft, military forces, or the citizens of a state abroad. Although all uses of force are not of the scope, duration, and intensity to constitute an armed attack, state activities that constitute a hostile act are a use of force that constitutes an armed attack, and a state that is the object of that hostile act has the right to use necessary and proportional force to respond in self-defense.

Consider the situation that is clearly an unlawful threat of the use of force. Before the virtual blockade, State B threatens to use armed military force to invade and occupy State A if it does not immediately cease subsidizing its currency to the detriment of the State B. Since the devaluation is not significant, State B does not have the right to threaten or use force in self-defense.

---

[240] Chairman of the Joint Chiefs of Staff Instruction 3121.01, *Standing Rules of Engagement for US Forces*, Oct. 1, 1994, as amended Dec. 22, 1994, Encl. A, paras. 5e, 5f, *reprinted in* OPERATIONAL LAW HANDBOOK, Chap. 8, at pp. 8-8 through 8-14, US Army Judge Advocate General's School (1997). Enclosure A to CJCSI 3121.01 is provided as APPENDIX E.

State A scoffs at the threat. To reinforce the seriousness of its unlawful threat, State B proves how easily it can control an occupied State A by jamming CNN and *all* international telecommunications that flow into State A for five minutes every hour on the hour. In the previous discussion on political and economic force, it was assumed that the jamming of telecommunications that flow into State A was unlawful under existing telecommunications agreements and other contractual obligations of State B, but that such unlawful jamming was not a use of force within the meaning of Article 2(4). Similarly, the unlawful jamming of all incoming telecommunications in this situation is not a use of force within the meaning of Article 2(4) or an armed attack, but, as an extension of the unlawful verbal threat to use force, it is an unlawful *threat* to use force within the meaning of Article 2(4).

Since the jamming activities of State B are only a *threat* to use force, State A does not have the right to use force in self-defense. This hypothetical, however, is an example of the special problems of conflict management caused by threats and the perceptions of threats that demonstrate hostile intent. In the context of State B's threats to use armed military force to invade and occupy State A, State B's jamming of all telecommunications that flow into State A could be subjectively perceived by State A as a demonstration of hostile intent that is a prelude to the threatened invasion and occupation. Under these circumstances, State A does have the lawful right to use force in anticipatory self-defense. As concluded in Chapter 3, *supra,* what constitutes a prohibited 'threat or use of force' is a question of fact that must be subjectively analyzed in each and every case in the context of all relevant law and circumstances.

Violations of International Law

A violation of international law is not *per se* a use of force and the unlawfulness of a violation follows from international norms other than

Article 2(4).[241] A violation, however, "which involves an exercise of power in the territorial domain but no use of arms" may constitute an unlawful use of force prohibited by Article 2(4).[242] While there appears to be no evidence in the working documents of the drafters of the Charter to support this view,[243] issues concerning a prohibited use of force have arisen under state practice when the territorial integrity of a state has been violated despite the fact that no armed military force occurs.

The Soviet government, for example, has asserted that military intelligence flights by reconnaissance aircraft over Soviet territory and territorial waters are *per se* acts of aggression.[244] The Soviet response to the 1960 U-2 flight by the United States over Soviet territory emphasizes that under modern conditions "the incursion of a single plane is 'an act of aggression' because it might carry a deadly load and, further, might cause those in charge of detection systems to order retaliation on the assumption that an attack had begun."[245] The United Nations Security Council disagreed, and concluded that the U-2 flight was a violation of Soviet airspace but not a use of force within the meaning of Article 2(4).[246] Also contrary to the Soviet position, one scholar concludes that military intelligence flights and tactics of psychological warfare may be deliberate and illegal intrusions, but are not necessarily a use of force.[247]

---

[241] THE CHARTER OF THE UNITED NATIONS: A COMMENTARY, *supra* note 70, at 112-13.

[242] US Naval War College, *International Law Studies, Collective Security under International Law,* p. 57; as discussed in BROWNLIE, INTERNATIONAL LAW, *supra* note 91, at 361-62.

[243] BROWNLIE, INTERNATIONAL LAW, *supra* note 91, at 362-63.

[244] *Id.* at 363.

[245] *Id.*

[246] NATIONAL SECURITY LAW 439 (John Norton Moore *et al.* eds., 1990).

[247] BROWNLIE, INTERNATIONAL LAW, *supra* note 91, at 364.

The consensus view is that while all nations have complete and exclusive sovereignty over their national airspace and never lose their inherent right of self-defense, customary international law requires a reasonable response to civilian and military aircraft that violate another state's airspace that is proportionate to the danger posed by the presence of the trespassing aircraft.[248]

The danger posed by the presence of a trespassing aircraft is a subjective determination that must be based upon the circumstances at the time of the actual or perceived attack.[249] Intrusions into another state's territory "may be regarded by the state which is their object as circumstantial evidence of an intention to attack or preparation for future attacks,"[250] thereby giving rise to concepts such as hostile act and hostile intent.[251] While international law is not completely clear in these cases of trespass, the right of self-defense may be invoked when a territorial state perceives a threat or use of force and has no alternative but to act to defend itself. All of the relevant circumstances must be subjectively analyzed to determine whether a threat or use of force, or a perception thereof, has occurred.

If the trespasser, for example, is an "unidentified fast aircraft which persists in a deliberate and deep penetration of airspace, it may be that, in view of the destructive power of even a single nuclear weapon carried by an aircraft, the territorial sovereign is justified in taking without

---

[248] *See* NATIONAL SECURITY LAW 439 (John Norton Moore *et al.* eds., 1990); COMMANDER'S HANDBOOK, *supra* note 109, at § 4.4; BROWNLIE, INTERNATIONAL LAW, *supra* note 91, at 363-64, 373-74.

[249] *See generally* COMMANDER'S HANDBOOK, *supra* note 109, at § 8.1.1.

[250] BROWNLIE, INTERNATIONAL LAW, *supra* note 91, at 364.

[251] *See id.*

any warning violent and immediate preventive measures."[252] In the context of a land invasion by armed troops, one scholar concludes that

> an incursion into the territory of another state constitutes an infringement of Art. 2(4), even if it is not intended to deprive that state of part of its territory and if the invading troops are meant to withdraw immediately after completing a temporary and limited operation ('in-and-out operations'). In other words, 'integrity' has to be read as 'inviolability', proscribing any kind of forcible trespassing.[253]

In contrast, it is "very doubtful if maritime trespassers in territorial waters may be subjected to violent measures unless they are engaged in an attack or are carrying aid to rebels."[254] The guidance from the International Court of Justice is that a trespasser must "be given reasonable opportunity to leave peacefully and measures against him should be preventive and not punitive."[255] Violations of international law may constitute a use of force within the meaning of Article 2(4) if they involve an exercise of power in the territorial domain of another state, even without the use of arms, but the response in self-defense must be necessary and proportional in view of the danger posed by the violation.

In the hypothetical scenario between States A and B, recall that as part of State B's unlawful threat to invade State A before the electronic blockade, it jams CNN and *all* international telecommunications that flow into State A for five minutes every hour on the hour. In previous discussions, it was concluded that the jamming of telecommunications

---

[252] *See id.* at 373.

[253] THE CHARTER OF THE UNITED NATIONS: A COMMENTARY, *supra* note 70, at 117.

[254] *See* BROWNLIE, INTERNATIONAL LAW, *supra* note 91, at 374.

[255] *See id.* at 364 n.3.

that flow *into* State A was not a use of force within the meaning of Article 2(4), although it was assumed that it violated telecommunications agreements and other contractual obligations of State B.

After five days of jamming external telecommunications flowing into State A, State B desires to strengthen its threat by completely jamming all electronic transmissions within, into, and out of State A for five minutes every hour on the hour. This constitutes an exercise of power in the territory of State A in violation of international law that constitutes a use of force within the meaning of Article 2(4), even in the absence of the context of the unlawful threat.

This violation of international law is a use of force, but it is not of the scope, duration, and intensity to constitute an armed attack. State A does, however, have the right of self-defense, and it could take necessary and proportionate measures to destroy or neutralize the source of the unlawful jamming. Moreover, such jamming could be considered by State A as an indication of hostile intent, *i.e.*, the beginning of an invasion by State B, thereby escalating the situation to actual hostilities if State A preemptively attacks under its right of anticipatory self-defense.

Unarmed, Non-Military Physical Force

Recognizing that unarmed, non-military physical force by a state can affect another state just as severely as the use of armed military force, the Article 2(4) prohibition on the use of force also covers "physical force of a non-military nature"[256] committed by any state agency.[257] Examples of such unarmed physical force include "the cross-frontier expulsion of populations, the diversion of a river by an up-stream state, the release of

---

[256] *See* THE CHARTER OF THE UNITED NATIONS: A COMMENTARY, *supra* note 70, at 113.

[257] BROWNLIE, INTERNATIONAL LAW, *supra* note 91, at 361.

large quantities of water down a valley, and the spreading of fire across a frontier."[258] It is also recognized that unarmed, non-military physical force "may produce the effects of an armed attack prompting the right of self-defense laid down in Art. 51."[259]

This principle places far-reaching restrictions on a state's activities in cyberspace that 'attack' the critical infrastructure of another state and cause destructive effects. A state can potentially cause significant property and economic damage, as well as human fatalities, in another state by utilizing the internet to cause

- flooding by opening the gates of a dam,
- train wrecks by switching tracks for oncoming trains,
- plane crashes by shutting down or manipulating air traffic control systems,
- large chemical explosions and fires by readjusting the mix of volatile chemicals at an industrial complex,
- a run on the banks or a massive economic crisis by crashing national stock exchanges,

and any number of other examples that are limited only by the imagination of the state actors. Almost any destructive state activity intentionally caused within the sovereign territory of another state is an unlawful use of force. The cross-border use of weapons and bombardment of the territory of one state by another state is one of the classic cases of a use of force within the meaning of Article 2(4).[260] The effect can be the same, if not more severe, as if the destruction were caused by conventional kinetic means of warfare. Accordingly, all state

---

[258] THE CHARTER OF THE UNITED NATIONS: A COMMENTARY, *supra* note 70, at 113. *See also id.* at 362-63.

[259] THE CHARTER OF THE UNITED NATIONS: A COMMENTARY, *supra* note 70, at 113.

[260] *See id.* at 670.

activities in cyberspace that intentionally cause almost *any* destructive effect within the sovereign territory of another state are probably an unlawful use of force.

Inadvertent damage done within the sovereign territory of another state will give rise to civil liability, but will very likely not be a use of force within the meaning of Article 2(4) that invokes a state's right to use force in self-defense. For example, consider the hypothetical of State A using cyberspace to attack State B in such a way that utilized the telecommunications infrastructure of State C. Then assume that this destructive activity of State A inadvertently caused severe damage to the information infrastructure of State C, and perhaps even significant consequent physical damage within State C. This hypothetical highlights why use of force issues must be subjectively analyzed, as previously discussed, in each and every case in the context of all relevant circumstances. While State A has caused a destructive act in State C, State A has not engaged in a use of force against State C within the meaning of Article 2(4), and State C does not have the right to use force in self-defense against State A. State A does, however, incur civil liability under international law for the damage it caused to State C.

Furthermore, even if such intentional destructive activities are an unlawful use of force that unequivocally invokes a victim state's right of self-defense, that right of self-defense does not necessarily justify a use of force in response. Similarly, not every destructive activity of this nature will reach the "intensity that enables them to be classified as 'armed attacks'."[261] The scope, duration, and intensity of the force must be analyzed to determine if an armed attack has occurred. For example, it would be an unlawful use of force but most likely not an armed attack if a state uses the internet to cause a single train crash. Furthermore, absent a continuing threat, a use of force in response would not be authorized under the victim state's right of self-defense. In contrast, it would very likely be considered an armed attack should a state cause dozens of such

---

[261] *Id.*

incidents, or if a state caused one major incident such as the complete and long-term crash of the New York Stock Exchange. In the case of such an armed attack, a victim state would be completely justified in using force in self-defense.

Protection of Nationals Abroad

Before the Charter's prohibition on the aggressive use of force, states and jurists universally accepted as lawful the use of force to protect the lives and property of nationals.[262] This position was justified on the theory that nationals of a state are an extension of the state, and that the purpose of the state is the protection of its nationals.[263] Today, there is considerable debate among jurists over whether the nonconsensual use of military force to protect nationals abroad is a use of force proscribed by Article 2(4), and if so, whether such use of force is authorized by Article 51.[264] One author concludes that only military units sent abroad can be the targets of an armed attack, and that diplomatic missions, individual nationals, and other external positions of a state cannot be the targets of an armed attack within the meaning of Article 51 which would invoke the right of a state to use force in self-defense.[265] The same author, however, modifies this seemingly absolute rule by noting that attacks on the "whole of the civilian marine or air fleet," as distinct from individual vessels or aircraft, may constitute an armed attack.[266] When applied to actual situations, this minority position is not an absolute prohibition as much as

---

[262] BROWNLIE, INTERNATIONAL LAW, *supra* note 91, at 289.

[263] *Id.*

[264] *See* THE CHARTER OF THE UNITED NATIONS: A COMMENTARY, *supra* note 70, at 124-26.

[265] *Id.* at 671.

[266] *Id.*

it is an evaluation of the level of the attack that has occurred and the effect it has had on the victim state.

State practice, however, resolves the debate in favor of the lawfulness of such rescue operations. The United States and other nations have repeatedly rescued their own nationals abroad since the Charter entered into force, and have declared that Article 2(4) "does not prohibit the use of armed force in order to rescue a state's own nationals whose lives or health are endangered in a foreign state, provided that the latter is not able or not willing to provide the required protection."[267] Notable examples include the protective measures taken in 1960 by Belgium in the Congo, the 1964 Belgian-United States rescue operation in the Congo, the 1976 Israeli raid on Entebbe, the 1980 attempt by the United States to rescue its diplomatic staff held hostage in Iran, the 1983 rescue operation by the United States in Grenada, and the 1989 intervention in Panama by the United States to protect its nationals.[268]

A growing number of noted jurists and scholars have justified the lawfulness of these military operations under one of three arguments.[269] Some maintain that a use of force on such a comparatively small scale does not infringe Article 2(4) because the territorial integrity and political independence of the states concerned are not affected.[270] Others recognize a case of conflicting obligations that justify these rescue operations through a weighing of protected interests.[271]

---

[267] *Id.* at 124.

[268] *Id.*

[269] *Id.* at 124-25.

[270] *Id.* at 125.

[271] *Id.*

The remaining and largest group of jurists conclude that the forcible protection of a state's own nationals is lawful self-defense under Article 51.[272] Their view is that if Article 51 of the Charter recognizes an inherent right of self-defense, then the content of that right must be determined by reference to customary international law.[273] Before the Charter, however, international law permitted states to use aggressive force freely, and references to self-defense were politically motivated without any particular legal denotation.[274] Accordingly, customary international law since the creation of the Charter's Article 2(4) prohibition has greater value than pre-Charter customary norms.

While one scholar concludes that the "forcible protection of nationals against the will of the territorial state is illegal,"[275] he acknowledges the divisiveness of the controversy among jurists. More importantly, he also concludes that state practice, supported by organs of the United Nations, demonstrates "considerable reluctance" to qualify armed rescue operations of a state's own nationals unlawful in any case—possibly giving rise to a new customary norm.[276] Another scholar concludes that three conditions for such lawful rescue operations exist in customary international law.[277] Based upon the principles espoused by Secretary of State Daniel Webster in the *Caroline* case, he concludes that

(*a*) an imminent threat of injury to the nationals;

---

[272] *Id.* at 125 & n.156.

[273] *See* BROWNLIE, INTERNATIONAL LAW, *supra* note 91, at 299.

[274] THE CHARTER OF THE UNITED NATIONS: A COMMENTARY, *supra* note 70, at 662-63.

[275] *Id.* at 125 (footnotes omitted).

[276] *Id.* at 125-26.

[277] BROWNLIE, INTERNATIONAL LAW, *supra* note 91, at 299.

(*b*) a failure or inability on the part of the territorial sovereign to protect them;
(*c*) that the measures of protection should be strictly confined to the object of protecting them against injury.[278]

A similar test was framed by the United States during the United Nations Security Council debate on the 1976 Entebbe incident. After reaffirming the principle of territorial sovereignty, the United States representative stated that

> Israel's action in rescuing the hostages necessarily involved a temporary breach of the territorial integrity of Uganda. Normally, such a breach would be impermissible under the Charter. However, there was a well established right to use limited force for the protection of one's own nationals from an imminent threat of injury or death in a situation where the State in whose territory they were located was either unwilling or unable to protect them. The right, flowing from the right of self-defence, was limited to such use of force as was necessary and appropriate to protect threatened nationals from injury.[279]

Two resolutions were presented to the Security Council after the raid on Entebbe—the United Kingdom-United States draft which condemned the hijacking in support of the Israeli rescue, and the Tanzania-Libya-Benin draft which condemned Israel's violation of Uganda's sovereignty and territorial integrity.[280] Only the first was brought to vote, and it failed to pass.[281] In contrast to the position of the United States, the delegate

---

[278] *Id.* at 299.

[279] THE INTERNATIONAL LEGAL SYSTEM 1287 (Joseph Modeste Sweeney *et al.* eds., 1981).

[280] *Id.* at 1285-86.

[281] *Id.*

from the Soviet Union stated during the Security Council debate that there are "no laws in the world, no moral or international laws, which could justify such action."[282]

The international custom of states since the creation of the Article 2(4) norm, demonstrated and reinforced by repeated state practice, is strong evidence that the nonconsensual use of limited military force that is necessary and proportional to protect nationals abroad is a use of force within the meaning of Article 2(4) and that such a use of force is lawful self-defense authorized by Article 51. Given the limited scope, duration, and intensity of nonconsensual rescue operations, they are not an armed attack. This conclusion is also supported by the writings of a growing number of scholars and jurists, which are also, *ipso facto*, indicia of the existence of a customary norm.

State activities in cyberspace would not likely constitute a rescue operation, but they very likely could be conducted in support of a rescue operation or other effort intended to protect a state's nationals from an imminent threat of injury or harm. For example, as previously discussed, it would be an unlawful use of force for a state to shut down the air traffic control system for an entire state without justification by crashing its computers via the internet. During a heliborne rescue operation, however, it would be a lawful act of self-defense for a rescuing state to manipulate the territorial state's computers via the internet to create a blind air corridor in which the territorial state is unable to detect the rescue aircraft. Similarly, it would be a lawful act of self-defense during a rescue operation to electronically jam anti-aircraft fire control radars either by overhead aircraft or via the internet. It would also be a lawful act of self-defense for a state to assist its nationals, who are in imminent threat of injury or harm, to escape surreptitiously from a state by manipulating that state's database via the internet to falsify travel documents, identity papers, passports, or other documents or databases that would facilitate their escape.

---

[282] *Id.* at 1288.

During the nonconsensual rescue operation that is a lawful use of force in self-defense, the territorial state has the right to use force in self-defense, but that use of force must be necessary and proportional. As previously discussed in the section concerning violations of international law, all nations have complete and exclusive sovereignty over their national airspace and never lose their inherent right of self-defense. Customary international law, however, requires a reasonable response to civilian and military aircraft that violate another state's airspace that is proportionate to the danger posed by the presence of the trespassing aircraft.[283]

Similarly, the International Court of Justice has held that a trespasser must "be given reasonable opportunity to leave peacefully and measures against him should be preventive and not punitive."[284] Based on these principles, if a territorial state was unaware of an ongoing rescue operation and perceived an attack on its territory, it would have the right to use armed force in self-defense to prevent the intrusion. In contrast, if circumstances were such that the territorial state was completely aware of the limited and humanitarian nature of the rescue operation, and the lack of any intent to do the territorial state any harm, its right to use necessary and proportional force in self-defense would not allow the territorial state to shoot down the rescue aircraft.

Indirect Armed Force

Article 2(4)'s prohibition on the use of force also includes the use of indirect armed force and it is widely accepted that indirect force can constitute an armed attack.[285] This Article 2(4) prohibition extends to the

---

[283] *See* COMMANDER'S HANDBOOK, *supra* note 109, at § 4.4]; BROWNLIE, INTERNATIONAL LAW, *supra* note 91, at 363-64, 373-74.

[284] *See* BROWNLIE, INTERNATIONAL LAW, *supra* note 91, at 364 n.3.

[285] THE CHARTER OF THE UNITED NATIONS: A COMMENTARY, *supra* note 70, at 113, 673.

participation of one state in the use of force by another state, for example, "by allowing parts of its own territory to be used for violent acts against a third state."[286] It also extends to a "state's participation in the use of force by unofficial bands organized in a military manner, such as irregulars, mercenaries, or rebels."[287]

The scope of this prohibition is clarified by the General Assembly's 1970 "Declaration on Principles of International Law Concerning Friendly Relations Among States In Accordance With the Charter of the United Nations," which provides that

> Every state has the duty to refrain from organising or encouraging the organisation of irregular forces or armed bands, including mercenaries, for incursion into the territory of another state. Every state has the duty to refrain from organising, instigating, assisting or participating in acts of civil strife or terrorist acts in another state or acquiescing in organized activities within its territory directed towards the commission of such acts, when the acts referred to in the present paragraph involve a threat or use of force.[288]

Although General Assembly resolutions are not binding on states, the International Court of Justice has held that these provisions reflect customary international law.[289] The prevailing view is that third states are "entitled to intervene in a civil war in support of the legitimate

---

[286] *Id.*; BROWNLIE, INTERNATIONAL LAW, *supra* note 91, at 374-75.

[287] THE CHARTER OF THE UNITED NATIONS: A COMMENTARY, *supra* note 70, at 113; BROWNLIE, INTERNATIONAL LAW, *supra* note 91, at 369-72.

[288] G.A. Res. 2625, U.N. GAOR, 25th Sess., *quoted in* THE CHARTER OF THE UNITED NATIONS: A COMMENTARY, *supra* note 70, at 114.

[289] THE CHARTER OF THE UNITED NATIONS: A COMMENTARY, *supra* note 70, at 115.

government, upon its request or at least with its consent, whereas giving support to rebels is said to be entirely prohibited."[290] What constitutes prohibited support remains unclear, but the International Court of Justice has held that the arming and training of irregular forces is a use of force, but merely supplying funds is not.[291]

There are two general circumstances where this principle may restrict state activities in cyberspace. First, and the most obvious, those state activities in cyberspace that either support or constitute a prohibited indirect use of force are themselves unlawful as well.[292] For example, it would be unlawful for one state to manipulate information in another state via cyberspace during peacetime in an effort to instigate a plot to overthrow the leader of that country. Suppose that State C wants to quietly have the president of State D overthrown. To do so, State C uses the internet to plant rumors that the president of State D is selling the majority of the food rations purchased for his military forces for his own private gain. State C then uses the internet to manipulate the State D's computer databases in such a way that most of the food rations are directed 'by the order of the President of State D' to a local coastal port, leaving little or no food for the military forces. State C then uses the internet to anonymously report the missing military food sitting at the port to the local press of State D. The military forces of State D are furious, and a military coup ensues. Absent the context of this occurring

---

[290] *Id.* at 116.

[291] *Id.* at 115.

[292] This may explain why most state activities of this nature are conducted by states as covert operations which attempt to conceal the identity of the state actor. A covert operation is so "planned and executed as to conceal the identity of or permit plausible denial by the sponsor . . . [which] differs from a clandestine operation in that emphasis is placed on concealment of identity of sponsor rather than on concealment of the operation." US DEP'T OF DEF., JOINT PUBLICATION 1-02, DICTIONARY OF MILITARY AND ASSOCIATED TERMS 129 (1998) [hereinafter JOINT PUB 1-02]. Available at http://www.dtic.mil/doctrine/ jel/new_pubs/jp1_02.pdf.

furious, and a military coup ensues. Absent the context of this occurring during an armed conflict between States C and D, State C's activities would be an unlawful use of indirect force, but it is very unlikely it would be considered of a scope, duration, and intensity that would constitute an armed attack.

Second, it would be unlawful under customary international law for a state to allow its information infrastructure to be used by another state or insurgents to engage in violent acts against a third state. It is an unlawful use of indirect force, for example, for State E to knowingly permit or acquiesce in the use of its internet gateways by State F during peacetime to commit unlawful uses of force against State G. Mere acquiescence would not constitute an armed attack, but using the information infrastructures to help arm or train insurgents would constitute an armed attack. If this occurs in the context of hostilities between States F and G, then State E has the duty under the law of neutrality to prevent the construction of telecommunications infrastructures by a belligerent on its territory and the use by a belligerent of telecommunications infrastructures that existed before hostilities for solely military purposes that are not open to the public for use.[293] As a neutral state during hostilities, State E is not obligated to forbid or restrict the use of its telecommunications infrastructure by the belligerents, but whatever restrictions are put into effect must be applied impartially to both belligerents.[294]

Armed Attack

Although a consensus on the meaning of armed attack is of the "utmost significance for the effectiveness of the rules of international law

---

[293] *See* Hague Convention (V) Respecting the Rights and Duties of Neutral Powers and Persons in Case of War on Land, Oct. 18, 1907, [hereinafter 1907 Hague Convention V, Neutral Powers] arts. 3-5, *reprinted in* ROBERTS & GUELFF, *supra* note 4, at 63.

[294] *Id.* at arts. 8-9.

on war prevention," the phrase is not defined anywhere in the Charter.[295] It is easy, however, to identify a few state activities universally accepted as an armed attack. One scholar, for example, identifies the "Iraqi panzer blitzkrieg against Kuwait ... [as] a paradigm armed attack" and concludes that if "this attack were not an armed attack..., then nothing would be."[296] This Iraqi attack fits squarely within the 1974 General Assembly Definition on Aggression resolution, and it was also condemned by numerous Security Council resolutions as an armed attack.[297]

While the General Assembly resolution on the Definition of Aggression[298] does not define armed attack, its list of acts of aggression are considered as examples of armed attack.[299] An analysis of the Definition of Aggression resolution by one scholar that provides a widely accepted view of what constitutes an armed attack is summarized as follows:

(a) *Invasion, bombardment and cross-border shooting.* These examples represent the classic cases of armed attacks, provided "that the military actions are on a certain scale and have a major effect, and are thus not to be considered mere frontier incidents."

(b) *Blockade.* An effective blocking of a state's ports or coasts by the armed forces of another state is an armed attack. The barring of passage for land-locked states to the open

---

[295] THE CHARTER OF THE UNITED NATIONS: A COMMENTARY, *supra* note 70, at 668.

[296] MOORE, CRISIS IN THE GULF, *supra* note 96 at 151.

[297] *Id.* at 22-23.

[298] *Definition of Aggression,* G.A. Res. 3314, U.N. GAOR, 29th Sess., U.N. Doc. A/RES/3314 (1974). G.A. Res. 3314 is provided as APPENDIX D.

[299] THE CHARTER OF THE UNITED NATIONS: A COMMENTARY, *supra* note 70, at 670.

sea across another state's territory has not been accepted as an armed attack.

(c) *Attack on the land, sea, or air forces or on the civilian marine and air fleets.* An armed attack occurs when the armed forces of one state attack the land, sea, or air forces, or the civilian marine and air fleets, of another state. The regular forces of a state, wherever they are, always have the right to defend themselves by military force.

(d) *Breach of stationing agreements.* An armed attack may occur when a state uses its armed forces within the territory of another state in contravention of the conditions provided for in the agreement, or any extension of their presence beyond the termination of the agreement; provided, however, that the breach of the terms of the agreement has the effect of an invasion or occupation.

(e) *Placing territory at another state's disposal.* The voluntary action of a state in allowing another state to use its territory for committing an armed attack is also an armed attack.

(f) *Participation in the use of force by militarily organized unofficial groups.* It is widely accepted that indirect force falls under the definition of armed attack. The sending of armed bands to use armed force in another state makes the armed bands a *de facto* state agent, thus the sending state has engaged in an armed attack. Similarly, 'substantial involvement' in the activities of an armed band may also constitute an armed attack.[300]

An armed attack may also occur when a state uses bacteriological, biological, and chemical agents against another state, even though they may not involve any explosive effect with shock waves and heat.[301] Although these 'weapons' may be used by the military in conjunction with

---

[300] *Id.* at 669-74.

[301] BROWNLIE, INTERNATIONAL LAW, *supra* note 91, at 362.

other traditional weapons, the use of these devices may constitute an armed attack when a state uses any of its civilian agencies to release the agents in another state.

One key point to take away from this summary is that an armed attack may occur when a use of force or an activity not traditionally considered an armed attack is used in such a way that it becomes tantamount in effect to an armed attack. Accordingly, state practice defining what is an armed attack within the meaning of Article 51 of the Charter places far-reaching restrictions on state activities in cyberspace.

This chapter has provided a number of examples of state activities in cyberspace that define what is and what is not an armed attack within the meaning of Article 51 of the Charter. For example, in the discussion of political and economic force, State B's virtual blockade of State A was determined to be a use of force within the meaning of Article 2(4) and a cyberspace equivalent of an armed attack. This is consistent with the conclusion of one scholar that in extreme cases where an impairment is equivalent to a "military invasion (*e.g.*, by cutting off all communication routes) an 'armed attack' can be taken to exist."[302]

In the discussion of threats of the use of force, State B's jamming of CNN and *all* other international telecommunications that flowed into State A for five minutes every hour on the hour, in the context of reinforcing an unlawful threat to invade, was determined to be an unlawful threat to use force within the meaning of Article 2(4). It was not, however, considered a use of force within the meaning of Article 2(4) or an armed attack.

In the discussion of violations of international law, State B strengthened its threat by completely jamming all electronic transmissions within, into, and out of State A for five minutes every hour on the hour.

---

[302] THE CHARTER OF THE UNITED NATIONS: A COMMENTARY, *supra* note 70, at 670.

This constituted an exercise of power in the territory of State A in violation of international law that constituted a use of force within the meaning of Article 2(4), but was not of the scope, duration, and intensity to constitute an armed attack. It was also noted that such jamming could be considered by State A as the beginning of an invasion by State B, thereby escalating the situation to actual hostilities if State A preemptively attacks under its right of anticipatory self-defense.

In the discussion of unarmed, non-military physical force, it was concluded that almost *all* state activities in cyberspace which that intentionally cause almost *any* destructive effect within the sovereign territory of another state are probably unlawful uses of force. It was concluded that if a state uses the internet to cause a single train crash in another state, it would be an unlawful use of force but not one of a scope, duration, and intensity to constitute an armed attack. In contrast, it would very likely be considered an armed attack should a state cause dozens of such incidents, or if a state caused one major incident such as the complete and long-term crash of the New York Stock Exchange.

In the discussion of the protection of nationals abroad, it was concluded that the nonconsensual use of limited military force that is necessary and proportional to protect nationals abroad is a use of force within the meaning of Article 2(4) and that such a use of force is lawful self-defense authorized by Article 51. Given the limited scope, duration, and intensity of nonconsensual rescue operations, however, they are not an armed attack. Accordingly, it would be a lawful act of self-defense for a rescuing state to manipulate the territorial state's computers via the internet to prevent the territorial state from detecting the rescue aircraft and to electronically jam anti-aircraft fire control radars. It is also a lawful act of self-defense for a state to assist its nationals, who are in imminent threat of injury or harm, to escape surreptitiously from a state by manipulating that state's database via the internet to falsify travel documents, identity papers, passports, or other documents or databases that would facilitate their escape.

Finally, in the discussion of indirect armed force, it was concluded that those state activities in cyberspace that either support or constitute a prohibited indirect use of force are themselves an unlawful use of force as well. For example, it would be an unlawful use of indirect force during peacetime for one state to manipulate information in another state via cyberspace in an effort to instigate a plot to overthrow the leader of that country. It is very unlikely, however, that it would be considered an armed attack. Similarly, it would be unlawful under customary international law for a state to allow its information infrastructure to be used by another state or insurgents to engage in violent acts against a third state. It is an unlawful use of indirect force, for example, for State E to knowingly permit or acquiesce in the use of its internet gateways by State F during peacetime to commit unlawful uses of force against State G. Mere acquiescence would not constitute an armed attack, however, using the information infrastructures to help arm or train insurgents in another state would constitute an armed attack.

The Future Development of "Use of Force"

Before returning to summarize the state of the law today, it would be useful to survey the horizon for possible vectors for its development. In a seminal 1998 article[303], Professor Michael Schmitt, then of the US Naval War College's Department of Oceans Law and Policy, identified three evolving trends in the law governing the use of force.[304] One such trend is the increasing pressure toward anticipatory self-defense, driven by the increasing speed of modern warfare and particularly the near-instantaneous operation of info-weapons.[305] The second is the role of

---

[303] Michael N. Schmitt, *Bellum Americanum: The US View of Twenty-First Century War and Its Possible Implications for the Law of Armed Conflict,* 19 MICH. J. INT'L. L. 1051 (1998). [hereinafter Schmitt, *Bellum Americanum*].

[304] *Id.* at 1069-1074.

[305] *Id.* at 1072. Schmitt explains:

> Even beyond information warfare, the reality of military operations in the next century will stress existing understandings of the line between a

non-state actors on the international scene, and the likelihood that a significant percentage of military operations in the future would be conducted against such groups.[306] That the entire structure of international law evolved with respect to *states'* rights and duties, and the uncertainty of the extent of a "host" state's complicity in the activities of a non-state malefactor, complicate targeting decisions under international law. The third trend, and the one most apropos of this discussion, is the extraordinarily complex question of consequences versus actions as the normative vector for the future development of the law.[307] He writes:

> In the current normative scheme the consequences of an act are often less important than its nature. For instance, a devastating economic embargo is not a "use of force" nor an "armed attack" justifying forcible self-defense, even though the embargo may result in enormous suffering. [footnote omitted] On the other hand, a relatively minor, armed incursion across a

---

> premature use of "defensive" force and valid self-defense. In tomorrow's high-tech battle the first shot may be the last. As weapons become more lethal, the incentive to strike first grows [footnote omitted] and the threshold for preemption in self-defense on the basis of apparent hostile intent drops precipitously. [footnote omitted]

Id.

[306] *Id.* at 1073-74. Schmitt states:

> Since the Peace of Westphalia, in 1648 and the rise of the nation-state, war has been the province and, until the turn of this century, the prerogative of states. When non-state actors have participated in organized violence, the normative paradigm has been that of international and domestic criminal law, not the law of armed conflict... If the twenty-first century national security threats are to come from non-state actors, then the law governing the resort to force is bound to evolve in a way that permits an effective defense against them...

Id.

[307] *Id.* at 1071-72 .

> border is both a use of force and an armed attack. [footnote omitted]. This contrary result derives from the law's use of "acts" as cognitive shorthand for what really matters—consequences. Acts are more easily expressed (to "use force" versus to cause a certain quantum and quality of harm) and more easily discerned than an effects-based standard, on the harm suffered. This cognitive shorthand does not work well in the age of information operations because information attacks, albeit potentially disastrous, may be physically imperceptible. Thus, as the nature of an hostile act becomes less determinative of its consequences, current notions of "lawful" coercive behavior by states, and the appropriate responses thereto, are likely to evolve accordingly.[308]

Professor Schmitt dilated on this trend in a 1999 article specifically addressing computer network attack and the means by which international law can contribute to preventing, or at least ameliorating the consequences of, such aggression.[309]

Central to Schmitt's analysis is that the consequences of a computer network attack, or any information operation, have *two* principal characteristics: a *quantitative* aspect, cited by the majority of commentators, which describes the quantum of damage, or the real-world consequences, of an operation, and a less-often acknowledged *qualitative* aspect, which describes the nature of the instrument used to produce the effect.[310] Schmitt explains:

> At least since the promulgation of the Charter, this use of force paradigm has been instrument based; determination of whether or not the standard has been

---

[308] *Id.*

[309] *Id.* at 885.

[310] *Id.* at 909-914.

> breached depends upon the type of coercive *instrument*—diplomatic, economic, or military—selected to attain the nationals objectives in question. The first two types of instruments might rise to the level of intervention, but they do not engage the normatively more flagrant act of using force.[311]

This two-prong test, then, would establish a much lower level of permissible damage done (the consequence) for military force than for economic or diplomatic force (the instrument). For this reason, the qualitative characterization of an information operation can matter as much as the quantitative level of harm inflicted.[312]

To understand why this is the case, it is useful to follow Schmitt's analysis of the origins of the Charter paradigm, particularly the choice to

---

[311] *Id.* at 909. Schmitt concludes his thought:

> However, despite the instrument-classing, in actual practice it does not follow that coercive acts involving armed force necessarily operate at counter-purposes with community values ... Even when they do, it is not always the case that they do greater violence thereto. For instance, a temporally and spatially limited border incursion is probably a lesser threat to international peace and security or the right of states to conduct their affairs free from outside interference than was the 1973-1974 Arab oil embargo. [footnote omitted] Yet, the prescriptive framework would proscribe the former, but not the latter. [footnote omitted].

Id. at 909-10.

[312] Kanuck identified this tension in a 1996 article:

> Over the years, the traditional Grotian paradigm succeeded in reforming itself to accommodate each new development; however, the time has finally come for a reconsideration that commences with the most basic tenets of international law. Interactions, and not physical territory, must become the basis of the new system; only then can aggression and intervention be redefined in terms of undesired effects and not merely the direct or indirect use of armed forces.

Sean P. Kanuck, *Information Warfare: New Challenges for Public International Law*, 37 HARV. INT'L L. J. 272, 292 (1996).

distinguish between military coercion on one hand and diplomatic and economic coercion on the other:

> In order to understand the distinction, one must first inquire into why the limitation exists at all...The prohibition on the use of force is designed to advance [the objectives of the international community] by restricting those acts which endanger them—uses of force. In fact, the international community is not directly concerned with the particular instrumentality used (force in this case), but rather the consequences of its use. However, it would prove extraordinarily difficult to quantify or qualify consequences in a normatively practical manner...The difficulty in looking to consequences themselves as criteria for calculating lawfulness led the Charter drafters to use prescriptive short-hand to achieve their goals. Because force represents a consistently serious menace to intermediate and ultimate objectives, the prohibition of resort to it is a relatively reliable instrument-based surrogate for a ban on deleterious consequences. It eases the evaluative process by simply asking whether force has been used, rather than requiring a far more difficult assessment of the consequences that have resulted...Any imprecision in this prescriptive short-hand is more than outweighed by its clarity and ease of application.[313]

The intellectual challenge, then, is to evaluate the *quantitative* consequences of a proposed information operation in terms of the *qualitative* criteria that distinguish military force from diplomatic or economic coercion.

By analyzing an information operation in this manner, two goals will be satisfied: first, its real-world destructive consequences would form the basis of the decision of whether or not to carry it out, and second, these consequences would be expressed in such way as to bring the

---

[313] Michael N. Schmitt, *Computer Network Attack and the Use of Force in International Law: Thoughts on a Normative Framework*, 37 COLUM. J. TRANSNAT'L L. 887 (1999) at 910-12.

analysis in line with the Charter drafters' principled distinction between military and other forms of coercion.[314]

Schmitt offers six such criteria, with the suggestion of a seventh:

1) *Severity*: Armed attacks threaten physical injury or destruction of property to a much greater extent than other forms of coercion. Physical well-being usually occupies the [lowest, most basic level] of the human hierarchy of need.

2) *Immediacy*: The negative consequences of armed coercion, or threat thereof, usually occur with great immediacy, while those of other forms of coercion develop more slowly. Thus, the opportunity for the target state or the international community to seek peaceful accommodation is hampered in the former case.

3) *Directness*: The consequences of armed coercion are more directly tied to the *actus reus* than in other forms of coercion, which often depend on numerous contributory factors to operate. Thus, the prohibition on force precludes negative consequences with greater certainty.

---

[314] *Id.* at 913-15. Schmitt elaborates:

> These consequence commonalities can serve as the ties between CNA [Computer Network Attack] and the prevailing instrument-based prescriptive short-hand. [footnote omitted] By this scheme, one measures the consequences of a computer network attack against the commonalities to ascertain whether they more closely approximate consequences of the sort characterizing armed force or whether they are better placed outside the use of force boundary. *This technique allows the force box to expand to fill lacunae (that became apparent upon the emergence of coercive possibilities enabled by technological advances) without altering the balance of the current framework—the growth is cast in terms of the underlying factors driving the existing classifications... The nature of the act's reasonably foreseeable consequences would be assessed to determine whether they resemble those of armed coercion* [emphasis added].

Id. at 915-16.

4) *Invasiveness*: In armed coercion, the act causing the harm usually crosses into the target state, whereas in economic warfare the acts generally occur beyond the target's borders. As a result, even though armed and economic acts may have roughly similar consequences, the former represents a greater intrusion on the rights of the target state and, therefore, is more likely to disrupt international stability.

5) *Measurability*: While the consequences of armed coercion are usually easy to ascertain (e.g., a certain level of destruction), the actual negative consequences of other forms of coercion are harder to measure. This fact renders the appropriateness of community condemnation, and the degree of vehemence contained therein, less suspect in the case of armed force.

6) *Presumptive Legitimacy*: In most cases, whether under domestic or international law, the application of violence is deemed illegitimate absent some specific exception such as self-defense. The cognitive approach is prohibitory. By contrast, most other forms of coercion—again in the domestic and international sphere—are presumptively lawful, absent a prohibition to the contrary. The cognitive approach is permissive. Thus, the consequences of armed coercion are presumptively impermissible, whereas those of other coercive acts are not (as a very generalized rule).[315]

---

[315] *Id.* at 914-15. Schmitt offers a seventh possible criterion:

> Arguably, *responsibility* is a seventh commonality. Armed coercion is the exclusive province of states; only they may generally engage in uses of force across borders, and in most cases only they have the ability to do so with any meaningful impact. By contrast, non-governmental entities are often capable of engaging in other forms of coercion (propaganda, boycotts, etc.). Therefore with armed coercion the likelihood of blurring the relative responsibility of the State, a traditional object of international prescription, and private entities, usually only the object of international administration, narrows. In sum, the consequences of armed coercion are more susceptible to being charged to the State actor than in the case of other forms of coercion. However, this is an issue

These factors, then, are the lens through which information operations may be analyzed to determine both the quantum of their real-world destructiveness and their qualitative conformance with existing Charter standards of permissible and impermissible coercion.[316] They are illustrated in two different formats on the pages that follow. Figure 7 displays Schmitt's cascading use of force analysis. Figure 8 portrays each factor, divided into three levels of intensity. These levels of intensity span the spectrum of actions within each factor; examples are provided for each level. The examples in the white areas will tend to drive the analysis below the Article 2(4) threshold; the examples in the black area will tend to drive the analysis above the threshold. The mid-range examples, both literally and metaphorically in the gray area, may be characterized either way, and, with closer evaluation, may drive the analysis in either direction. The cumulative level of intensity of each of the seven factors determines the overall level of forcefulness, which is either above or below the Article 2(4) threshold.

> of assessing State responsibility, not lawfulness. It is a practical challenge, not a normative one.

Id. at 915, fn. 81.

[316] Schmitt concludes:

> It may appear tortuous to use the prescriptive shorthand (instrument-based classification as a point of departure, rather than simply ask to what degree the consequences of computer network attack threaten shared community values. One might simply look no further than the severity of the consequences. [footnote omitted] Indeed, at conferences and among those who have considered the subject in any depth, there is a tendency to take this stance when struggling with the dilemma of how to account for non-kinetically based harm with a system designed to regulate kinetic activities. The flaw in doing so lies in the fact that it calls for a new normative architecture altogether to handle such actions, an architecture that amounts to more than interpretive dilation of the use of force standard. It would constitute a new standard.

Id. at 917.

Conclusion

There are no clear definitions for a use of force or an armed attack under all circumstances, but this chapter has provided a number of examples of state activities in cyberspace that define the contours of what is and what is not a use of force or an armed attack within the meaning of Articles 2(4) and 51 of the Charter of the United Nations. Since these issues are very fact-dependent, the following definitions derived from this chapter's analysis remain very general:

Use of Force. A state activity that threatens the territorial integrity or political independence of another state constitutes an unlawful use of force within the meaning of Article 2(4) of the Charter of the United Nations.

Armed Attack. A use of force that rises to a certain scope, duration, and intensity threshold constitutes an armed attack within the meaning of Article 51 of the Charter of the United Nations that invokes a state's inherent right of self-defense.

The best way, however, to accurately predict what may be considered a use of force and an armed attack within the meaning of Articles 2(4) and 51 is by studying state practice. In addition to traditional, universally accepted examples such as an armed cross-border invasion, an armed attack may occur when a use of force or an activity not traditionally considered an armed attack is used in such a way that it becomes tantamount in effect to an armed attack. Figure 9 completes the development of a linear model for a heuristic use of force analysis by adding some of the examples discussed in this chapter to the spectrum of interstate relations. This figure should serve as a useful reference tool that summarizes all of the key principles discussed in Chapters 3 through 6.

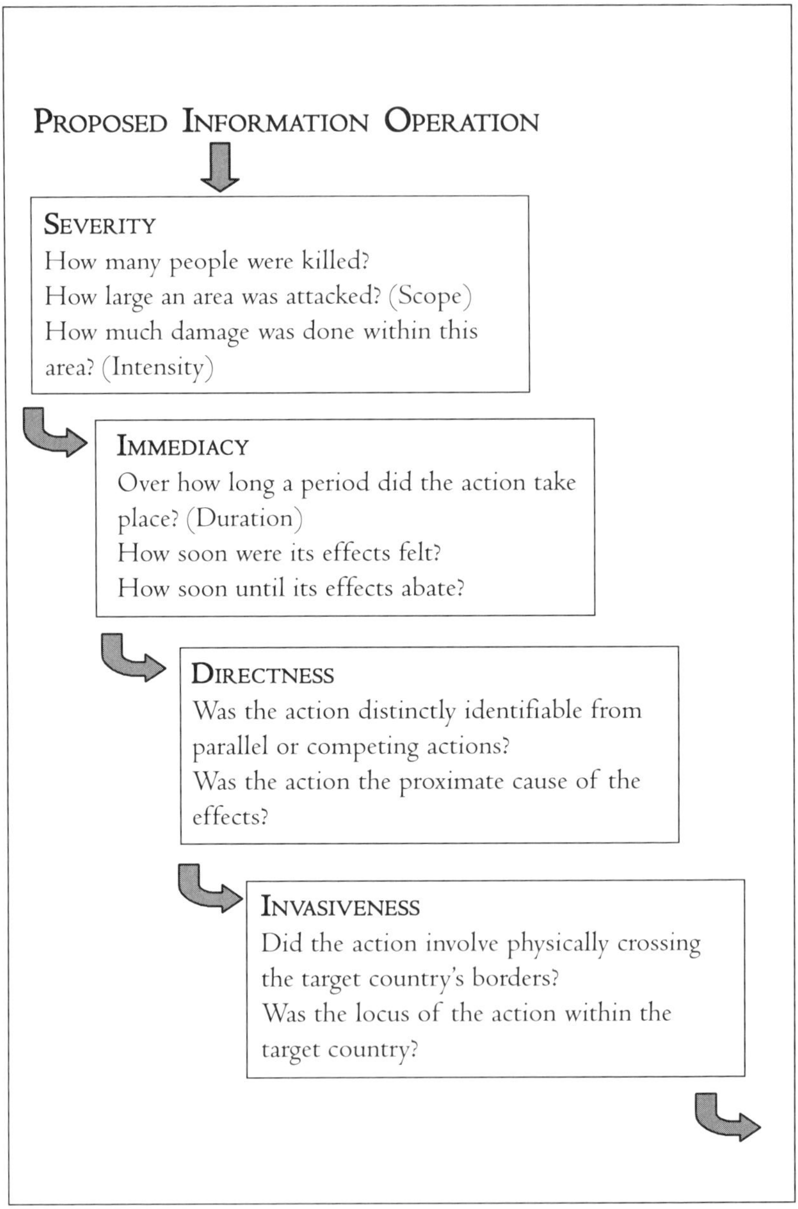

Figure 7. The Schmitt Use of Force Analysis

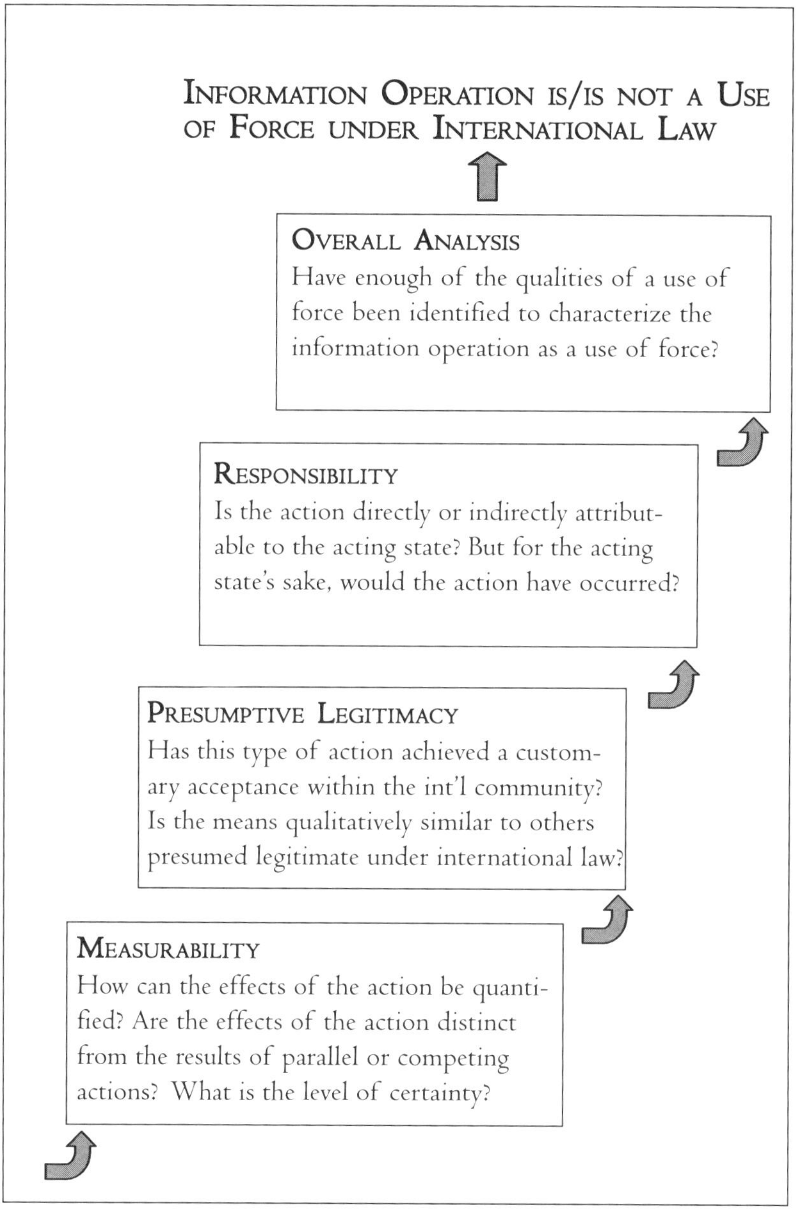

Figure 7. continued

Severity	Immediacy
People Killed; Severe Property Damage	**Seconds to Minutes**
People Injured; Moderate Property Damage	Hours to Days
People Unaffected; No Discernable Property Damage	Weeks to Months

Directness	Invasiveness
Action Sole Cause of Result	**Border Physically Crossed; Action Has Point Locus**
Action Identifiable as One Cause of Result, and to an Indefinite Degree	Border Electronically Crossed; Action Occurs Over Diffuse Area
Action Played No Identifiable Role in Result	Border Not Crossed; Action Has No Identifiable Locus in Target Country

Figure 8. Schmitt Analysis Factors

MEASURABILITY	PRESUMPTIVE LEGITIMACY	RESPONSIBILITY	OVERALL ANALYSIS
Effects Can Be Quantified Immediately by Traditional Means (BDA, etc.) with High Degree of Certainty	Action Accomplished By Means of Kinetic Attack	Responsibility for Action Acknowledged by Acting State; Degree of Involvement Large	Use of Force Under Article 2(4)
Effects Can Be Estimated by Rough Order of Magnitude with Moderate Certainty	Action Accomplished in Cyberspace but Manifested by a "Smoking Hole" in Physical Space	Target State Government Aware of Acting State's Responsibility; Public Role Unacknowledged; Degree of Involvement Low	Arguably Use of Force or Not
Effects Cannot Be Separated from Those of Other Actions; Overall Certainty is Low	Action Accomplished in Cyberspace and Effects Not Apparent in Physical World	Action Unattributable to Acting State; Degree of Involvement Low	Not a Use of Force Under Article 2(4)

Figure 8. continued

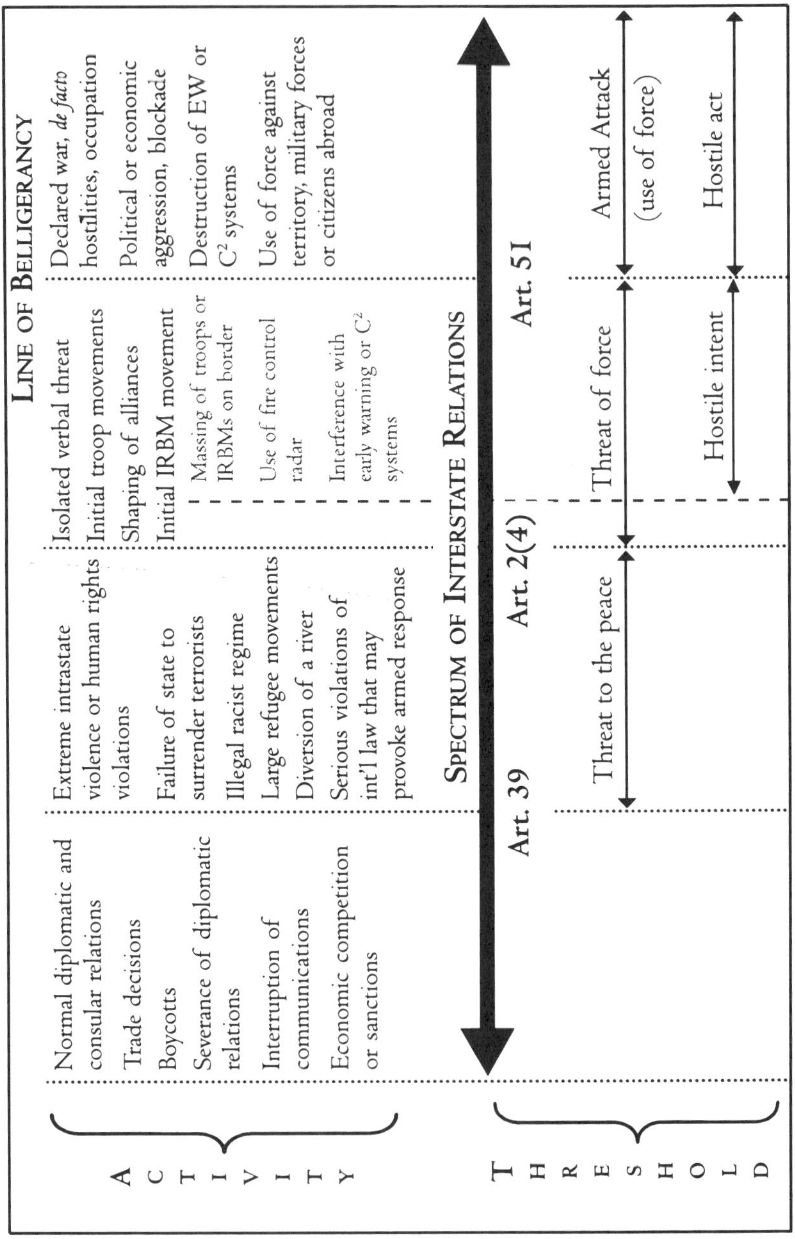

Figure 9. Use of Force Analysis

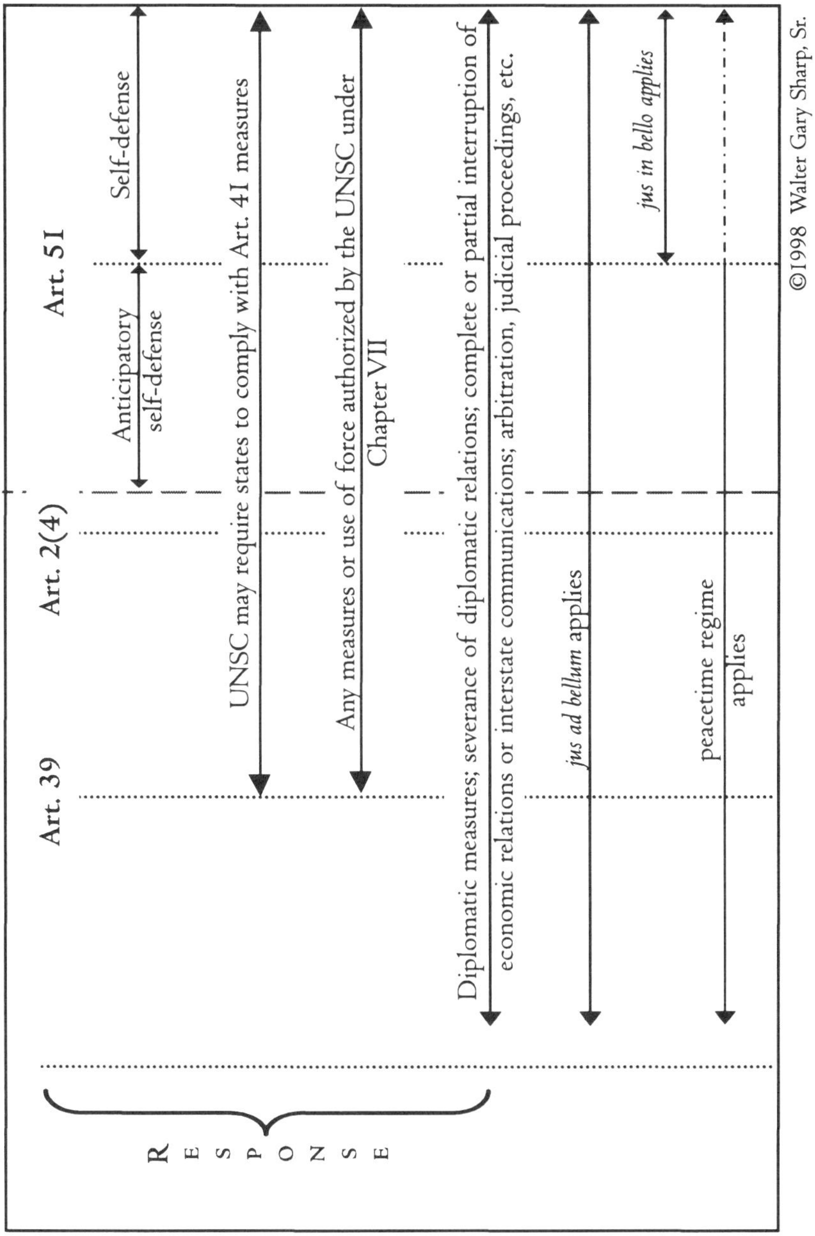

Figure 9. continued

PART III. *JUS IN BELLO*

◆

CHAPTER 7: THE LAW OF ARMED CONFLICT†

The law of information conflict begins with the broader law of armed conflict. That body of law, also known as the law of war or the jus in bello, *governs the behavior of both belligerents and neutrals during hostilities. After a brief historical review, this chapter will concentrate on the four major principles of the law of armed conflict: discrimination, necessity, proportionality, and chivalry. Discrimination is the requirement to distinguish combatants and military objectives from noncombatants (civilians and certain protected persons) and civilian objects, and to attack only the former. Necessity limits the quantum of force used to that required for mission accomplishment and force protection, and forbids any additional use of force (destruction for the sake of destruction). In addition, the subsidiary doctrine of "unnecessary suffering" limits the type of force used, placing off-limits certain inhumane means and methods of war (such as chemical and biological weapons), or weapons which cause superfluous injury, such as exploding bullets or transparent rounds which cannot be located in the body by medical personnel. Proportionality is* not *the requirement that one side can use only as much force as was used against it, and* not *the requirement to employ the same type of weapon as the attacking force, but rather the requirement to balance the concrete and direct military advantage expected from attack against the probable collateral damage that will be done to civilians and their property. Finally, chivalry, far from being an anachronistic throwback to the Middle Ages, is a modern doctrine that permits legitimate ruses of war, but prohibits treacherous acts of perfidy. This distinction is of particular interest in the law of information conflict, in that the majority of information operations are ruses of war and will demand especially close scrutiny under this doctrine. This chapter heavily emphasizes the academic, theoretical aspects of the customary international law of armed conflict, the better to provide a firm foundation for the following chapter's review of the key treaties in*

---

† The introduction to this chapter is drawn substantially from work by Prof. Walter Gary Sharp, Sr., for Aegis Research Corporation.

the field, and with them, a more speculative approach to their application to the law of information conflict.

Over two thousand years ago, Cicero wrote *silent enim leges inter arma, i.e.*, the "laws are inoperative in war."[317] During 20th century military operations, however, nothing could be further from the truth. Colonel Raymond Ruppert, the senior military lawyer in the field during the 1991 Persian Gulf War, described the war as the "most legalistic war we've ever fought."[318] Indeed, the Persian Gulf War was, *per capita*, the largest-ever wartime deployment of lawyers, and the first reservists called to active duty in support of the war were six international lawyers.[319]

Professor John Norton Moore[320] has defined the importance of the role of law during contemporary armed conflict in more depth than any other scholar. From his prescient 1973 Foreign Affairs article on the role of law in the management of national security affairs[321] to his definitive 1992 study on the rule of law during the Persian Gulf War[322] and his dramatic 1997 work on defining a new paradigm of war avoidance centered on democracy, deterrence, and the rule of law,[323] Professor

---

[317] See this famous quote from Cicero's work *Pro Milone*, ch. IV, sec. xi, (44-43 B.C.), THE OXFORD DICTIONARY OF QUOTATIONS 151 (3rd ed. 1980).

[318] Steven Keeva, *Lawyers in the War Room*, A.B.A. J., Dec. 1991, at 52.

[319] *See id.* at 54.

[320] Walter L. Brown Professor of Law and Director of the Center for National Security Law at the University of Virginia School of Law.

[321] John Norton Moore, *Law and National Security*, 51 Foreign Aff. 408, 408-421 1973). [hereinafter *Law and National Security*].

[322] *See generally* MOORE, CRISIS IN THE GULF, *supra* note 96.

[323] John Norton Moore, Toward a New Paradigm: Enhanced Effectiveness in United Nations Peacekeeping, Collective Security, and War Avoidance, 37 VA. J. INT'L L. 811, 811-890 (1997).

Moore has clearly described the fundamentally important value of adhering to the rule of law during armed conflict. Some policy-makers, warfighters, journalists, and scholars incorrectly believe that international law is constraining, and that it prevents the United States from winning wars and accomplishing other international objectives.[324] In sharp contrast, however, Professor Moore aptly notes that

> International law may also help to elaborate the national interest in a variety of operational settings. For example, international law has distilled from centuries of experience a substantial body of norms for the conduct of hostilities. Not to comply with these is to risk breakdowns in military discipline, brutalization of participants with resultant social costs on return to civilian life, unnecessary escalation or continuation of conflict, reciprocal mistreatment of nationals, domestic loss of support and unnecessary destruction of human and material resources.[325]

Indeed, the law of armed conflict developed as self-imposed, advantageous restraints designed to mitigate the savagery of war,[326] which were never intended to impede achieving the goals of hostilities.[327] Official United States doctrine echoes this view. The US Department of the Navy handbook for commanders on the law of naval operations, for example, concludes that the law of armed conflict complements and supports "the principles of warfare embodied in the military concepts of objective, mass, economy of force, surprise, and security."[328] In sum, adherence to the rule

---

[324] *See, e.g.*, Keeva, *Lawyers* at 55.

[325] Moore, *Law and National Security*, *supra* note 321, at 408, 416.

[326] *See* MORRIS GREENSPAN, THE MODERN LAW OF LAND WARFARE 4 (1959) [hereinafter GREENSPAN, THE MODERN LAW OF LAND WARFARE].

[327] *See* COMMANDER'S HANDBOOK, *supra* note 109, at § 5.2.

[328] *Id.*

of law during armed conflict should be a major political objective of all nation states because it strengthens system-wide deterrence and international peace.[329]

Introduction to the Law of Armed Conflict

In early times war was waged "with all the unalleviated cruelty of which human fiendishness is capable."[330] To mitigate this savagery, reciprocal restraints on the conduct of war were believed advantageous and became common as the means and methods of warfare became more destructive.[331] Restraints on the conduct of armed conflict were developed through the practice of almost all societies over thousands of years—from the classical era of the Greeks and Romans to the Middle Ages to the twentieth century.[332] During the fourth century B.C., for example, the Hindu civilization of India produced a book of rules governing land warfare that prohibited the use of poisoned or barbed weapons, and the killing of those belligerents who are asleep, naked, disarmed, or grievously wounded.[333] Similarly, the customs of the ancient Hebrews, set forth in Deuteronomy, required them to spare trees of the field and permitted them to kill only their immediate enemies.[334]

Some of these self-imposed restraints became so universally observed they gradually became legally binding customs of war, and offenders were punishable as war criminals.[335] These customs ultimately

---

[329] *See* MOORE, CRISIS IN THE GULF, *supra* note 96, at 304, 338.

[330] *See* GREENSPAN, THE MODERN LAW OF LAND WARFARE, *supra* note 326.

[331] *See id.* at 4.

[332] *See* ROBERTS & GUELFF, *supra* note 4, at 1-2; FRIEDMAN, *supra* note 63, at 3.

[333] *See* FRIEDMAN, *supra* note 63, at 3.

[334] *See id.*

[335] *See* GREENSPAN, THE MODERN LAW OF LAND WARFARE, *supra* note 326, at 4.

evolved into customary international law[336] and are collectively referred to as the law of armed conflict, the law of war, or the *jus in bello*. They were not codified until the seventeenth century when a Dutch lawyer, Hugo Grotius, compiled and recorded these customs of war in his seminal three-volume work, DE JURE BELLI AC PACIS LIBRI TRES (ON THE LAW OF WAR AND PEACE IN THREE BOOKS).[337] His work is the "most comprehensive attempt to bring together both classical and medieval

---

[336] Arend and Beck describe the process by which customary international law develops:

> [F]or a norm of customary international law to exist, there must be two elements. First there must be state practice. States must comply with the putative rule in their actions. For instance, if there were to be a norm that diplomats were not to be arrested by the host country, states would have to demonstrate a practice of not arresting diplomats. Second, the proposed rule must be perceived to be law. Not only must states actually engage in a practice, they must do so because they believe that the practice is required by law. Thus, in the example given above, states must not only refrain from arresting diplomats, they must do so because they believe that it is *unlawful* to do so.

ANTHONY CLARK AREND and ROBERT J. BECK, INTERNATIONAL LAW AND THE USE OF FORCE: BEYOND THE UN CHARTER PARADIGM 6 (1993).

[337] *See* FRIEDMAN, *supra* note 63, at 14; GREENSPAN, THE MODERN LAW OF LAND WARFARE, *supra* note 326, at 4.

Hugo Grotius (1583-1645) was a "Dutch jurist, humanist, and statesman, whose legal writings laid the foundation for modern international law. ...[H]e began the practice of law [in 1599]. In 1607 he was appointed attorney general of the province of Holland. [His]... first published work on international law, MARE LIBERUM (THE FREE SEA, 1609), challenged the right of any nation to claim any part of the open sea as exclusively its own. Such a claim, Grotius argued, was against natural law and the basic law of humanity. He took the same line of argument in his DE JURE BELLI AC PACIS (ON THE LAW OF WAR AND PEACE, 1625): that war violates natural law, which applies to the conduct of nations and of individuals. His contention was that war can be condoned only if it is for a righteous cause and conciliation has failed, and he called for humanitarian limits on such warfare."

HUGO GROTIUS, MICROSOFT ENCARTA '95 (CD-ROM Multimedia Encyclopedia, 1994).

thought on war" and is considered the "starting point for the study and development of the modern law of war."[338]

The practice of codifying the law of armed conflict in binding international agreements began in the nineteenth century with the 1856 Paris Declaration on Maritime War,[339] and codification accelerated at the turn of the twentieth century with the Hague Peace Conferences of 1899 and 1907.[340] The law of armed conflict has generally developed in two regimes: the Hague regulations, which govern the means and methods of warfare, and the Geneva conventions, which govern the protection of victims of war.[341] Codification of the law of armed conflict peaked, however, just before World War I, and the international community has not since been able to adopt conventions as wide-ranging as were adopted at the two Hague Peace Conferences at the turn of the century.[342]

Most conventions adopted in the following decades simply clarified and adapted the principles of earlier agreements to the changes in the conduct of armed conflicts and military technology.[343] After World War I, for example, only four prominent conventions were adopted: a 1922 treaty that codified rules on submarine warfare, a 1925 treaty that prohibited the use of poisonous gases and bacteriological agents in war, and two 1929 Geneva conventions that addressed protections for the wounded and sick in the field and prisoners of war.[344] Most other

---

338 FRIEDMAN, *supra* note 63, at 14.

339 *See* ROBERTS & GUELFF, *supra* note 4, at 2-4.

340 *See id.* at 3.

341 *See generally id.*

342 *See* SCHINDLER & TOMAN, ARMED CONFLICTS, *supra* note 5, at vii.

343 *See id.* at vii-viii.

344 *See id.* at viii.

attempts to advance the law of armed conflict failed.[345] Similarly, the end of World War II saw a period of neglect of the law of armed conflict.[346] The only significant treaties that immediately followed World War II were the four Geneva Conventions of 1949 and a 1954 Geneva convention for the protection of cultural property.[347]

The international community renewed its interest in the law of armed conflict in the mid-1960s as a result of the Viet Nam war and a number of non-international armed conflicts that were ongoing worldwide.[348] Extended efforts by the United Nations and the International Committee of the Red Cross yielded two extensive protocols to the four Geneva Conventions of 1949, significantly advancing the law of armed conflict.[349] Nevertheless, many deficiencies remained.[350] For example, two scholars observe that

- the breadth of the existing rules is very complex (the four Geneva Conventions of 1949 and their two Protocols alone contain over 600 articles);
- the dissemination of the rules is inconsistent;
- many rules are open to various interpretations;
- there are no specific rules for many new weapons;
- there are no rules regarding the application of the law of armed conflict to hostilities when U.N. forces are engaged

---

[345] *See id.*

[346] *See id.*

[347] *See id.*

[348] *See id.* at ix.

[349] *See id.* at ix-x.

[350] *See id.* at x.

or when multinational armed forces are involved in an armed conflict;
- the maritime warfare conventions are outdated;
- economic warfare remains outside the scope of the law of armed conflict; and,
- the rights and duties of belligerents and neutrals are highly uncertain.[351]

Despite these deficiencies, there is much evidence that the law of armed conflict has influenced the conduct of hostilities by bringing about a degree of acceptance and observance of certain general principles.[352] Indeed, with the creation of the International Criminal Tribunal for the Former Yugoslavia in 1993 by the Security Council and the adoption of the Rome Statute of the International Criminal Court in 1998, the international community has clearly demonstrated its expectation that belligerents conduct hostilities in accordance with the law of armed conflict.

The law of armed conflict is, by definition, that international law which applies to state activities *during* international armed conflict.[353] The issue of when such conflict exists is a separate issue addressed by a

---

[351] *See id.*

[352] *See* ROBERTS & GUELFF, *supra* note 4, at 14.

[353] There is a small portion of the law of armed conflict that does apply to internal armed conflicts which occur within the territory of a single state. Common Article 3 of the four Geneva Conventions of 1949, for example, sets forth a few minimum standards that shall apply in the "case of armed conflict not of an international character occurring in the territory" of a state. *See* 1949 Geneva Convention No. I, *supra* note 148, *reprinted in id.* at 171. A protocol to the four Geneva Conventions of 1949 relating to non-international armed conflicts developed and supplemented the brief provisions of Common Article 3. *See* ROBERTS & GUELFF, *supra* note 4, at 13. *See also* Protocol Additional to the Geneva Conventions of 12 August 1949, and Relating to the Protection of Victims of Non-International Armed Conflicts, Dec. 12, 1977, 1125 U.N.T.S. 609 [hereinafter 1977 Geneva Protocol II], *reprinted in id.* at 449.

international armed conflict exists between states in each of three factual conditions: the declaration of war, the occurrence of "any other armed conflict" between two or more contracting parties, even if the state of war is not recognized by one of them, and in all cases of partial or total occupation, even if met with no armed resistance.[354]

Despite the emphasis in codifying the law of armed conflict over the last century, it is very important to note that customary international law continues to exist independent of treaty law, which serves to clarify and implement the general principles of the customary law of armed conflict.[355]

---

[354] *See* 1949 Geneva Convention No. IV, *supra* note 148, *reprinted in* ROBERTS & GUELFF, *supra* note 4, at 272; 1949 Geneva Convention No. IV Commentary, *supra* note 152, at 20.

[355] *See* ROBERTS & GUELFF, *supra* note 4, at 2-4. This principle is codified in the preamble of the 1907 Hague Convention IV, which provides that

> in cases not included in the Regulations adopted by them, the inhabitants and the belligerents remain under the protection and the rule of the principles of the law of nations, as they result from the usages established among civilized peoples, from the laws of humanity, and the dictates of the public conscience.
>
> Hague Convention IV Respecting the Laws and Customs of War on Land, Oct. 18, 1907, preamble, 36 Stat. 2295, 1 Bevans 643 [hereinafter 1907 Hague Convention IV, Customs of War on Land], *reprinted in* ROBERTS & GUELFF, *supra* note 4, at 44. Similarly, Article 2 of the 1977 Geneva Protocol I provides that
>
> [i]n cases not covered by this Protocol or by other international agreements, civilians and combatants remain under the protection and authority of the principles of international law derived from established custom, from the principles of humanity and from the dictates of public conscience.

Protocol Additional to the Geneva Conventions of 12 August 1949, and Relating to the Protection of Victims of International Armed Conflicts, art. 1(2), Dec. 12, 1977, 1125 U.N.T.S. 3 [hereinafter 1977 Geneva Protocol I], *reprinted in* ROBERTS & GUELFF, *supra* note 4, at 389.

Indeed, a significant portion of the law of armed conflict continues to exist in unwritten customary principles.[356] Accordingly, customary international law plays a very important role in determining the legality of means and methods of warfare under the law of armed conflict.

Although belligerents are lawful targets, the most fundamental customary tenet of the contemporary law of armed conflict is that the right of belligerents to adopt means of injuring the enemy is not unlimited.[357] From this tenet flow several interrelated principles of customary international law. There are numerous ways to organize and describe the law of armed conflict. Here, we recount the basic concepts before adding greater complexity.

Discrimination

Discrimination is the customary rule of international law, which requires that belligerents distinguish between combatants and noncombatants, avoid targeting civilians and their property, and take all reasonable precautions against injuring civilians or damaging their property in the course of striking military targets.[358]

---

[356] *See* ROBERTS & GUELFF, *supra* note 4, at 4.

[357] *See id.* This principle is codified in Article 22 of the 1907 Hague Convention IV, Customs of War on Land, Annex (Regulations), *supra* note 355, at 643 *reprinted in* ROBERTS & GUELFF, *supra* note 4, at 48.

[358] COMMANDER'S HANDBOOK, *supra* note 109, at § 8.1. The Commander's Handbook captures this rule as three interrelated proscriptions:

1. The right of belligerents to adopt means of injuring the enemy is not unlimited. [footnote omitted]

2. It is prohibited to launch attacks against the civilian population as such. [footnote omitted]

Combatants and other military objectives are lawful Targets. Combatants are

> Those persons who have the right under international law to participate directly in armed conflict during hostilities. Combatants, therefore, include all members of the regularly organized armed forces of a party to a conflict (except medical personnel, chaplains, civil defense personnel, and members of the armed forces who have acquired civil defense status), as well as irregular forces who are under responsible command and subject to military discipline, carry their arms openly, and otherwise distinguish themselves clearly from the civilian population. [footnote omitted][359]

Military objectives have been authoritatively defined as:

> Combatants and those objects which, by their nature, location, purpose, or use, effectively contribute[360] to the enemy's

---

> 3. Distinctions must be made between combatants and noncombatants, to the effect that noncombatants be spared as much as possible. [footnote omitted]

Id. See also FM 27-10, *supra* note 154, and US DEP'T OF THE AIR FORCE, AFP 110-31, INTERNATIONAL LAW—THE CONDUCT OF ARMED CONFLICT AND AIR OPERATIONS § 5-3 and 11-2 (1993) [hereinafter AFP 110-31]. The authoritative articulation of this principle under international law is found in article 48 of Protocol I to the Geneva Conventions: "In order to ensure respect for and protection of the civilian population and between civilian objects, the Parties to the conflict shall at all times distinguish between the civilian population and combatants and civilian objects and military objectives and accordingly shall direct their operations only against military objectives."

1977 Geneva Protocol I, *supra* note 355, at art. 48.

[359] COMMANDER'S HANDBOOK, *supra* note 109, at § 5.3.

[360] Prof. Schmitt notes that [o]bjects which make an effective contribution are those that are by nature beneficial to the military effort: weapons, aircraft, communications, etc." Schmitt, *Bellum Americanum, supra* note 360, at 1075-76, *citing* INTERNATIONAL COMMITTEE

> warfighting or war sustaining capability and whose total or partial destruction, capture, or neutralization would constitute a definite military advantage[361] to the attacker under the circumstances at the time of the attack. [footnote omitted] Military advantage may involve a variety of considerations, including the security of the attacking force.[362]

TO THE GENEVA CONVENTIONS OF 12 AUGUST, 1949 at 635 (Yves Sandoz, *et al.*, eds., 1987) [hereinafter RED CROSS COMMENTARY]. This dichotomy is somewhat blurred in the Commander's Handbook, which, as quoted *supra*, bases the effectiveness of the contribution on "nature, location, purpose, *or* use." [emphasis added] COMMANDER'S HANDBOOK, *supra* note 109, at § 5.3.

[361] Professor Schmitt explains that "[d]efinite military advantage refers to objects which contribute by virtue of their location, such as bridges and buildings used for shelter." Schmitt, *Bellum Americanum*, *supra* note 303, at 1075-76 *citing* RED CROSS COMMENTARY, *supra* note 360, at 635. Although Schmitt describes the Red Cross Commentary as requiring a less restrictive disjunctive test for targeting ("effective contribution" *or* "military advantage"), and the Commander's Handbook advances a more restrictive conjunctive test ("effective contribution" *and* "military advantage"), this is largely a distinction without a difference, as it is difficult to conceive of a target offering one but not the other.

[362] COMMANDER'S HANDBOOK, *supra* note 109, at § 8.1.1. The Commander's Handbook goes on to give a thorough roster of lawful targets:

> Proper targets for naval attack include such military objectives as enemy warships and military aircraft, naval and military auxiliaries, naval and military bases ashore, warship construction and repair facilities, military depots and warehouses, petroleums/oils/lubricants (POL) storage areas, docks, port facilities, harbors, bridges, airfields, military vehicles, armor, artillery, ammunition stores, troop concentrations and embarkation points, lines of communication and other objects used to conduct or support military operations.

Id. The authors continue, listing targets which, though not immediately threatening or even inherently military, are nonetheless lawful:

> Proper naval targets also include geographic targets, such as a mountain pass, [footnote omitted] and buildings and facilities that provide administrative and personnel support for military and naval operations such as barracks, communications and command and control facilities, headquarters buildings, mess halls, and training areas.

Noncombatants and civilian objects are unlawful targets. Noncombatants are

> Those individuals who do not form a part of the armed forces and who otherwise refrain from the commission or direct support of hostile acts. In this context, noncombatants and, generally, the civilian population, are synonymous. The term noncombatants may, however, also embrace certain categories of persons who, although members of or accompanying the armed forces, enjoy special protected status, such as medical officers, corpsmen, chaplains, technical (i.e., contractor) representatives, and civilian war correspondents... The term is also applied to armed forces personnel who are unable to engage in combat because of wounds, sickness, shipwreck, or capture. [footnote omitted][363]

Civilian objects consist of

> All civilian property and activities other than those used to support or sustain the enemy's warfighting capability [footnote omitted]. . . the intentional destruction of food, crops, livestock, drinking water, and other objects indispensable to the survival of the civilian population, for the specific purpose of denying the civilian population of their use, is prohibited.[364]

> Proper economic targets for naval attack include enemy lines of communication, rail yards, bridges, rolling stock, barges, lighters, industrial installations producing war-fighting products, and power generation plants. Economic targets of the enemy that indirectly but effectively support and sustain the enemy's war-fighting capability may also be attacked. [footnote omitted]

Id.

[363] COMMANDER'S HANDBOOK, *supra* note 109, at § 5.3. Chapter 11 of the Commander's Handbook goes into much greater detail about noncombatants and their immunities. *Id.* at §§ 11.1-11.8.

[364] *Id.* at § 8.1.2.

Graphically, then, this distinction may be represented as a two-region chart with a vertical division, as shown below.

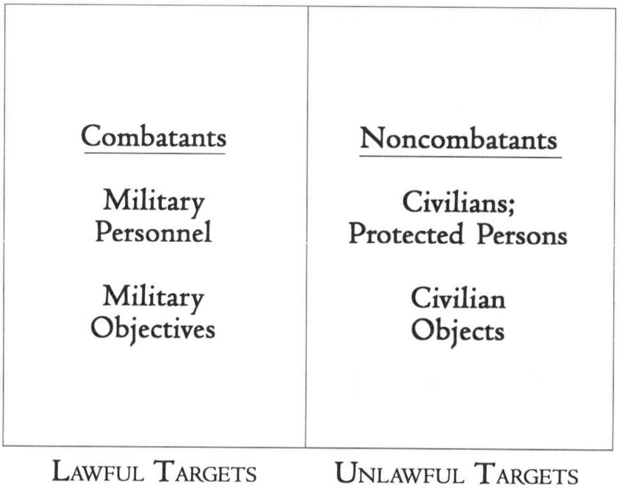

Figure 10. Discrimination

Using these distinctions as its basis, the doctrine of indiscriminate effect[365] limits the use of weapons in an inherently indiscriminate manner. According to the Commander's Handbook,

> weapons which by their nature are incapable of being directed against specifically military objectives, and therefore that put noncombatants at equivalent risk, are forbidden due to their indiscriminate effect. [footnote omitted] A few weapons, such as poisoned projectiles, are unlawful, no matter how employed.

---

[365] 1977 Geneva Protocol I, *supra* note 355, at arts. 51 para 4(b) & 51 para 5.

> [footnote omitted][366] Others may be rendered unlawful by alteration, such as by coating ammunition with a poison.[367] Still others may be unlawfully employed, such as by setting armed contact naval mines adrift so as to endanger innocent as well as enemy shipping.[368] And finally, any weapon may be set to an unlawful purpose when it is directed against noncombatants[369] and other protected persons and property.[370]

Although the concept of indiscriminate effect[371] and the academic distinction between the armed forces and the civilian populace is clear in

---

[366] While this is an example of unnecessary suffering, a weapon could be unlawful due to its inherently indiscriminate nature, such as a SCUD missile with no guidance package, launched purely as a terror weapon. *See* section on Unnecessary Suffering, *infra*.

[367] This is a case similar to the one above. A poisonous coating would violate the principle of unnecessary suffering, but a weapons modification, such as the removal of the guidance package from a SCUD missile, would violate the principle of indiscriminate effect. *See* section on Unnecessary Suffering, *infra*.

[368] This is an example of purely indiscriminate effect, based not on the quality of the weapon, but the complete negligence in its targeting.

[369] This final example is also purely indiscriminate, but based on the wilful, and not merely negligent, targeting of noncombatants.

[370] COMMANDER'S HANDBOOK, *supra* note 109, at § 9.1.

[371] The Commander's Handbook elaborates on the concept:

> Weapons that are incapable of being controlled (i.e., directed at a military target) are forbidden as being indiscriminate in their effect. [footnote omitted] Drifting armed contact mines and long-range unguided missiles (such as the German V-1 and V-2 rockets of Word War II) fall into this category. A weapon is not indiscriminate simply because it may cause incidental or collateral civilian casualties, provided such casualties are not foreseeably excessive in light of the expected military advantage to be gained. [footnote omitted] An artillery round that is capable of being directed with a reasonable degree of accuracy at a military target is not an indiscriminate weapon simply because it may miss its mark or inflict collateral damage. Conversely, uncontrolled balloon-borne bombs, such as those released by the Japanese against the west coast of the

theory, their application to real-world targeting issues can become hazy and complex. Nowhere is this more true than in the world of information operations, where "shooters" rarely "see" their targets, where interconnected, dual-use infrastructures support civilian and military customers, and where unintentional secondary and tertiary ripple effects can cause results unforeseeable, or at least unforeseen, by the attacker. Separating military targets from civilian sites presents a special problem in intelligence collection, legal analysis, and operational execution.

The nature of modern weapons and the nature of modern adversaries are shaping the evolution of the rule of discrimination at the beginning of the 21st century. As the rapidly advancing technology of First World militaries allows them to discriminate between military objectives and civilian objects with more precision than ever before, the failure of Third World adversaries to abide by the customary rule of discrimination is retarding this advance. On the offensive, such outlaw countries, or the subnational groups which they support, often engage in operations whose sole purpose is to spread the terror arising from attacks on civilians. They spurn the very concept of discrimination as a tool used by militarily powerful nations to keep future conflicts on a favorable, qualitatively symmetrical and quantitatively asymmetrical basis.

On the defensive, these same nations frequently resort to intentionally placing civilians near military objectives, or *vice versa.*[372] This,

---

> United States and Canada inn World War II lack that capability of direction and are , therefore, unlawful. [footnote omitted]

Id. at § 9.1.2.

[372] Schmitt notes:

> The use by Saddam Hussein of civilians and cultural sites as shields is well known. [footnote omitted] Indeed, since the conflict ended, Iraqi civilians have flooded potential targets on numerous occasions to protect them in the face of threatened air attacks against which the Iraqi military would likely prove impotent.

of course, is a violation of international law.[373] While the forced co-location of civilian objects does not render a military objective unlawful *per se*, the attacking force does assume the additional responsibility of including the newly introduced civilians in its calculation of proportionality (discussed *infra*).[374] Thus skewed, the proportionality calculation may very well render a previously vulnerable target effectively off limits.

Even without the intentional use of human shields, the problem remains in a somewhat attenuated form. As mentioned earlier in this text, the increasing interconnectedness of information systems vital to a nation's critical infrastructure, and the especial reliance placed by developing countries on the dual use of such systems, renders discrimination far more complex in cyberspace than in physical space. There are two keys to following the just-discussed prescriptions of international law. First, an attacker must gather finely sifted intelligence so that competent decisions can be made regarding which computers, or indeed, which programs within

---

Schmitt, *Bellum Americanum*, *supra* note 303, at 1077-78. In a footnote, he continues:

> After attacking Kuwait, the Iraqis used Western and Kuwaiti hostages to shield their military sites from Coalition air attacks. The non-Kuwaiti civilians were eventually released in December 1990 when the tactic resulted in near universal condemnation... Other examples included the dispersal of helicopters to residential areas, placing surface-to-air missiles in a school in a populated area of Kuwait City, and placement of fighter aircraft next to the Temple of Ur.

Id., fn. 110, at 1078.

[373] Using civilians in this way constitutes a grave breach of the Geneva Conventions and Protocol I. 1949 Geneva Convention IV, *supra* note 148, at arts. 29 & 147 and, 1977 Geneva Protocol I, *supra* note 355, at arts. 75 para 2(c) & 85 para 2. States Parties to Protocol I are obligated to "endeavour to remove the civilian population, individual civilians and civilian objects under their control from the vicinity from military objectives," 1977 Geneva Protocol I, *supra* note 355, at art. 58, para a.

[374] 1977 Geneva Protocol I, *supra* note 355, at art. 51, paras 7-8.

computers, are lawful targets. Second, information weapons sufficiently precise to attack specifically selected cybernetic targets must be developed. While a brute-force approach would remain perfectly lawful for an attack against a stand-alone system, a more discriminating attack would have to be mounted against a system of systems, if the attacker wishes to comply with international law and prevent unanticipated secondary and tertiary consequences far more destructive than the primary attack.

Necessity

The umbrella term "necessity" actually covers two interrelated concepts, "military necessity" and "unnecessary suffering." The concept of military necessity, more quantitative than qualitative, demands that:

> Only that degree and kind of force, not otherwise prohibited by the law of armed conflict, required for the partial or complete submission of the enemy with a minimum expenditure of time, life, and physical resources may be applied.[375]

---

[375] COMMANDER'S HANDBOOK, *supra* note 109, at § 5.2. The Commanders' Handbook goes on to quote a seminal elaboration on this definition:

> Military necessity has been invoked by the defendants as justifying the killing of innocent members of the population and the destruction of villages and towns in the occupied territory. Military necessity permits a belligerent, subject to the laws of war, to apply any amount and kind of force to compel the complete submission of the enemy with the least possible expenditure of time, life, and money. In general, it sanctions measures by an occupant necessary to protect the safety of his forces and to facilitate the success of his operations. It permits the destruction of life of armed enemies and other persons whose destruction is incidentally unavoidable by the armed conflicts of the war; it allows the capturing of armed enemies and others of peculiar danger, but it does not permit the killing of innocent inhabitants for the purposes of revenge or the satisfaction of a lust to kill. The destruction of property to be lawful musty be imperatively demanded by the necessities of war. Destruction as en end in itself is a violation of international law. There must be some reasonable connection between the destruction of property and the overcoming of the enemy forces. It is lawful to destroy railways, lines of communication, or any other property that

That is, only that quantum of force required for lawful mission accomplishment may be employed; any amount above that level is unlawful.

Central to the concept of military necessity is that it is not superseded by exigent circumstances, but, rather, was specifically designed to cover military exigencies:

> It is an essence of war that one or the other side must lose and the experienced generals and statesmen knew this when they drafted the rules and customs of land warfare. In short, these rules and customs of warfare are designed specifically for all phases of war. They comprise the law for such emergency. To claim that they can be wantonly—and at the sole discretion of any one belligerent—disregarded when he considers his own situation to be critical, means nothing more or less than to abrogate the laws and customs of war entirely.[376]

Accepting, however, that the principle of military necessity is at all times operative, the commander is given wide latitude in assessing the situation before him. As Hanseman points out:

> Obviously, commanders cannot always tell just how much force they will need to conclude an attack successfully, and there is no requirement that they know. They can make estimates, and even add a little safety margin. What they

---

> might be utilized by the enemy. Private homes and churches even may be destroyed if necessary for military operations. It does not admit the wanton devastation of a district or the willful infliction of suffering upon its inhabitants for the sake of suffering alone.

The Hostages Case (United States v. List et al.), 11 TWC 759, 1253-54 (1950), *quoted in id.*

[376] *The Krupp Trial (Trial of Alfred Felix Alwyn Krupp von Bohlen und Halbach and Eleven Others)*, 10 LRTWC 139 (1949), *quoted in* COMMANDER'S HANDBOOK, *supra* note 109, at § 5.2.

> cannot do is deliberately apply more force than they believe will possibly be necessary, in hopes of killing rather than capturing enemy combatants or civilians.[377]

Based on the advice of his staff and his own training and experience, the commander is charged with making a determination that is reasonable in the light of the information he has or should have at the time of the targeting decision. The requirement of the doctrine of military necessity is principally one of the quantity of force to be expended against a target.

The complementary doctrine of "unnecessary suffering" is primarily qualitative in nature, but it does have a quantitative aspect. This latter aspect is depicted in Figure 11, on the following page, as simply that quantum of force above the free-floating threshold of military necessity.

In addition to this straightforward quantitative component, there is a slightly more complex qualitative component.[378] This aspect is rooted in the famous Martens Clause:

> Until a more complete code of laws has been issued, the high Contracting Parties deem it expedient to declare that, in cases not included in the regulations adopted by them, the inhabitants and belligerents remain under the protection and the rule of principles of the laws of nations, as they result from

---

[377] Robert G. Hanseman, Capt, USAF, *The Realities and Legalities of Information Warfare*, 42 A.F.L. REV 173, 182-83 (1997). Capt Hanseman also cites AFI 14-207, which requires United States Air Force targeting specialists to advise USAF commanders on the type and amount of force to apply to any given target. AFI 14-207, AIR FORCE TARGETING (25 Nov. 1993) at § 6.3.

[378] Some commentators refer to this qualitative component of unnecessary suffering as the principle of "humanity," while others use the term as a synonym for the principle of proportionality, discussed *infra*. To avoid confusion here, the term "humanity" will be applied in neither context.

the usages established among civilized peoples, from the laws of humanity, and from the dictates of public conscience.[379]

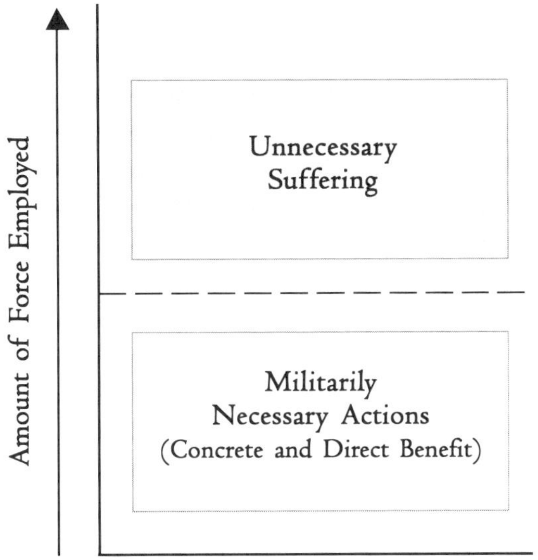

Figure 11. Military Necessity and Unnecessary Suffering (Quantitative)

This "savings clause" from the early Hague law has assumed an independent status as a statement of the irreducible standards below which behavior, even under the stresses of combat, may not sink.

---

[379] 1907 Hague Convention IV, Customs of War on Land, *supra* note 355, at 631.

As Prof. Schmitt points out:

> Putting this cumulative [quantitative] component of humanity [unnecessary suffering] aside, what remains are *ab initio* prohibitions on the methods and means of warfare that are not so much inhumane as inhuman. We intuitively recognize them as wrongful regardless of the context in which they occur. To some extent, they are acts which violate the "dictates of public conscience," [footnote omitted] acts that civilized people *just don't do.*[380]

These limitations were first applied very narrowly, to poisons, burning agents, or bullets of a certain size, shape, or composition.[381] Over time, however, approaches to limiting unnecessary suffering has become more broadly categorical.[382] Schmitt discusses this trend toward banning weapons toward which there is broad, public antipathy (*i.e.*, chemical weapons, biological weapons, and landmines), and states that an unintended consequence of this humanitarian impulse will be to take certain nonlethal

---

[380] Schmitt, *Bellum Americanum, supra* note 303, at fn. 130, at 1084.

[381] Such early treaties included the St. Petersburg Declaration Renouncing the Use, in Time of War, of Explosive Projectiles Under 400 Grammes Weight, December 11, 1868, [hereinafter St. Petersburg Declaration] *reprinted in* ROBERTS & GUELFF, *supra* note 4, at 30-3, the Hague Declaration 2 Concerning Asphyxiating Gases, July 29, 1899 [hereinafter 1899 Hague Declaration 2, Asphyxiating Gases], *reprinted in* ROBERTS & GUELFF, *supra* note 4, at 35, and the Hague Declaration 3 Concerning Expanding Bullets, July 29, 1899 [hereinafter 1899 Hague Declaration 3, Expanding Bullets], *reprinted in* ROBERTS & GUELFF, *supra* note 4, at 40. The development of the law through these and many other treaties is examined in greater depth in Chapter 8, *infra*.

[382] The Commander's Handbook summarizes:

> Antipersonnel weapons are designed to kill or disable enemy combatants and are lawful notwithstanding the death, pain, and suffering they inflict. Weapons that are designed to cause unnecessary suffering or superfluous injury are, however, prohibited because the degree of pain or injury, or the certainty of death they produce is needlessly or clearly disproportionate to the military advantage to be gained by their use. Poisoned projectiles and small arms ammunition intended to cause superfluous injury or unnecessary suffering fall into this category.

COMMANDER'S HANDBOOK, *supra* note 109, at § 9.1.2.

weapons off the table, leaving only traditional lethal weapons at the commander's disposal.[383]

The doctrine of unnecessary suffering prohibits both force that is quantitatively beyond the upper limit of military necessity and that is qualitatively distasteful.[384] For our purposes, we can reduce this distinction to a slightly more complex version of Figure 12, pictured on the next page. The vertical axis represents the permissible and impermissible quanta of force, while the horizontal axis distinguishes the lawful from the unlawful weapons.

---

[383] Schmitt, *Bellum Americanum, supra* note 303, at 1084-85. Schmitt sees this trend continuing into the future, affecting the development and deployment of conventional and information weapons:

> Recall just some of the weapons imagined for the twenty-first century—acoustic weapons that induce vomiting, microwaves that cause the human body to heat up, and electromagnetic pulses that will cause an airplane to fall to the earth because its engine shuts down. Such weapons may be humane in certain circumstances, but there is little doubt that many individuals will react to them viscerally as *inhuman*. Given the current trend in humanity-based conventions, we can expect many of these weapons to be targeted for prohibition, regardless of the possibilities the offer for proportionate use.

Id. at 1085. The author believes that Prof. Schmitt's analysis can be advanced by distinguishing weapons that target human beings (*i.e.*, the first two weapons cited above) from weapons that target equipment (*i.e.*, the last weapon cited above). If an aircraft is "shot down" by EMP rather than a missile, and the pilot is afforded an even better chance of ejecting than if he had taken a missile hit, then the current humanitarian trend would quite probably favor the development and deployment of such a system. The principle criterion appears to be the gruesomeness with which the proposed weapon acts on the human being targeted.

[384] In addition to the examples cited above, the use of projectiles that are difficult to locate by x-ray or the naked eye (such as clear plastic bullets) are prohibited by the Convention if the victim's medical treatment were made more dangerous and complex. Convention on Prohibitions or Restrictions on the Use of Certain Conventional Weapons Which May be Deemed to be Excessively Injurious or to Have Indiscriminate Effects, Protocol I, 10 April 1981 [hereinafter Indiscriminate Effects], *reprinted in* ROBERTS & GUELFF, *supra* note 4, at 479.

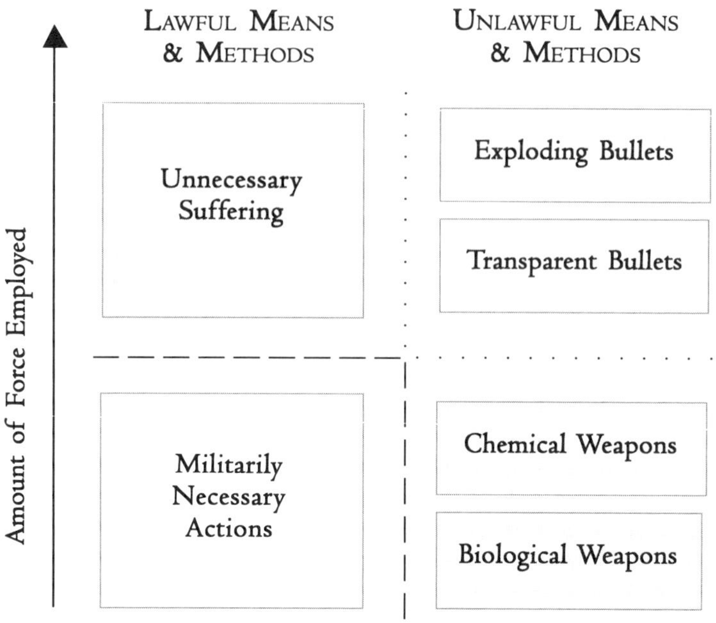

Figure 12. Military Necessity and Unnecessary Suffering (Quantitative and Qualitative)

Proportionality

Just as the principle of necessity prohibits the use of superfluous force against combatants, the complementary principal of proportionality limits the effects of an attack against noncombatants.[385] While the principle of discrimination prohibits attacks against civilians *per se*, "it is not unlawful to cause *incidental* injury to civilians, or collateral damage to civilian objects,

---

[385] *See generally* William J. Fenrick, *The Rule of Proportionality and Protocol I in Conventional Warfare*, 98 MIL. L. REV. 91 (1982) and Judith G. Gardam, *Proportionality and Force in International Law*, 87 AM. J. INT'L L. 391 (1993).

during an attack upon a legitimate military target."[386] The principle of proportionality prohibits attacks

> which may be expected to cause incidental loss of civilian life, injury to civilians, damage to civilian objects, or a combination thereof, which would be excessive in relation to the concrete and direct military advantage anticipated.[387]

This prohibition is expressed as a balancing test between the anticipated "concrete and direct"[388] military advantage on one hand, and the expected civilian losses on the other.[389] As the value of a potential target increases, so does the level of permissible collateral damage. Figure 13 on the next page makes this balancing test clear:

---

[386] COMMANDER'S HANDBOOK, *supra* note 109, at § 8.1.2.1.

[387] 1977 Geneva Protocol I, *supra* note 355, at arts. 51, para 5(b) and 57, para 2(a)(iii).

[388] This term refers to "the advantage anticipated from the specific military operation of which the attack is a part taken as a whole and not from isolated or particular parts of the operation." BOTHE, PARTSCH & SOLF, NEW RULES FOR VICTIMS OF ARMED CONFLICTS [hereinafter BOTHE, PARTSCH & SOLF, NEW RULES], *quoted in* COMMANDER'S HANDBOOK, *supra* note 109, at arts. 51, para 5(b) and 57, para 2(a)(iii). By this standard, an attack need not stand on its own in the proportionality analysis, but may be viewed as an integral part of a much larger effort. This would permit attacks that, in and of themselves, provide too little "concrete and direct" military advantage to justify the damage done to surrounding civilian objects.

[389] Hanseman points out:

> The military advantage gained by attacking military structures is generally minuscule compared to the resulting loss of human life and culture, and thus runs afoul of proportionality. This is why attacking hospitals, schools, religious structures and other cultural institutions is banned [under the independent doctrine of discrimination], unless the enemy is taking advantage of the situation by hiding military assets there.

Robert G. Hanseman, *The Realities and Legalities of Information Warfare*, 42 A.F. L. REV. 173 (1997) at 182. In a sense, the principle of discrimination is a codification of this repeated experience of proportionality calculations against these categories of targets.

Figure 13. Proportionality

A corollary of this principle, explicitly stated in Article 57 of Protocol I,[390] is that "commanders must take all reasonable[391] precautions, taking into

[390] 1977 Geneva Protocol I, *supra* note 355, at arts. 51, para 5(b) and 57, para 2(a)(iii).

[391] Protocol I requires that belligerents take all *feasible* precautions, setting a very high standard, closer to "possible" than "practical. *Id* at arts. 51, para 5(b) and 57, para 2(a)(iii).

This was interpreted by NATO countries (Belgium, the Netherlands, Italy, and the United Kingdom) to actually mean "that which is practical, or practically possible, taking into account all circumstances at the time, including those relevant to the success of military operations," creating a somewhat lower, if more reasonable, standard. BOTHE, PARTSCH & SOLF, NEW RULES, *supra* note 388, *quoted in* COMMANDER'S HANDBOOK, *supra* note 109, at § 8.1.2.1, fns. 17 and 18.

The International Committee of the Red Cross [ICRC] has, understandably, elaborated the highest standard in its Commentary on Protocol I:

> [T]he identification of the objective, particularly when it is located at a great distance, should be carried out with great care. Admittedly, those who plan or decide upon such an attack will base their decision on information given them, and they cannot be expected to have personal knowledge of the objective to be attacked and of its exact nature. However, this does not detract from their responsibility, and in case of doubt, even if there is only a slight doubt, they must call for additional information and if need be give orders for further

account military and humanitarian considerations, to keep civilian casualties and damage to the minimum consistent with mission accomplishment and the security of the force."[392]

These decisions must be based on an "honest and reasonable estimate of the facts,"[393] available at the time the decision is made. Finally,

> the commander must decide, in light of all the facts known or reasonably available to him, [footnote omitted] including the need to conserve resources and complete the mission successfully, whether to adopt an alternative method of attack,

---

> reconnaissance... The evaluation of the information obtained must include a serious check of its accuracy.

RED CROSS COMMENTARY, *supra* note 360, at 680-81.

[392] COMMANDER'S HANDBOOK, *supra* note 109, at § 8.1.2.1. P. J. O'Rourke recounts a recent example of this doctrine in action:

> Colonel Snell . . . all but emitted a visible glow of pride in Operation Just Cause. What made Snell so happy, however, was not the brave fighting his troops had done or the modest casualties they'd suffered doing it but something it never occurred to me a professional soldier would prize. "Where my soldiers fought," said Snell, "their restraint was remarkable." Even in the heat of combat, he said—and the fight for the Commanceia had lasted the whole first night of the operation—his troops had used "proportional response." That is, they didn't soften up positions with mile-wide artillery barrages or huge air strikes. To take out individuals firing at them, they relied on snipers instead of M16 fusillades. They used ninety-millimeter recoilless rifles to punch holes, right where they wanted them, in cinder block buildings. "Minimum collateral damage—minimum injury or death of innocents," said Snell with a smack of satisfaction a Mongol khan might have used with the words "Built a pyramid of human skulls."

P.J. O'ROURKE, PARLIAMENT OF WHORES (1994) at 170-71.

[393] COMMANDER'S HANDBOOK, *supra* note 109, at § 8.1.2.1.

> if reasonably available, to reduce civilian casualties and damage. [footnote omitted][394]

Schmitt cites the three ways in which proportionality is most often (unintentionally, but possibly negligently) violated:

1) a lack of full knowledge as to what is being hit;

2) the inability to surgically craft the amount of force being applied to the target; and,

3) the inability to ensure the weapon strikes precisely the right point.[395]

---

[394] *Id.* Variables in the attack include "choice of attack platforms, weaponeering, fusing of ordnance, time of attack, and angle of approach to the target." *Id.*, at § 8.1.2.1, fn. 20.

[395] Schmitt, *Bellum Americanum, supra* note 303, at 1080. Schmitt gives an example of each of the three:

> With regard to the first, consider the Al-'Amariyah Bunker incident. [footnote omitted] Some 300 noncombatants were killed during the Persian Gulf War when US aircraft destroyed an Iraqi command and control bunker unaware that civilians had entered it during the night.
>
> As to weapons availability and capability, extended gaps along the continuum of force remain. For instance, because non-lethals are absent from the inventories of most militaries, forces sent into a crowd or perimeter defense situation have nothing to resort to between warning shots and the use of deadly force.
>
> Finally, in terms of accuracy—and despite the morbidly spectacular film of PGM [Precision Guided Munition] strikes during the Persian Gulf War—the reality is that many weapons continue to lack fully reliable precision guidance. Today, for instance, fighter-bombers still "toss," "drop," or "dive bomb" the majority of their weapons, which in most cases are unguided general purpose bombs.

Id. at 1080-81.

The calibration of force has been almost completely refined with modern technology, and the ability to steer toward a selected target is scarcely less advanced. The lack of complete intelligence, however, is and will remain a problem. From the Al-'Amariyah bunker in Baghdad to the Chinese Embassy in Belgrade, examples abound of high-tech weapons being guided with flawless precision to an inappropriate target. As weapon-mounted sensors continue to advance fire-and-forget weapons from "smart" to "brilliant," an important dichotomy will become more clear: weapons may become almost totally reliable at identifying and striking inherently military targets, such as tanks, artillery pieces, or surface-to-air missile batteries, but when it comes to identifying targets whose lawfulness is not patent (such as an office building housing intelligence personnel, a warehouse containing military supplies, or a radio tower broadcasting genocidal programming), then even the most reliable on-board computer targeting system will be no better than the targeting intelligence which was collected, analyzed, and downloaded into its memory.

This has an even greater application in cyberspace, where the difference between the power distribution codes for an early warning radar and an intensive care unit may be no more than a few ones and zeros in the same computer. In addition, the opportunity for unintended consequences (discussed in the section *supra* on the principle of discrimination) further complicates the targeting picture. The principle of proportionality in cyberspace may call for a fidelity and granularity of intelligence collection and analysis beyond current demonstrated capabilities, and may also call for modeling and simulation of attacks to give the commander at least minimal assurances that his attack will hit his target and not produce unintentional reverberating effects.

Chivalry

The law of chivalry has informed the development of the principles of discrimination, necessity, and proportionality, from the proper protections accorded noncombatants to the amount of damage that may be done in "just" pursuit of a military objective. There are other, more anachronistic, remnants: some of the codified distinctions between the handling of officers and enlisted men after capture, and even the post-WW II survival of the doctrine of "parole." As strongly as the law of chivalry is woven into the fabric of the modern law of war, it remains most intact in the distinction between lawful ruses and treacherous perfidy.[396]

The criteria for discriminating lawful ruses from unlawful perfidy have never been more important than now: information operations, the modern military effort to limit the physical violence of war by affecting the analysis options of the opposing commander though a variety of means, is virtually a synonym for "ruses of war." Precisely what in the realm of information operations is permissible and what is not, given the virtually unlimited technical capabilities on the horizon, may be the principal legal question of operational military lawyers in the next century.

Before addressing the modern rule of law, it is important to understand its historical context. This desire to be successful in war while

---

[396] Professor Schmitt sets the standard in treachery:

> [T]he term "treachery," a critical component in the current law of armed conflict, is designated as a breach of confidence by an assailant. However, one must be careful not to define treachery too broadly. Use of stealth or trickery, for instance, is not precluded, and will not render an otherwise lawful killing [unlawful] ... Treachery exists only when the victim possessed an affirmative reason to trust the assailant. Thinking in terms of ruses and perfidy is useful in understanding this distinction: ruses are planned to mislead the enemy, for example, by causing him to become reckless or choose a particular course of action. By contrast, perfidy involves an act designed to convince the enemy that the actor is entitled to protected status under the law of war, with the intent of betraying that confidence. Treachery, as construed by early scholars, is thus broader than the concept of perfidy; nevertheless, the same basic criteria that are used to distinguish lawful ruses from unlawful ruses can be applied to determinations of treachery. [footnotes omitted]

Michael N. Schmitt, *State-Sponsored Assassination in International and Domestic Law,* 17 YALE J. INT'L LAW 609, 617 (1992).

preserving one's immortal soul was a prime motivator for much chivalrous conduct.

To begin with, the entire chivalric system, as a complex network of personal obligations, depended on the trustworthiness of knights:

> Loyalty, meaning the pledged word, was chivalry's fulcrum. The extreme emphasis given to it derived from the time when a pledge between lord and vassal was the only form of government. A knight who broke his oath was charged with "treason" for betraying the order of knighthood. The concept of loyalty did not preclude treachery of the most egregious trickery as long as no knightly oath was broken. When a party of armed knights gained entrance to a walled town by declaring themselves allies and then proceeded to slaughter the defenders, chivalry evidently was not violated, no oath having been made to the burghers.[397]

This doctrine slowly evolved from the end of the Middle Ages to the beginning of the modern era. Balthazar Ayala, a sixteenth century commentator, looked beyond the existence of an actual contract or agreement between the two parties. Instead, Ayala proposed that a general duty to avoid treachery existed, and that such treachery could arise in the absence of a specific understanding between the aggressor and the defender. They key for Ayala was a distinction between permissible "trickery," similar to our understanding ruses of war, and impermissible "frauds and snares," which we would equate with perfidy. The distinction between the two was to be made in the mind of the reasonable defender.[398]

---

[397] BARBARA TUCHMAN, A DISTANT MIRROR: THE CALAMITOUS 14TH CENTURY at 64 (1978). Prestwich echoes Tuchman: "Where no prior agreement was involved . . . surprise and guile might be considered perfectly legitimate. [footnote omitted] Low cunning was not itself dishonorable; what brought shame was perjury of an oath promising to abstain from certain acts." MICHAEL PRESTWICH, ARMIES AND WARFARE IN THE MIDDLE AGES: THE ENGLISH EXPERIENCE (1996) at 128.

[398] BALTHAZAR AYALA, THREE BOOKS ON THE LAW OF WAR AND THE DUTIES CONNECTED WITH WAR AND ON THE MILITARY DISCIPLINE 84 (1582), *reprinted in* 2(2)

Alberico Gentili wrote in the early seventeenth century. He echoed the chivalric belief that treacherous killing was not worthy of an honorable warrior, but he extended the concept of treachery to include not just those who took advantage of a reasonably placed trust in safety, but also those who encouraged or commissioned violations of that trust.[399] Under Gentili's model, it would be permissible for a warrior to stealthily make his way into an enemy's camp to kill him as he sleeps, but not to hire a one of the enemy's own soldiers to perform the same act. Removed from the equation is the question of whether or not the enemy combatant is noble or not, and the question of whether or not the enemies had any previous agreement about such actions.

Hugo Grotius also wrote in the early seventeenth century. His work was pivotal in moving away from Aquinas' "just war" concept as license for any subsidiary means and toward the modern concept of examining only the means employed by each side, regardless of how hostilities commenced.[400] By removing the presumption of legality for acts perpetrated by the "just" party in a conflict, Grotius eliminated the catchall doctrine that permitted atrocious behavior by those claiming the mandate of heaven.

Later commentators, such as Vattel and Lieber, would sharpen and refine this doctrine, but the post-Grotius understanding of the distinction between ruses of war and perfidy is virtually indistinguishable from the modern standard.

THE CLASSICS OF INTERNATIONAL LAW 84-85 (J. Bate trans. 1912). *See also* Schmitt, *State-Sponsored Assassination* at 614.

[399] ALBERICO GENTILI, DE IURE BELLI LIBRE TRES (The Classics of International Law No. 16) 168 (John Rolfe trans., 1933) (1612). *See also* Schmitt, *State-Sponsored Assassination* at 615.

[400] HUGO GROTIUS, THE LAW OF WAR AND PEACE (1625), *quoted in* FRIEDMAN, *supra* note 63, at 16, 39 [footnote omitted], *quoted in* Schmitt, *State-Sponsored Assassination* at 615.

That standard is codified in Article 37 of Protocol I:

> It is prohibited to kill, injure or capture an adversary by resort to perfidy. Acts inviting the confidence of an adversary to lead him to believe he is entitled to, or is obligated to accord, protection under the rules of international law applicable in armed conflict, with intent to betray that confidence, shall constitute perfidy.[401]

The use of "stratagems,"[402] the term of art encompassing both ruses of war and perfidy, is addressed by the three principal reference manuals on the operational aspects of the law of war.[403] The army

---

[401] 1977 Geneva Protocol I, *supra* note 355, at arts. 51, para 5(b) and 57, para 2(a)(iii). As to what would constitute perfidious acts, Protocol I offers these four examples:

1. feigning a desire to negotiate under a truce or surrender flag;
2. feigning incapacitation by wounds or sickness;
3. feigning civilian, noncombatant status
4. feigning protected status by the use of signs, emblems or other uniforms of the United Nations, neutral states, or other states not party to the conflict.

Id. Offering a bounty, equivalent to a ransom, is treacherous. So, too, is launching an attack in civilian clothes, unless the enemy is not deceived by or does not rely upon the civilian clothing worn by the attacker. Schmitt, *State-Sponsored Assassination* at 635-39. *See also* 1907 Hague Convention IV, Customs of War on Land, *supra* note 355, at art. 23 para F.

[402] FM 27-10, *supra* note 154, at 22-23.

[403] *See id.* at 22-23; COMMANDER'S HANDBOOK, *supra* note 109, at §§ 12.1-12.10; and, AFP 110-31, *supra* note 358, at 8-1 to 8-5. For an excellent distillation of these rules, *see* CAPT Ashley J. Roach, USN, *Ruses and Perfidy: Deception During Armed Conflict*, 23 TOL. L. REV. 395 (1992). *See also* Hall, *False Colors and Dummy Ships: The Use of Ruse in Naval Warfare*, NAV. WAR COLL. REV., Summer 1989.

manual, FM 27-10, explains the general rule very concisely: "48. Stratagems Permissible . . . Ruses of war and the employment of measures necessary for obtaining information about the enemy and the country are considered permissible."[404] The manual goes on to restate the requirement for good faith in all dealings with the enemy, excepting those which the enemy can reasonably foresee and against which can reasonably defend.[405]

The key question here is the breadth of the measure against which the enemy is expected to protect himself. If the area is broadly drawn, defining one such area as "protect all computer systems from all intrusion," then the enemy is on constructive notice against almost any conceivable computer network attack (CNA). If, however, one defines the duty to protect as applying to more narrowly defined systems or against more narrowly defined techniques, then this aspiration toward good faith may be violated. As an example, suppose a nation sought to conduct psychological operations, perhaps including false statements about an enemy dictator's inappropriate relationship with his collie, and, by using a completely new technology, inserted this information into the CNN data stream reaching every TV set in the enemy country.[406] If a

---

[404] FM 27-10, *supra* note 154, at 22.

[405] Paragraph 49 reads:

> Absolute good faith with the enemy must be observed as a rule of conduct; but this does not prevent measures such as using spies and secret agents, encouraging defection or insurrection among the enemy civilian population, corrupting enemy civilians and soldiers by bribes, or inducing the enemy's soldiers to desert, surrender, or rebel. In general, a belligerent may resort to those measures for mystifying or misleading the enemy against which the enemy ought to take measures to protect himself.

Id. at 22.

[406] Such an operation would not be mounted by the United States, which scrupulously adheres to a policy of disseminating selected facets of the truth in its psychological operations. The United States would mount such an operation only if the dictator did, in fact, have such a relationship with his collie.

broad definition of "psychological operations" is accepted, then any psychological operation would be one against which the enemy should defend himself. If a narrow definition is chosen, perhaps limited to known technologies, then the duty of good faith would be violated because the enemy would not be protecting his civilian television system from "attack." The same may be said of other nontraditional targets or techniques.

On the other hand, since such targets and techniques are being widely discussed in the open press, one could argue that the enemy is on constructive notice to protect any system now susceptible to attack. This type of notice would satisfy the rule's requirement for good faith in these operations.

FM 27-10 goes on to address treachery and perfidy: "Ruses of war are legitimate so long as they do not involve treachery or perfidy on the part of the belligerent resorting to them. They are however, forbidden if they contravene any generally accepted rule."[407] The paragraph closes with the reason behind the rule: "Treacherous or perfidious conduct in war is forbidden because it destroys the basis for a restoration of peace short of the complete annihilation of one belligerent by the other."[408]

---

[407] FM 27-10, *supra* note 154, at 22. Paragraph 50 continues:

> The line of demarcation between legitimate ruses and forbidden acts or perfidy is sometimes indistinct, but the following examples indicate the correct principles. It would be an improper practice to secure an advantage of the enemy by deliberate lying or misleading conduct which involves a breach of faith, or when there is a moral duty to speak the truth. For example, it is improper to feign surrender so as to secure an advantage over the opposing belligerent thereby. So similarly, to broadcast to the enemy that an armistice has been agreed upon when such is not the case would be treacherous. On the other hand, it is a perfectly proper ruse to summon a force to surrender on the ground that it is surrounded and thereby induce such surrender with a small force.

Id.

[408] *Id.* at 22.

This explanation is completely in keeping with the Article 37 definition from Protocol I.

Paragraph 51 of the army manual is particularly helpful, in that it provides a long list of permissible ruses:

> Among legitimate ruses may be counted surprises, ambushes, feigning attacks, retreats, or flights, simulating quiet and inactivity, use of small forces to simulate large units, transmitting false or misleading radio or telephone messages, deception of the enemy by bogus orders purporting to have been issued by the enemy commander, making use of the enemy's signals and passwords, pretending to communicate with troops or reinforcements which have no existence, deceptive supply movements, deliberate planting of false information, use of spies and secret agents, moving landmarks, putting up dummy guns and vehicles or laying dummy mines, erection of dummy installations and airfields, removing unit identifications from uniforms, use of signal deceptive measures, and psychological warfare activities.[409]

---

[409] *Id.* at 22-23. The Commander's Handbook and AFP 110-31 provide service-specific examples of permissible ruses. The naval manual lists "camouflage, deceptive lighting, dummy ships and other armament, decoys, simulated forces, feigned attacks and withdrawals, ambushes, false intelligence information, electronic deceptions, and utilization of enemy codes, passwords, and countersigns." COMMANDER'S HANDBOOK, *supra* note 109, at arts. 51, para 5(b) and 57, para 2(a)(iii). The Commander's Handbook also lists "allowing false messages to fall into enemy hands," as was done by British naval intelligence during Operation Mincemeat in WWII. *Id.* AFP 110-31 lists use of aircraft decoys, staging air combats, imitation of enemy signals, use of flares and fires, camouflage use, and operational ruses, such as switched raids. AFP 110-31, *supra* note 358, at 8-2. Of these, the most important for the law of information conflict is the "imitation of enemy signals," a category under which many information operations could be accurately placed. The Air Force Manual goes on to explain:

> No objection can be made to the use by friendly forces of the signals or codes of the adversary. The signals or codes used by enemy aircraft or by enemy ground installations in contact with their aircraft may properly be employed by friendly forces to deceive or mislead an adversary. However, misuse of distress signals or

Among these numerous examples, it is important to point out that lawful "camouflage" can and does include disguising a military objective as a civilian object.[410] This is not as contradictory as it may seem at first blush; the true distinction, for the purposes of likenesses that may be appropriated for ruses of war, is between protected symbols, discussed *infra*, and ordinary civilian objects. Therefore, it would be lawful to disguise a military warehouse as a baby milk factory, but *not* as a hospital.

The remainder of the section on stratagems is devoted to the improper use of identifying devices.[411] The general rule is that "[I]t is

---

> distinctive signals internationally recognized as reserved for the exclusive use of medical aircraft would be perfidious.

Id.

[410] The Commander's Handbook states that "it is a common practice, not prohibited by [Protocol I], to disguise a military object to appear to be a civilian object." COMMANDER'S HANDBOOK, *supra* note 109, at § 12.1.1.

[411] *See* FM 27-10, *supra* note 154, at 23. *See also* COMMANDER'S HANDBOOK, *supra* note 109, at §§ 12.2-12.5. Although land and ground forces are prohibited from using neutral flags, insignia, or uniforms, naval forces are permitted considerable latitude before hostilities commence:

> Under the customary international law of naval warfare, it is permitted for a belligerent warship to fly false colors and disguises its outward appearance in other ways in order to deceive the enemy into believing the vessel is of neutral nationality or is other than a warship. However, it is unlawful for a warship to go into action without first showing her true colors. [footnote omitted] Use of neutral flags, insignia, or uniforms during an actual armed engagement at sea is, therefore, forbidden.

COMMANDER'S HANDBOOK, *supra* note 109, at § 12.3.1. For more on the modern trend toward relaxing the traditional requirements for combatancy, *see* Schmitt, *Bellum Americanum, supra* note 303, at 1078-79. Schmitt argues that the increasing transparency of the modern battlefield is driving third-world adversaries to perfidiously disguises for men and facilities, accelerating a trend which began in the 1960s with nonuniformed guerillas waging "wars of national liberation." *Id.*

especially forbidden to make improper use of a flag of truce, of the national flag, or of the military insignia and uniform of the enemy, as well as the distinctive badges of the Geneva Convention."[412] These insignia are protected because they directly induce the reliance of the enemy, and endanger those who legitimately rely on such symbols for protection.[413] Paragraph 53 prohibits misuse of flags of truce,[414] and paragraph 55 prohibits the misuse of Geneva Convention symbols, such as the emblem of the Red Cross.[415] Paragraph 54 does the same for national flags,

---

[412] *See* FM 27-10, *supra* note 154, at 23.

[413] *See* COMMANDER'S HANDBOOK, *supra* note 109, at §§ 12.6 and 12.7. For the same reason, feigning distress in order to attack a rescuing enemy, and falsely claiming noncombatant status, are perfidious and punishable as war crimes. *Id.* *See also* BOTHE, PARTSCH & SOLF, NEW RULES, *supra* note 388, at 207, *citing* 10 WHITEMAN, DIGEST OF INTERNATIONAL LAW (Whiteman ed. 1973):

> It would be a legitimate ruse to use the electronic transponder aboard a combatant aircraft to respond with the code used for identifying friendly aircraft (IFF), but it would be perfidious to use for this purpose the electronic signal established under annex I, Art. 8, [Protocol I] for the exclusive use of medical aircraft. Similarly, the use of distress signals established under the Radio Regulations of the International Telecommunications Union is prohibited under the second sentence of Art. 38, para. 1 [of Protocol I] and might be violative of Art. 37 [of Protocol I].

Id.

[414] Paragraph 53 states:

> Flags of truce must not be used surreptitiously to obtain military information or merely to obtain time to effect a retreat or secure reinforcements or to feign a surrender in order to surprise an enemy. An officer receiving them is not on this account absolved from the duty of exercising proper precautions with regard to them.

FM 27-10, *supra* note 154, at 23.

[415] *Id.* Paragraph 55 prohibits using personnel, vehicles, and facilities displaying these symbols "for cloaking acts of hostility." *Id.*

insignia, and uniforms.[416] The underlying rationale is to avoid dilution of the absolute nature of these symbols when encountered, reinforcing the legal protections they bestow and the legal obligations they exact. The use of these protected symbols in information operations would be included in the general prohibition.

In a recent article, an author states: "analogy strongly weighs against sending a logic bomb disguised as e-mail from the International Committee of the Red Cross (ICRC) or even from "Microsoft Software Support—where such a message might be permissible without perfidious labels."[417] The flaw in this statement is that it incorrectly equates two information attacks that are quite different under international law. The first, a false message from the ICRC, is clearly perfidious, in that it delivers a weapon under the protection of the Red Cross symbol—an action directly analogous to delivering a car bomb in an ambulance. The second, however, is just as clearly lawful, in that Microsoft Corporation enjoys no protected status under international law. A message from Microsoft would be no different from a message from any other firm with which a belligerent is doing business. The analogy here would be a commando team emplacing a bomb in enemy headquarters while disguised in the coveralls of a local plumbing company.

Espionage in wartime (intelligence collection behind enemy lines) is governed by two legal principles, one defining who, under international law, actually is a spy, and the other regulating a spy's treatment after capture.

---

[416] *Id.* Paragraph 54 states: "It is certainly forbidden to employ them during combat, but their use at other times is not forbidden." *Id.*

[417] Mark R. Schulman, *Discrimination in the Laws of Information Warfare*, 37 COLUM. J. TRANSNAT'L L. 939, 959 (1999).

As for the first,

> [a] spy is someone who, while in territory under enemy control or the zone of operations of a belligerent force,[418] seeks to obtain information while operating under a false claim of noncombatant or friendly forces status with the intention of passing that information to an opposing belligerent. [footnote omitted] Members of the armed forces who penetrate enemy-held territory[419] in civilian attire or enemy uniform to collect intelligence are spies. [footnote omitted] Conversely, personnel conducting reconnaissance missions behind enemy lines while properly uniformed are not spies. [footnote omitted] Crewmembers of warships and military aircraft engaged in intelligence collection missions in enemy waters or airspace are not spies unless the ship or aircraft displays false civilian, neutral, or enemy markings. [footnote omitted][420]

The second, on the treatment of captured spies, is equally succinct:

> Spying during armed conflict is not a violation of international law. Captured spies are not, however, entitled to prisoner-of-war status. [footnote omitted] The captor nation may try and punish spies in accordance with its national law. [footnote omitted] Should a spy succeed in eluding capture and return to friendly territory, liability to punishment terminates. If subsequently captured during some other military operation, the former spy cannot be tried or punished for the earlier act of espionage.[421]

---

[418] There is no clear consensus on the application of the term "zone of operations of a belligerent force" to cyberspace. Until clearly distinguishable norms arise, it is probably best to apply, as closely as possible, the standards of the physical world by analogy.

[419] *See id.*

[420] COMMANDER'S HANDBOOK, *supra* note 109, at § 12.8.

[421] *Id.* at § 12.8.1.

The liability of an intelligence officer later captured behind enemy lines appears to include only those activities physically conducted by that office during that mission. Previous cyber "entries" into and "exits" from the enemy country were not contemplated in the formulation of this rule of law, and appear not to provide the basis for any additional liability. However, not all information operations are mounted from outside enemy borders, and those that require physical entry to enemy territory, and which can be plausibly characterized as intelligence operations, will expose the operator to traditional liability.

Although superficially complex, the basic relationship of these concepts can be displayed in a relatively simple chart, shown below.

	Military Appearance	Civilian Appearance	Protected Appearance
Hostilities	Acceptable	Prohibited	Prohibited
Pre-Hostilities	Acceptable	Acceptable	Prohibited

Figure 14. Ruses of War and Perfidy

In summary, whatever use is made of the enemy commander's information environment, the techniques employed must resemble, or at least be analogous to, the paragraph 51 list of permissible ruses, or the naval or air force equivalents. Although there are exceptions due to specific circumstances, the general rule is that military forces may masquerade as civilians or civilian objects until engaging in hostilities, but they may not use protected symbols (the Red Cross, the U.N. flag) in any way which would induce the enemy's detrimental reliance or dilute the effect of those symbols and the safety of those depending on their protection.

CHAPTER 8: TREATIES ON THE LAW OF ARMED CONFLICT[†]

Having examined the theoretical bases for the law of information conflict in the previous chapter, it is now appropriate to turn to the historical application of these principles to conflict through the development of treaty law. It is in treaties that existing international norms are codified and made specific. Treaties serve another purpose, in that they may advance a new rule of law, and rely on the signatories obeying the new standard based purely on the treaty obligation. Such new standards may be too far in advance of the international community, and so may never win the support necessary to become a customary norm applicable to all nations, and not merely the treaty signatories. Usually, though, treaty law has advanced at roughly the same pace as customary international law. The twenty-nine treaties or international agreements listed here cover all major facets of the law of armed conflict. The accompanying commentary discusses how the legal standard in each may be applicable to the emerging law of information conflict. Just as the previous chapter was intended to clearly convey existing standards, this chapter is intended to spur thought and discussion of the future evolution of those standards.

[†] The concept for this chapter was originated by Prof. Walter Gary Sharp, Sr., for Aegis Research Corporation.

I. 1856 Paris Declaration Respecting Maritime Law[422]

The Paris Declaration began the 19th Century trend of codifying the customary international law of arms. Immediately following the Crimean War, the Declaration captured the agreed-upon norms for that war as the new European, and then international, standard.[423] Schindler and Toman note that "[i]t is the outcome of a *modus vivendi* which was adopted between France and Great Britain in 1854 and was originally intended for the Crimean War only."[424] The Declaration is not surprising that the first such codification was in the area of maritime operations. The law of the sea was the first international law, and it was a natural place to begin the internationalization of the law of armed conflict. Also helpful was the existence of a moribund doctrine, privateering, over which the international community could easily forge a consensus for prohibition.

"1. Privateering is, and remains, abolished;"[425]

The naval equivalent of using private actors to accomplish forcible government ends in cyberspace is privateering. It has fallen into desuetude because formerly private functions, such as transport and armed patrol, have been absorbed by traditional naval or quasi-naval services.

---

[422] Paris Declaration Respecting Maritime Law, April 16, 1856, *reprinted in* ROBERTS & GUELFF, *supra* note 4, at 23 [hereinafter Paris Declaration]. *See also* SCHINDLER & TOMAN, ARMED CONFLICTS, *supra* note 5, at 787-790.

[423] ROBERTS & GUELFF, *supra* note 4, at 23.

[424] SCHINDLER & TOMAN, ARMED CONFLICTS, *supra* note 5, at 787. Schindler and Toman go on to state that "[a]ll the belligerents furthermore proclaimed that they would not issue letters of marque."

[425] Paris Declaration, *supra* note 422, at art. 1. *See also* SCHINDLER & TOMAN, ARMED CONFLICTS, *supra* note 5, at 788.

The nature of privateering in cyberspace, given its overwhelmingly civilian, commercial focus, remains open.

> "2. The neutral flag covers enemy's goods, with the exception of contraband of war;"[426]

First granted by France during the Crimean War,[427] this protection ensured that the property of nonbelligerent nations in the custody of belligerents would remain safe, or at least as safe as practicable. The cyberspace analog to this guarantee would be the inviolability of information belonging to nonbelligerents, stored in the systems of belligerent nations. Such distinctions can be maintained only in "boarding" scenarios, in which cargo or information can be properly sorted, and not in "sinking" scenarios, where the entire cargo of a vessel or memory of a computer is destroyed.

> "3. Neutral goods, with the exception of contraband of war, are not liable to capture under enemy's flag;"[428]

This protection was first offered by the British during the Crimean War.[429] This guarantee complimented the French provision, in that it safeguarded the nonbelligerent vessels carrying at least some cargo of a belligerent nation. Attacks on neutral vessels, and by extension on neutral information systems containing enemy information, would similarly be prohibited under this doctrine, if applied.

---

[426] Paris Declaration, *supra* note 422, at art. 3. *See also* SCHINDLER & TOMAN, ARMED CONFLICTS, *supra* note 5, at 788.

[427] ROBERTS & GUELFF, *supra* note 4, at 23.

[428] Paris Declaration, *supra* note 422, at art. 3. *See also* SCHINDLER & TOMAN, ARMED CONFLICTS, *supra* note 5, at 788.

[429] ROBERTS & GUELFF, *supra* note 4, at 23.

> "4. Blockades, in order to be binding, must be effective, that is to say, maintained by a force sufficient really to prevent access to the coast of the enemy."[430]

The concept of an "information embargo" or an "information blockade" is an attractive one to a military planner, but, given the increasingly distributed nature of information systems, is one increasingly difficult to impose. The main lesson of this doctrine is that any such blockade must be effective in the real world in order to have legal effect, and not based merely on the statement of a power unable to impose such a quarantine.

> "Convinced that the maxims which they now proclaim cannot but be received with gratitude by the whole world, the undersigned Plenipotentiaries doubt not that the efforts of their Governments to obtain the general adoption thereof will be crowned with full success."[431]

This passage has less importance for its substance than for its tone. It reveals the unbridled optimism and naïveté of one of the first attempts to codify a civilizing rule of law in warfare. The whole world may have to continue to withhold its gratitude, delaying the coronation of full success, until information operators are able to address these and other competing interests (*i.e.*, the encryption debate).

---

[430] Paris Declaration, *supra* note 422 at art. 4. *See also* SCHINDLER & TOMAN, ARMED CONFLICTS, *supra* note 5, at 788.

[431] *See generally* Paris Declaration, *supra* note 422. *See also* SCHINDLER & TOMAN, ARMED CONFLICTS, *supra* note 5, at 788.

2. 1868 St. Petersburg Declaration Renouncing the Use, in Time of War, of Explosive Projectiles Under 400 Grammes Weight[432]

Roberts and Guelff tell us that "[t]he St. Petersburg Declaration has been regarded as the first major international agreement prohibiting the use of a particular weapon in warfare."[433] Reisman and Antoniou concur, stating that "[t]he Declaration of St. Petersburg ...was one of the earliest international efforts to develop explicit limitations on how war is to be conducted."[434] They go on to note that in the United States, the *Lieber Instructions* of 1863 were one of the first attempts at codifying the laws of war.[435]

---

[432] St. Petersburg Declaration, *supra* note 381, *reprinted in* ROBERTS & GUELFF, *supra* note 4, at 29. *See also* THE LAWS OF WAR: A COMPREHENSIVE COLLECTION OF PRIMARY DOCUMENTS ON INTERNATIONAL LAWS GOVERNING ARMED CONFLICT 35-36, 48, Reisman and Antoniou eds., [hereinafter REISMAN & ANTONIOU]; SCHINDLER & TOMAN, ARMED CONFLICTS, *supra* note 5, at 101-102.

[433] ROBERTS & GUELFF, *supra* note 4, at 29.

[434] REISMAN & ANTONIOU, *supra* note 432, at 35. Schindler and Toman offer further clarification:

> It had its origin in the invention, in 1863, by Russian military authorities of a bullet which exploded on contact with a hard substance and whose primary object was to blow up ammunition wagons. In 1867 the projectile was so modified as to explode on contact with a soft substance.

SCHINDLER & TOMAN, ARMED CONFLICTS, *supra* note 5, at 101.

[435] REISMAN & ANTONIOU, *supra* note 432. Reisman and Antoniou state:

> In 1863 the United States attempted to codify the laws of war for Union forces fighting in the Civil War. Called the *Lieber Instructions*, they were prepared by Professor Francis Lieber of Columbia College, revised by a board of Army officers, and then issued by President Lincoln. As the first formal, fairly comprehensive codification of the laws of war, the *Lieber Instructions* had a significant impact upon future codifications.

Outlawing the use of exploding bullets against personnel was part of a rich legal tradition of such prohibitions at the national level, stretching back thousands of years.[436] This declaration implements the principle of preventing unnecessary suffering, later captured in Article 23(e) of the Regulations annexed to the 1899 Hague Convention II, *infra*.[437] The desire to limit the amount of damage done or pain inflicted to the minimum necessary to remove a soldier from combat inspired this declaration and its progeny.

> Considering that the progress of civilization should have the effect of alleviating as much as possible the calamities of war;
>
> That the only legitimate object which States should endeavour to accomplish during war is to weaken the military forces of the enemy;
>
> That for this purpose it is sufficient to disable the greatest possible number of men;
>
> That this object would be exceeded by the employment of arms which uselessly aggravate the sufferings of disabled men, or render their death inevitable;

---

[436] ROBERTS & GUELFF, *supra* note 4, at 29. Roberts and Guelff explain:

> Attempts to prohibit the use of particular weapons in warfare have been made in various civilizations over a long period of time. For example, in ancient times, the Laws of Manu (the greatest of the ancient Hindu codes) prohibited Hindus from using poisoned arrows; and the Greeks and Romans customarily observed a prohibition against using poison or poison weapons. During the Middle Ages the Lateran Council of 1132 declared that the crossbow and arbalest were 'unchristian' weapons.

Id.

[437] 1907 Hague Convention IV, Customs of War on Land, Annex art. 23 (e), *supra* note 355, *reprinted in* ROBERTS & GUELFF at 29. 1989). *See also* REISMAN & ANTONIOU, *supra* note 432; SCHINDLER & TOMAN, ARMED CONFLICTS, *supra* note 5, at 83.

> That the employment of such arms would, therefore, be contrary to the laws of humanity...[438]

Although there have been numerous attempts throughout history to limit the employment of weapons, this declaration is generally considered the first to prohibit, internationally, a specific type of weapon.[439] Schindler and Toman concur, stating that "[t]he Declaration of St. Petersburg is the first formal agreement prohibiting the use of certain weapons in war."[440] The weight limit of 400 grams was chosen to distinguish between exploding bullets and artillery shells, the former being more reprehensible because they actually exploded *inside* the human body.[441]

This desire to prohibit new and frightening weapons is a common, though usually ineffective, response to military technological advancement. This is less likely to be the case in cyberspace, as the new means and methods of warfare are often indistinguishable from existing and widely employed civilian technology. The realization that these efficient systems can be used to attack the basic infrastructures of our

---

[438] St. Petersburg Declaration, *supra* note 381, *reprinted in* ROBERTS & GUELFF, *supra* note 4, at 30-31. *See also* REISMAN & ANTONIOU, *supra* note 432; SCHINDLER & TOMAN, ARMED CONFLICTS, *supra* note 5, at 102.

[439] ROBERTS & GUELFF, *supra* note 4, at 29.

[440] SCHINDLER & TOMAN, ARMED CONFLICTS, *supra* note 5, at 101.

[441] ROBERTS & GUELFF, *supra* note 4, at 29. *See also* REISMAN & ANTONIOU, *supra* note 432, at 48. The text states:

> "The contracting parties engage, mutually, to renounce, in case of war among themselves, the employment, by their military or naval forces, of any projectile of less weight than 400 grammes, which is explosive, or is charged with fulminating or inflammable substances."

REISMAN & ANTONIOU, *supra* note 432, at 48.

civilization is only slowly dawning. In this case, the frightening aspect is not the means, but the end.

> The Contracting or Acceding Parties reserve to themselves to come hereafter to an understanding whenever a precise proposition shall be drawn up in view of future improvements which science may effect in the armament of troops, in order to maintain the principles which they have established, and to conciliate the necessities of war with the laws of humanity.[442]

This declaration applies the principle of unnecessary suffering to a particular type of weapon, and finds the entire type violative of the principle. This closing passage leaves open the possibility that science would produce other weapons requiring limitation or prohibition. As the law or arms control developed, several more types of weapons (*i.e.*, chemical and biological) were categorically prohibited. Although the majority of information weapons are not, in themselves, particularly violent or destructive, their downstream consequences, some perhaps unintended, could subject some information weapons to such a categorical ban.

---

[442] *See* St. Petersburg Declaration, *supra* note 381, *reprinted in* ROBERTS & GUELFF, *supra* note 4, at 31. *See also* SCHINDLER & TOMAN, ARMED CONFLICTS, *supra* note 5, at 102.

3. 1899 Hague Declaration 2 Concerning Asphyxiating Gases[443]

This declaration follows closely in the footsteps of the previous one on exploding bullets. It is the progenitor of the later, and somewhat more complex, Chemical Weapons Convention. Schindler and Toman tell us that "[t]his Declaration gives expression, with regard to a particular instrument of warfare, to the customary rules prohibiting the use of poison and of material causing unnecessary suffering," and cite a number of legal precedents to support this statement.[444]

> "The contracting Powers agree to abstain from the use of projectiles the sole object of which is the diffusion of asphyxiating or deleterious gases."[445]

While this declaration offers no unique insights into the law of information conflict, it is important for two reasons: first, it continued the tradition of single-weapon prohibitions, and second, it may be the only international agreement in history whose operative text is captured in a single sentence.

---

[443] 1899 Hague Declaration 2, Asphyxiating Gases, *supra* note 381, *reprinted in* ROBERTS & GUELFF, *supra* note 4, at 35. *See also* REISMAN & ANTONIOU, *supra* note 432, at 57; SCHINDLER & TOMAN, ARMED CONFLICTS, *supra* note 5, at 105-107.

[444] SCHINDLER & TOMAN, ARMED CONFLICTS, *supra* note 5, at 105.

[445] 1899 Hague Declaration 2, Asphyxiating Gases, *supra* note 381, *reprinted in* ROBERTS & GUELFF, *supra* note 4, at 35. *See also* REISMAN & ANTONIOU, *supra* note 432, at 57; SCHINDLER & TOMAN, ARMED CONFLICTS, *supra* note 5, at 105.

4. 1899 Hague Declaration 3 Concerning Expanding Bullets[446]

This declaration addressed a new technological problem not covered by the ban on exploding bullets: bullets which, due to their composition and shape, expanded or flattened upon entering the body, transferring a high percentage of their kinetic energy to the surrounding tissue. These bullets had impressive "stopping power," making them attractive to armies at the turn of the century. Known as "dum-dum" bullets (named for the first versions, which were produced at the British arsenal in Dum-Dum, India),[447] they drew the attention of humanitarian activists for their apparent similarities to exploding bullets.

> "The contracting Parties agree to abstain from the use of bullets which expand or flatten easily in the human body, such as bullets with a hard envelope which does not entirely cover the core or is pierced with incisions."[448]

This declaration, almost as brief as its asphyxiating-gases predecessor, is also an example of single-weapon prohibition. The most instructive thing about this agreement is the record of objections while it was being negotiated:

> Great Britain objected to the proposed prohibition on the grounds that the dumdum bullet did not expand in such a

---

[446] 1899 Hague Declaration 3, Expanding Bullets, *supra* note 381, *reprinted in* ROBERTS & GUELFF, *supra* note 4, at 39. *See also* REISMAN & ANTONIOU, *supra* note 432, at 49; SCHINDLER & TOMAN, ARMED CONFLICTS, *supra* note 5, at 109-111.

[447] ROBERTS & GUELFF, *supra* note 4, at 39. *Cf.*, SCHINDLER & TOMAN, ARMED CONFLICTS, *supra* note 5, at 109, where Schindler and Toman state the Declaration "prohibit(s) the use of weapons which inflict unnecessarily cruel wounds."

[448] *See* 1899 Hague Declaration 3, Expanding Bullets, *supra* note 381, *reprinted in* ROBERTS & GUELFF, *supra* note 4, at 40. *See also* REISMAN & ANTONIOU, *supra* note 432, at 49; SCHINDLER & TOMAN, ARMED CONFLICTS, *supra* note 5, at 109.

> manner as to create wounds of exceptional cruelty and that, in fact, the wounds were less severe than those produced by certain rifles which had been in use. The USA objected to the proposed declaration on three grounds: first, a prohibition based on the specification of details of construction might not be useful because of the possibility that a bullet might be made to expand in such a regular manner as to simply assume the form of a larger calibre; second, such an expanding bullet might be the most humane way of increasing the shocking power of the bullet; and third, the bullet was actually being used by the British Army, and any condemnation should follow from real evidence against its effects, rather than from implication draw from its design.[449]

These objections could be applied to the law of information conflict. The British objection captured the concern that the most inhumane weapons were not being outlawed, merely the easiest to categorize. The British preferred that prohibitions attach to the weapons or methods actually causing the most unnecessary suffering. The American objections bear directly on problems in information operations: first, that a technique may be too broadly defined, and a prohibition based upon such a definition may inappropriately restrict certain means and methods; second, that an information weapon could be mistakenly prohibited when, in fact, it could be the most humane way to accomplish a military mission (resulting in a return to more traditional kinetic solutions); and third, that the effect of a weapon in the real world is the more appropriate basis for evaluating its effects—not extrapolation from untested technical schematics and projections. Collectively, these objections show the difficulty that the law will have in staying ahead of or simply even with technology. Restrictions based on the core principles of military necessity and proportionality are far more likely to remain useful in the development of the law of information conflict.

---

[449] ROBERTS & GUELFF, *supra* note 4, at 39.

5. 1907 Hague Convention IV Respecting the Laws and Customs of War on Land[450]

This convention grew out of the 1899 Hague Convention II Respecting the Laws and Customs of War on Land, not reproduced here due to its substantial similarity to the convention *infra* and the 1907 Convention's role as the successor document.[451] The Convention itself is almost entirely procedural; the regulations governing land warfare, the true substance of the Convention, are captured in the Annex. For those cases not covered by the Annex, the famous Martens Clause states that the Convention is not exhaustive or exclusive, and that customary international law will continue to fill any gaps.[452] The principles espoused by the Convention continue to form the foundation of the law of armed conflict and, by extension, the law of information conflict.[453]

---

[450] 1907 Hague Convention IV, Customs of War on Land, *supra* note 355, *reprinted in* ROBERTS & GUELFF, *supra* note 4, at 43. *See also* REISMAN & ANTONIOU, *supra* note 432, at 25, 36-38, 41-42, 47-48, 81, 94, 130-131, 150-151 and 232-233; SCHINDLER & TOMAN, ARMED CONFLICTS, *supra* note 5, at 63-98.

[451] ROBERTS & GUELFF, *supra* note 4, at 43.

[452] *Id.* The Convention states:

> Until a more complete code of laws of war has been issued, the high contracting Parties deem it expedient to declare that, in the cases not included in the Regulations adopted by them, the inhabitants and the belligerents remain under the protection and the rule of the principles of the law of nations, as they result from the usages established among civilized peoples, from the laws of humanity, and the dictates of the public conscience.

1907 Hague Convention IV, Customs of War on Land, *supra* note 355, *reprinted in* ROBERTS & GUELFF, *supra* note 4, at 45. *See also* SCHINDLER & TOMAN, ARMED CONFLICTS, *supra* note 5, at 70.

[453] Schindler and Toman note that "[t]he provisions of the two Conventions on land warfare, like most of the substantive provisions of the Hague Conventions of 1899 and 1907, are considered as embodying rules of customary international law. As such they are

> "The contracting Powers shall issue instructions to their armed land forces which shall be in conformity with the Regulations respecting the laws and customs of war on land. Annexed to the present Convention."[454]

This Convention contains nine articles and an annex, but only article I, quoted above, and the annex contain nonprocedural material. Article I requires signatory nations to issue (and, presumably, implement) standing orders that conform to the rules set out in the annex. This annex is divided into three sections: "On Belligerents," "Hostilities," and "Military Authority over the Territory of the Hostile State." Section I is further subdivided into three chapters: "The Qualifications of Belligerents," "Prisoners of War," and "The Sick and Wounded." Of these, only the first chapter is implicated by developments in the law of information conflict.

> The laws, rights, and duties of war apply not only to armies, but also to militia and volunteer corps fulfilling the following conditions:
>
> 1. To be commanded by a person responsible for his subordinates;
>
> 2. To have a fixed distinctive emblem recognizable at a distance;
>
> 3. To carry arms openly; and

---

also binding on states which are not formally parties to them." SCHINDLER & TOMAN, ARMED CONFLICTS, *supra* note 5, at 63.

[454] 1907 Hague Convention IV, Customs of War on Land, *supra* note 355, at art. I *reprinted in* ROBERTS & GUELFF, *supra* note 4, at 46. *See also* REISMAN & ANTONIOU, *supra* note 432, at 38; SCHINDLER & TOMAN, ARMED CONFLICTS, *supra* note 5, at 71.

4. To conduct their operations in accordance with the laws and customs of war.[455]

Reisman and Antoniou tell us that "[b]oth customary and conventional law have sought, at least until 1977, to maintain the distinction between combatants and noncombatants with as much clarity as possible."[456] In cyberspace, the identity of bad actors is difficult to verify. Lines of responsibility are blurred, no identifying uniform is visible, the "arms" wielded are not carried openly, and it is nearly impossible to confirm that opponents are complying with the existing law of war. Only the first factor above is clarified when a military unit, vice hired hackers, is responsible for the virtual use of force; the other three factors remain very cloudy.

> "The right of belligerents to adopt means of injuring the enemy is not unlimited."[457]

This is the most basic principle of the law of armed conflict. It applies to the law of information conflict with equal strength.

> In addition to the prohibitions provided by special Conventions, it is especially forbidden—
>
> (a) To employ poison or poisoned weapons;

---

[455] 1907 Hague Convention IV, Customs of War on Land, *supra* note 355, at art. I *reprinted in* ROBERTS & GUELFF, *supra* note 4, at 46. *See also* REISMAN & ANTONIOU, *supra* note 432, at 41 *and* SCHINDLER & TOMAN, ARMED CONFLICTS, *supra* note 5, at 75.

[456] REISMAN & ANTONIOU, *supra* note 432, at 41.

[457] 1907 Hague Convention IV, Customs of War on Land, *supra* note 355, Annex at art. 22 *reprinted in* ROBERTS & GUELFF, *supra* note 4, at 52. 1989). *See also* REISMAN & ANTONIOU, *supra* note 432, at 47 *and* SCHINDLER & TOMAN, ARMED CONFLICTS, *supra* note 5, at 82.

> (b) To kill or wound treacherously individuals belonging to the hostile nation or army;
>
> (c) To kill or wound an enemy who, having laid down his arms, or having no longer means of defence, has surrendered at discretion;
>
> (d) To declare that no quarter will be given;
>
> (e) To employ arms, projectiles, or material calculated to cause unnecessary suffering; [footnote omitted]
>
> (f) To make improper use of a flag of truce, of the national flag or of the military insignia and uniform of the enemy, as well as the distinctive badges of the Geneva Convention;
>
> (g) To destroy or seize property, unless such destruction or seizure be imperatively demanded by the necessities of war;
>
> (h) To declare abolished, suspended, or inadmissible in a court of law the rights and actions of the nationals of the hostile party.
>
> A belligerent is likewise forbidden to compel the nationals of the hostile party to take part in the operations of war directed against their own country, even if they were in the belligerent's service before the commencement of the war.[458]

Article 23 is one of the seminal sources of the law of war, codifying centuries of customary law. Each of its provisions should be examined in turn for its application to the law of information conflict.

---

[458] 1907 Hague Convention IV, Customs of War on Land, *supra* note 355, Annex at art. 23 *reprinted in* ROBERTS & GUELFF, *supra* note 4, at 52-53. *See also* REISMAN & ANTONIOU, *supra* note 432, at 47 *and* SCHINDLER & TOMAN, ARMED CONFLICTS, *supra* note 5, at 82-83.

Paragraph (a)'s restriction on poisons, most recently codified in the Chemical Weapons Convention, is inappropriate, unless one extends the prohibition metaphorically to the various "cyber poisons" available to those attacking an enemy computer system.

More problematic is paragraph (b), which captures the central principle of the law of chivalry. "Treachery" is generally understood to involve the violation of a trust or legal protection in an attack (*i.e.*, hiring an opponent's own guards to assassinate him, or violating a temporary truce or other agreement with an adversary to secure military advantage). The potential for treachery in the virtual world is vastly greater than in the real world, and information operations are predicated on some degree of surprise, confusion, and hidden identity. It is also generally agreed within the critical infrastructure protection community that the greatest threat is a knowledgeable but disgruntled insider, the prototypical treacherous means of attack.

Paragraphs (c) and (d) forbid harming those who have surrendered or, more importantly, have "no longer a means of defense."[459] It is easy to imagine a military unit, materially intact, being brought to a position of complete defenselessness by the corruption or manipulation of its information systems. Under paragraph (c), such an apparently undamaged military unit would no longer be a lawful target. In effect, this might present the choice between fighting conventionally and fighting electronically. The same would apply to part or all of an enemy nation similarly affected.[460]

---

[459] 1907 Hague Convention IV, Customs of War on Land, *supra* note 355, Annex at art. 23 *reprinted in* ROBERTS & GUELFF, *supra* note 4, at 52. 1989). *See also* REISMAN & ANTONIOU, *supra* note 432, at 47 *and* SCHINDLER & TOMAN, ARMED CONFLICTS, *supra* note 5, at 83.

[460] 1907 Hague Convention IV, Customs of War on Land, *supra* note 355, Annex at art. 23, *reprinted in* ROBERTS & GUELFF, *supra* note 4, at 52. *See also* REISMAN & ANTONIOU, *supra* note 432, at 47 *and* SCHINDLER & TOMAN, ARMED CONFLICTS, *supra* note 5, at 83.

Paragraph (e) captures the principle of humanity, outlawing unnecessary suffering. This requirement is balanced against the military necessity of the contemplated use of force, and is a factually driven decision based on all the information available to the commander at the time. This prohibition, however, is particularly relevant to the law of information conflict for two reasons: first, that the intended target of information operations could, perhaps in most cases, be a civilian infrastructure system; and second, that information attacks, especially against such large, interdependent systems, would have almost incalculable ripple effects cascading throughout the target nation. The ability to determine, with any degree of accuracy, such effects is extremely problematic. Further complicating the picture is that the nations most likely to mount such attacks on civilian targets are the same nations that have demonstrated little respect for the rule of law in or out of war.[461]

Paragraph (f) could prohibit those information operations employing distinctive visual signatures (*i.e.*, graphic user interfaces, etc.) of the opposing military. Obviously, information operations mounted under the guise of the Red Cross or similar symbols would be illegal.[462]

Paragraph (g) returns to the issues discussed under paragraphs (c) and (d), and suggest that the seizure of civilian intellectual property would require a finding of military necessity.[463]

---

[461] 1907 Hague Convention IV, Customs of War on Land, *supra* note 355, Annex at art. 23 *reprinted in* ROBERTS & GUELFF, *supra* note 4, at 52. *See also* REISMAN & ANTONIOU, *supra* note 432, at 47 *and* SCHINDLER & TOMAN, ARMED CONFLICTS, *supra* note 5, at 83.

[462] 1907 Hague Convention IV, Customs of War on Land, *supra* note 355, Annex at art. 23 *reprinted in* ROBERTS & GUELFF, *supra* note 4, at 52. *See also* REISMAN & ANTONIOU, *supra* note 432, at 47 *and* SCHINDLER & TOMAN, ARMED CONFLICTS, *supra* note 5, at 83.

[463] 1907 Hague Convention IV, Customs of War on Land, *supra* note 355, Annex at art. 23 *reprinted in* ROBERTS & GUELFF, *supra* note 4, at 52. *See also* REISMAN & ANTONIOU, *supra* note 432, at 47 *and* SCHINDLER & TOMAN, ARMED CONFLICTS, *supra* note 5, at 83.

Paragraph (h) provides an additional legal basis—beyond simple self-defense—for the protection of nationals abroad. No information operation in wartime could legitimately target the legal rights of a foreign national in the host nation.[464]

The closing paragraph of the article protects such foreign nationals from being compelled to act against their own nation. The interconnectedness of information infrastructures will provide numerous opportunities for belligerent nations to attempt to recruit or blackmail foreign nationals into the equivalent of taking up arms against their own nation.

> "Ruses of war and the employment of measures necessary for obtaining information about the enemy and the country are considered permissible."[465]

"Ruses of war" are distinct from treachery and perfidy in that they violate no pre-existing legal duty of disclosure or protection. Information operations, more than any other category of military activity, involve extensive ruses of war. For this reason, absolute clarity of analysis in distinguishing such ruses from treachery and perfidy is vital.

> "The attack or bombardment, *by whatever means*, of towns, villages, dwellings, or buildings which are undefended is prohibited."[466] [emphasis added]

---

[464] 1907 Hague Convention IV, Customs of War on Land, *supra* note 355, Annex at art. 23 *reprinted in* ROBERTS & GUELFF, *supra* note 4, at 52. *See also* REISMAN & ANTONIOU, *supra* note 432, at 47 *and* SCHINDLER & TOMAN, ARMED CONFLICTS, *supra* note 5, at 83.

[465] *See* 1907 Hague Convention IV, Customs of War on Land, *supra* note 355, Annex *reprinted in* ROBERTS & GUELFF, *supra* note 4, at 53. *See also* REISMAN & ANTONIOU, *supra* note 432, at 48 *and* SCHINDLER & TOMAN, ARMED CONFLICTS, *supra* note 5, at 83.

[466] 1907 Hague Convention IV, Customs of War on Land, *supra* note 355, Annex at art. 25 *reprinted in* ROBERTS & GUELFF, *supra* note 4, at 53. *See also* REISMAN & ANTONIOU, *supra* note 432 *and* SCHINDLER & TOMAN, ARMED CONFLICTS, *supra* note 5, at 83–84.

The concept of an undefended or "open" city dates back to the days of chivalry, and this article is the first of several which bring into modern times the law of siege, one of the major subdivisions of the law of chivalry. This doctrine is, metaphorically at least, applicable to modern information operations conducted against vulnerable civilian targets.

> "The officer in command of an attacking force must, before commencing a bombardment, except in cases of assault, do all in his power to warn the authorities."[467]

This convention of siege warfare seems almost quaint today, as any individual attack on a municipal target would almost certainly be characterized as an "assault" requiring the element of surprise, and therefore exempt from the rule of notification. However, even subnational groups of terrorists issue such "warnings" to a target nation before launching a campaign of discrete assaults.

> In sieges and bombardments all necessary steps must be taken to spare, as far as possible, buildings dedicated to religion, art, science, or charitable purposes, historic monuments, hospitals, and places where the sick and wounded are collected, provided they are not being used at the time for military purposes.
>
> It is the duty of the besieged to indicate the presence of such buildings or places by distinctive and visible signs, which shall be notified to the enemy beforehand.[468]

---

[467] 1907 Hague Convention IV, Customs of War on Land, *supra* note 355, Annex at art. 26 *reprinted in* ROBERTS & GUELFF, *supra* note 4, at 53. *See also* REISMAN & ANTONIOU, *supra* note 432, at 81 *and* SCHINDLER & TOMAN, ARMED CONFLICTS, *supra* note 5, at 84.

[468] 1907 Hague Convention IV, Customs of War on Land, *supra* note 355, Annex at art. 27 *reprinted in* ROBERTS & GUELFF, *supra* note 4, at 53. *See also* REISMAN & ANTONIOU, *supra* note 432, at 94 *and* SCHINDLER & TOMAN, ARMED CONFLICTS, *supra* note 5, at 84.

The interconnectedness of information infrastructures renders the old concept of discretely targetable buildings less helpful. Today's information systems are inextricably linked, mixing lawful and unlawful targets as never before. A computer routing military communications is obviously a legitimate target; if, however, the same computer is used for a country's civilian ambulance dispatch service, the picture is less clear. It is possible that using a computer for military purposes legitimately places it on a target list, regardless of its subsidiary functions. Such a doctrine would militate for information weapons of such precision that the military functions of any dual-use system could be isolated and surgically removed, leaving the peaceful applications in place. As with the development of precision-guided munitions, the development of such information weapons may serve to raise the legal standard of such discrimination.

The second paragraph suggests that the defending nation is under an obligation to place banners and other warnings on specified information systems, warning potential attackers of the protected nature of the information or connectivity within. Of course, false use of such warnings, employed as a cheap defense of lawful targets, would be a violation of international law.

> "The pillage of a town or place, even when taken by assault, is prohibited."[469]

In the Middle Ages, pillage was the principal means of feeding and rewarding one's troops; the prospect of booty taken in battle was the prime motivator for common and noble warriors. Since then, the norm of protecting civilians from military depredations has evolved to become customary international law. In the case of information operations, it is still relevant: no cyber force attacking, for example, a financial institution

---

[469] 1907 Hague Convention IV, Customs of War on Land, *supra* note 355, Annex at art. 28 *reprinted in* ROBERTS & GUELFF, *supra* note 4, at 53. *See also* REISMAN & ANTONIOU, *supra* note 432 at 81, *and* SCHINDLER & TOMAN, ARMED CONFLICTS, *supra* note 5, at 84.

could first reroute a few million dollars to a personal or even national account. Similarly, hackers hired by one country to conduct an information operation against a financial institution could not be legally employed on a commission basis, allowing them to keep some percentage of the funds misrouted.

> "A person can only be considered a spy when, acting clandestinely or on false pretences, he obtains or endeavors to obtain information in the zone of operations of a belligerent, with the intention of communicating it to a hostile party."[470]

Those information operations directed more toward the gathering of information than its destruction or manipulation fall under the category of "spying." Such operations would, by their very nature, be mounted clandestinely, and with the intent of retrieving information for the hostile party. The "zone of operations" limitation is rendered far less important by the global reach and interconnectedness of today's information systems.

---

[470] 1907 Hague Convention IV, Customs of War on Land, *supra* note 355, Annex at art. 29 *reprinted in* ROBERTS & GUELFF, *supra* note 4, at 53. *See also* REISMAN & ANTONIOU, *supra* note 432, at 42 1994) *and* SCHINDLER & TOMAN, ARMED CONFLICTS, *supra* note 5, at 84-85. Article 29 continues:

> Thus, soldiers not wearing a disguise who have penetrated into the zone of operations of the hostile army, for the purpose of obtaining information, are not considered spies. Similarly, the following are not considered spies: Soldiers and civilians, carrying out their mission openly, entrusted with the delivery of despatches intended either for their own army or for the enemy's army. To this class belong likewise persons sent in balloons for the purpose of carrying despatches and, generally, of maintaining communications between the different parts of an enemy or territory.

ROBERTS & GUELFF, *supra* note 4, at 53.

6. 1907 Hague Convention V Respecting the Rights and Duties of Neutral Powers and Persons in Case of War on Land[471]

Neutrality can be permanent (applying to states which opt for such status in a treaty or other instrument) or situational (applying to all but the belligerent states in any given conflict).[472] Because this latter status applies to so many states so often, the contours of the doctrine are vitally important to the law of armed conflict. The U.N. Charter has modified, but not replaced, these rights and duties of neutral states.[473] Reisman and Antoniou agree, noting that "[t]he Industrial Revolution brought on an expansion of world trade and increasing interdependence between states, and, consequently, it became harder and harder to isolate and contain wars between a few states. Even small wars began to threaten the commercial relationships upon which the rest of the international community depends."[474] This convention, then, provides the basis for the modern law of neutrality.

> "Belligerents are forbidden to move troops or convoys of either munitions of war or supplies across the territory of a neutral power."[475]

---

[471] 1907 Hague Convention V, Neutral Powers, *supra* note 293, *reprinted in* ROBERTS & GUELFF, *supra* note 4, at 61. *See also* REISMAN & ANTONIOU, *supra* note 432, at 134-137 *and* SCHINDLER & TOMAN, ARMED CONFLICTS, *supra* note 5, at 941-949.

[472] ROBERTS & GUELFF, *supra* note 4, at 61.

[473] *Id.* The only practical alteration would be in the case of a U.N. Security Council resolution, adopted under its Chapter VII powers, directing a neutral state to take or forgo a specific action. *Id.*

[474] REISMAN & ANTONIOU, *supra* note 432, at 133.

[475] 1907 Hague Convention V, Neutral Powers, *supra* note 293, at art. 2, October 18, 1907, *reprinted in* ROBERTS & GUELFF, *supra* note 4, at 63. *See also* REISMAN & ANTONIOU,

Neutrality law developed under circumstances analogous to the modern "borderless" commerce of today:

> The concept of neutrality in war emerged with the early development of international maritime law. The rapid growth and increasing importance of international trade in the eighteenth and nineteenth centuries, which led maritime states to seek a means of resisting belligerent interference with neutral trade, became the foundation for the contemporary development of neutrality.[476]

Now, the new "maritime" states are those whose technology allows them unimpeded access to the new ocean of cyberspace. These states seek the same protection that their predecessors did two centuries ago: to limit, to the extent practicable, the effect of hostilities to those nations actually engaged in conflict. Today, there may be an additional desire to limit the adverse effects of conflict even more narrowly, to the aggressors who initiated the conflict, and not to the defending belligerent.

> Belligerents are likewise forbidden to—
>
> (a) Erect on the territory of a neutral power a wireless telegraphy station or other apparatus for the purpose of communicating with belligerent forces on land or sea;
>
> (b) Use of any installation of this kind established by them before the war on the territory of a neutral Power for purely military purposes, and which has not been opened for the service of public messages.[477]

---

supra note 432, at 134 *and* SCHINDLER & TOMAN, ARMED CONFLICTS, *supra* note 5, at 942.

[476] ROBERTS & GUELFF, *supra* note 4, at 61.

[477] 1907 Hague Convention V, Neutral Powers, *supra* note 293, at art. *reprinted in* ROBERTS & GUELFF, *supra* note 4, at 63. *See also* REISMAN & ANTONIOU, *supra* note 432, at 134 *and* SCHINDLER & TOMAN, ARMED CONFLICTS, *supra* note 5, at 942-943.

While it is not likely that belligerents could or would erect dedicated military communications facilities on the territory of a neutral power, it is very likely that a large part of any belligerent's communications will be routed through the global web, consisting of computer and communications sites in countries around the world. These sites appear to escape the clause's prohibition because they are not dedicated to one type of communication (military) or one nation's communications (a belligerent's).

> A neutral Power is not called upon to forbid or restrict the use on behalf of the belligerents of telegraph or telephone cables or of wireless telegraphy apparatus belonging to it or to companies or private individuals.[478]
>
> Every measure of restriction or prohibition taken by a neutral Power in regard to the matters referred to...must be impartially applied by it to both belligerents.
>
> A neutral Power must see to the same obligation being observed by companies or private individuals owning telegraph or telephone cables or wireless telegraphy apparatus.[479]

These articles place no affirmative obligation on neutral states to be anything more than evenhanded in their relations with both sides of a conflict. The neutral states, however, are responsible for more than mere acts of state—they must also ensure that all private actors on their territory are equally evenhanded.

---

[478] 1907 Hague Convention V, Neutral Powers, *supra* note 293, at art. 8 *reprinted in* ROBERTS & GUELFF, *supra* note 4, at 64. *See also* REISMAN & ANTONIOU, *supra* note 432, at 135 *and* SCHINDLER & TOMAN, ARMED CONFLICTS, *supra* note 5, at 943.

[479] 1907 Hague Convention V, Neutral Powers, *supra* note 293, at art. 9 *reprinted in* ROBERTS & GUELFF, *supra* note 4, at 64.

7. 1907 Hague Convention VI Relating to the Status of Enemy Merchant Ships at the Outbreak of Hostilities[480]

This Convention was the first to codify the international rules regarding the treatment of a belligerent nation's merchant ships in foreign ports at the outbreak of hostilities.[481] Although the Convention was an attempt to capture the customary international law of the time, it was also largely aspirational. Its orderly scheme for the movement of shipping immediately after the start of a war was never realized in subsequent conflicts. The Convention demonstrates the ever-present tension between the desire to capture law as it exists, ensuring a high degree of acceptance and compliance, and the desire to advance international law with aspirational goals, raising the standard but risking lower compliance, if not outright rejection.

> Anxious to ensure the security of international commerce against the surprises of war, and wishing, in accordance with modern practice, to protect as far as possible operations undertaken in good faith and in process of being carried out before the outbreak of hostilities, have resolved to conclude a Convention to this effect . . .[482]

Roberts and Guelff explain the background of the Convention:

---

[480] Hague Convention VI Relating to the Status of Enemy Merchant Ships at the Outbreak of Hostilities, October 18, 1907 [hereinafter 1907 Hague Convention VI, Merchant Ships] *reprinted in* ROBERTS & GUELFF, *supra* note 4, at 71. *See also* REISMAN & ANTONIOU, *supra* note 432, at 73-74 *and* SCHINDLER & TOMAN, ARMED CONFLICTS, *supra* note 5, at 791-796.

[481] ROBERTS & GUELFF, *supra* note 4, at 71.

[482] 1907 Hague Convention VI, Merchant Ships, *supra* note 480, *reprinted in* ROBERTS & GUELFF, *supra* note 4, at 72. *See also* SCHINDLER & TOMAN, ARMED CONFLICTS, *supra* note 5, at 791.

> International law once recognized that, during peacetime, states could perform an act of reprisal (technically called 'embargo') whereby an injured state could prevent the ships of a state committing an unlawful act from leaving its ports in order to compel the offending state to make reparations for the act committed. When war seemed imminent, the opposing states could impose such an embargo on enemy merchant ships in port to facilitate capture and confiscation of such ships once war had broken out. However, the use of such reprisals during peacetime would now run up against the UN Charter's requirement to settle disputes by peaceful means and its prohibition on the threat or use of force.[483]

This application of this principle to merchant ships may violate the Charter, but its application to information embargoes immediately before the outbreak of hostilities may not. A nation's essential information in commerce—including the vast majority of financial reserves, which exist in tangible form only 10% of the time—is frequently at risk overseas when a conflict begins. This freezing of information assets may be a far more effective weapon against nations initiating hostilities than the seizure of mere tangible assets, such as merchant ships.

---

[483] ROBERTS & GUELFF, *supra* note 4, at 71. In addition to the Convention's apparent incompatibility with the Charter paradigm, there is a second problem:

> Many states failed to ratify the Convention... During the First World War, observance of the Convention was far from uniform. Great Britain regarded reciprocity and a high degree of uniformity of practice as essential, and therefore denounced the convention in 1925. France followed suit in 1939. At the outbreak of the Second World War, Great Britain and France did not grant a period of grace to enemy merchant ships in port, and enemy merchant ships (in port or on the high seas) were liable to seizure and confiscation. Other states adopted the same practice. In view of this state practice, it is unlikely that the favourable treatment of belligerent merchant ships in enemy ports at the outbreak of hostilities can be regarded as part of customary international law.

Id.

8. 1907 Hague Convention VII Relating to the Conversion of Merchant Ships into Warships[484]

The principle of discrimination provides for different treatment of combatants and noncombatants. The conversion, during hostilities, of noncombatant merchant vessels into warships complicates the application of this principle. In the late 19th and early 20th Century, a significant number of merchant ships were capable of such reconfiguration; this Convention was drafted with a view toward systematizing the process and removing treacherous surprise from the equation. Schindler and Toman note the difficulty inherent in classifying these conversions:

> The signatories of the present Convention, concluded at the Second Hague Peace Conference, although agreeing on the general principle, were unable to settle all the questions that gave rise to the controversy. In particular, they could not agree on the question whether the conversion might be performed in a harbour of the converting State only or also on the high seas (see preamble). Nor could they solve the question whether or not it was permissible to reconvert into a merchantman, before the end of the war, a vessel which during the war had been converted into a warship.[485]

Once again, the well thought out process embodied in the Convention was rarely employed in the real world. The concern for avoiding perfidious behavior, though, was preserved in future conventions and in state practice.

---

[484] 1907 Hague Convention VI, Merchant Ships, *supra* note 480, *reprinted in* ROBERTS & GUELFF, *supra* note 4, at 79. *See also* REISMAN & ANTONIOU, *supra* note 432, at 75 *and* SCHINDLER & TOMAN, ARMED CONFLICTS, *supra* note 5, at 797-802.

[485] SCHINDLER & TOMAN, ARMED CONFLICTS, *supra* note 5, at 797.

> "Whereas it is desirable, in view of the incorporation in time of war of merchant ships in the fighting fleet, to define the conditions subject to which the operation may be effected . . ."[486]

The modern trend toward building advanced weapon and sensor suites into warships had been thought to obviate the quaint practice of simply adding a gun or two to a merchant ship and thus declaring it a warship. However, the 1982 Falklands War saw Great Britain press merchant ships into service not only as combat transports, but even, in one case, as an aircraft carrier for Harrier jump-jets. This ingenious use of existing civilian assets renewed the question of the legal process of the transition from merchant ship to warship. Roberts and Guelff provide background:

> There has traditionally been a clear distinction between the treatment, in time of war, of enemy warships and enemy merchant ships. Warships could be attacked and destroyed, or captured with the title of the ship immediately passing to the capturing state. Merchant ships, on the other hand, were generally immune from attack and destruction, and although they could be captured, title to a ship could only pass after adjudication in the prize courts of the capturing state. However, if a merchant vessel refused to stop, actively resisted search and seizure, directly assisted its own state's warships or attacked enemy warships, any immunity from attack and destruction was forfeited.[487]

---

[486] 1907 Hague Convention VI, Merchant Ships, *supra* note 480, *reprinted in* ROBERTS & GUELFF, *supra* note 4, at 80. *See also* SCHINDLER & TOMAN, ARMED CONFLICTS, *supra* note 5, at 798.

[487] ROBERTS & GUELFF, *supra* note 4, at 79. There are two problems with the Convention itself, and a subsequent problem with state practice:

> The regime established by the Convention is regarded as unsatisfactory because it did not resolve the issues of whether conversion may be

"A merchant ship converted into a war-ship can not have the rights and duties accruing to such vessels unless it is placed under the direct authority, immediate control, and responsibility of the Power whose flag it flies."[488]

"Merchant ships converted into war-ships must bear the external marks which distinguish the warships of their nationality."[489]

"The commander must be in the service of the State and duly commissioned by the competent authorities. His name must figure on the list of the officers of the fighting fleet."[490]

---

performed on the high seas and whether a converted merchant ship may convert back to a merchant ship before the termination of the war.

During the First and Second World Wars, belligerents employed the practice of converting merchant ships into warships, but controversy remained over the place of conversion, the legitimacy of re-conversion, and the status of merchant ships which have not been openly converted. Moreover, the actions of belligerents during both wars challenged the traditional immunity of merchant ships from attack and destruction. To the extent that the noncombatant status of merchant ships is undermined, the practical relevance of a distinction between combatant and noncombatant ships is obviously reduced.

Id. 79-80.

[488] 1907 Hague Convention VI, Merchant Ships, *supra* note 480, *reprinted in* ROBERTS & GUELFF, *supra* note 4, at 80. *See also* REISMAN & ANTONIOU, *supra* note 432, at 75 *and* SCHINDLER & TOMAN, ARMED CONFLICTS, *supra* note 5, at 798.

[489] 1907 Hague Convention VI, Merchant Ships, *supra* note 480, at art. 2 *reprinted in* ROBERTS & GUELFF, *supra* note 4, at 81. *See also* REISMAN & ANTONIOU, *supra* note 432, at 75 *and* SCHINDLER & TOMAN, ARMED CONFLICTS, *supra* note 5, at 798.

[490] 1907 Hague Convention VI, Merchant Ships, *supra* note 480, at art. 3 *reprinted in* ROBERTS & GUELFF, *supra* note 4, at 81. *See also* REISMAN & ANTONIOU, *supra* note 432, at 75 *and* SCHINDLER & TOMAN, ARMED CONFLICTS, *supra* note 5, at 798.

> "The crew must be subject to military discipline."[491]

> "Every merchant ship converted into a war-ship must observe in its operations the laws and customs of war."[492]

> "A belligerent who converts a merchant ship into a warship must, as soon as possible, announce such conversion in the list of war-ships."[493]

These criteria, though less relevant to today's merchant ships, are more relevant to ostensibly civilian computer operations of a belligerent nation. Were a belligerent to "convert" such an operation toward a military purpose, the criteria outlined in the Convention still provide an excellent reference for the appropriate indicia of a transition. The failure to meet of the criteria would highlight the inherently clandestine nature of cyberspace in general and such operations in particular.

---

[491] 1907 Hague Convention VI, Merchant Ships, *supra* note 480, at art. 4 *reprinted in* ROBERTS & GUELFF, *supra* note 4, at 81. *See also* REISMAN & ANTONIOU, *supra* note 432, at 75 *and* SCHINDLER & TOMAN, ARMED CONFLICTS, *supra* note 5, at 798.

[492] 1907 Hague Convention VI, Merchant Ships, *supra* note 480, at art. 5 *reprinted in* ROBERTS & GUELFF, *supra* note 4, at 81. *See also* REISMAN & ANTONIOU, *supra* note 432, at 75, *and* SCHINDLER & TOMAN, ARMED CONFLICTS, *supra* note 5, at 798.

[493] 1907 Hague Convention VI, Merchant Ships, *supra* note 480, at art. 6 *reprinted in* ROBERTS & GUELFF, *supra* note 4, at 81. *See also* REISMAN & ANTONIOU, *supra* note 432, at 75 *and* SCHINDLER & TOMAN, ARMED CONFLICTS, *supra* note 5, at 798.

9. 1907 Hague Convention VIII Relative to the Laying of Automatic Submarine Contact Mines[494]

Naval mines were employed at least as early as the 16th Century,[495] but they were not a widespread threat to merchant shipping until the advent of modern manufacturing processes and mass production in the 19th Century. At that time, it was recognized that a belligerent power would be able to threaten, intentionally or not, the rest of the world's shipping in a given port. To address this indiscriminate threat, the Convention was adopted. Although advancing technology has marginalized the importance of contact mines, the Convention's underlying principles, especially that of notification to all threatened parties, remain robust.

> "Inspired by the of the freedom of sea routes, the common highway of all nations . . ."[496]

Cyberspace is the new "common highway of all nations," and new information weapons bear a strong metaphorical resemblance to contact mines in ports and the sea lanes. Specifically, there are close parallels in the areas of precision, discrimination, the conduct of neutrals, the requirements of war termination, and the "perfection" of weapons to allow compliance.

---

[494] Hague Convention VIII Relative to the Laying of Automatic Submarine Contact Mines, October 18, 1907, [hereinafter 1907 Hague Convention VIII, Contact Mines] *reprinted in* ROBERTS & GUELFF, *supra* note 4, at 85. *See also* REISMAN & ANTONIOU, *supra* note 432, at 56, 141 *and* SCHINDLER & TOMAN, ARMED CONFLICTS, *supra* note 5, at 803-809.

[495] ROBERTS & GUELFF, *supra* note 4, at 85.

[496] *See* 1907 Hague Convention VIII, Contact Mines, *supra* note 494, *reprinted in* ROBERTS & GUELFF, *supra* note 4, at 86. *See also* SCHINDLER & TOMAN, ARMED CONFLICTS, *supra* note 5, at 804.

> It is forbidden—
>
> 1. To lay unanchored automatic contact mines, except when they are so constructed as to become harmless one hour at most after the person who laid then ceases to control them;
>
> 2. To lay anchored automatic contact mines which do not become harmless as soon as they have broken loose from their moorings;
>
> 3. To use torpedoes which do not become harmless when they have missed their mark.[497]

This article stresses the need for precision in controlling what could be a dangerously random weapon. The law of information conflict should similarly require that information weapons become disabled at the time they are no longer under positive control, or soon after, and that when an information weapon which misses its intended target it not be permitted to continue roaming cyberspace, searching for another victim.

> "It is forbidden to lay automatic contact mines off the coast and ports of the enemy, with the sole object of intercepting commercial shipping."[498]
>
> When anchored automatic contact mines are employed, every possible precaution must be taken for the security of peaceful shipping.

---

[497] 1907 Hague Convention VIII, Contact Mines, *supra* note 494, at art. 1 *reprinted in* ROBERTS & GUELFF, *supra* note 4, at 87. *See also* REISMAN & ANTONIOU, *supra* note 432, at 56 *and* SCHINDLER & TOMAN, ARMED CONFLICTS, *supra* note 5, at 804.

[498] 1907 Hague Convention VIII, Contact Mines, *supra* note 494, at art. 2 *reprinted in* ROBERTS & GUELFF, *supra* note 4, at 87. *See also* REISMAN & ANTONIOU, *supra* note 432, at 56 *and* SCHINDLER & TOMAN, ARMED CONFLICTS, *supra* note 5, at 804.

> The belligerents undertake to do their utmost to render these mines harmless within a limited time, and, should they cease to be under surveillance, to notify the danger zones as soon as military exigencies permit, by a notice addressed to ship owners, which must also be communicated to the Governments through the diplomatic channel.[499]

The requirement to discriminate between lawful and unlawful targets is one of the basic doctrines of the law of armed conflict, and would be foundational to the law of information conflict. This article forbids only the targeting of *solely* commercial targets, setting a very low standard for the deployment of these weapons. Presumably, even a small amount of potentially military traffic in and out of a port would justify the use of moored contact mines under this Convention. Similarly, under this treaty, such a standard in cyberspace would allow the disruption of a great deal of civilian message traffic, provided there were *some* military communications intermixed.

> "Neutral Powers which lay automatic contact mines off their coasts must observe the same rules and take the same precautions as are imposed on belligerents."[500]

Given the potential for damage to innocent third parties, even the defensive use of such weapons must meet the standard set for belligerent nations. This rule would appear to span the entire spectrum of conflict under the law of information conflict.

---

[499] 1907 Hague Convention VIII, Contact Mines, *supra* note 494, at art. 3 *reprinted in* ROBERTS & GUELFF, *supra* note 4, at 87. *See also* REISMAN & ANTONIOU, *supra* note 432, at 56 *and* SCHINDLER & TOMAN, ARMED CONFLICTS, *supra* note 5, at 804.

[500] 1907 Hague Convention VIII, Contact Mines, *supra* note 494, at art. 4 *reprinted in* ROBERTS & GUELFF, *supra* note 4, at 87. *See also* REISMAN & ANTONIOU, *supra* note 432, at 141 *and* SCHINDLER & TOMAN, ARMED CONFLICTS, *supra* note 5, at 805.

> At the close of war, the contracting Powers undertake to do their utmost to remove the mines which they have laid, each Power removing its own mines.

> As regards anchored automatic contact mines laid by one of the belligerents off the coast of the other, their position must be notified to the other party by the Power which laid them, and each Power must proceed with the least possible delay to remove the mines in its own waters.[501]

Contact mines, like latent information weapons, have the potential to do damage long after a conflict ends. The regime envisioned in this Convention would appear to be applicable under the law of information conflict: that the party which laid the weapons remove them from areas to which it has lawful access at the end of hostilities, and notify the adversary of weapons remaining on its territory or within its information systems.

> The contracting Powers which do not at present own perfected mines of the pattern contemplated in the present Convention, and which, consequently could not at present carry out the rules laid down in [this Convention], undertake to convert the materiel of their mines as soon as possible, so as to bring it into conformity with the foregoing requirements.[502]

In the Gulf War, two high-tech US Navy ships were knocked out of the war by mines. Since the end of WWII, three-fourths of US ships damaged in combat were damaged by mines. An aggressive power today

---

[501] 1907 Hague Convention VIII, Contact Mines, *supra* note 494, at art. 5 *reprinted in* ROBERTS & GUELFF, *supra* note 4, at 87. *See also* REISMAN & ANTONIOU, *supra* note 432, at 56 *and* SCHINDLER & TOMAN, ARMED CONFLICTS, *supra* note 5, at 805.

[502] 1907 Hague Convention VIII, Contact Mines, *supra* note 494, at art. 6 *reprinted in* ROBERTS & GUELFF, *supra* note 4, at 88. *See also* SCHINDLER & TOMAN, ARMED CONFLICTS, *supra* note 5, at 805.

may develop the technology to cause severe damage to high-tech systems long before it develops the technological or the moral capacity to distinguish between lawful and unlawful targets. Advances in the precision of weapons (both information and kinetic) generally favor the forces of civilization.

10. 1907 Hague Convention IX Concerning Bombardment by Naval Forces in Time of War[503]

Naval bombardment differs qualitatively from land bombardment, in that one of the principal justifications for land bombardment, immediate reduction for immediate seizure, is absent.[504] The military necessity of protecting one's own forces weighs less heavily on a naval commander lobbing shells on a coastal target from outside the range of whatever coastal artillery may exist. Because of that, the equation that balances military necessity against unnecessary suffering is altered, and only the "big picture" operational or strategic justifications remain. Schindler and Toman agree, noting "[a]lthough Article I of the present Convention confirms the principle that undefended ports and towns may not be bombarded, Article 2 allows bombardment by naval forces of

---

[503] Convention (IX) Concerning Bombardment by Naval Forces in Time of War, October 18, 1907, [hereinafter 1907 Hague Convention IX, Bombardment] *reprinted in* ROBERTS & GUELFF, *supra* note 4, at 93. *See also* REISMAN & ANTONIOU, *supra* note 432, at 81-82, 94 *and* SCHINDLER & TOMAN, ARMED CONFLICTS, *supra* note 5, at 811-817.

[504] ROBERTS & GUELFF, *supra* note 4, at 93. Roberts and Guelff explain:

> In land warfare, the customary principle regarding bombardment (later codified in Article 25 of the Regulations annexed to both 1899 Hague Convention II and 1907 Hague Convention IV) prohibited the bombardment by land forces of undefended targets. The principle was based on the notion that an undefended target was open to immediate entry and occupation, and bombardment would only cause unnecessary destruction.
>
> In naval warfare, it was recognized that enemy coastal targets which were defended could be bombarded by naval forces, whether such forces were acting in cooperation with a besieging army or independently. However, the question as to whether or not undefended coastal targets could be bombarded by naval forces remained controversial. Unlike in land warfare, the entry and occupation of undefended coastal targets was relatively rare: however, the objective of eliminating an enemy's military resources remained.

Id.

military objectives in undefended towns. This new rule eventually became applicable in air warfare, too."[505]

This Convention was developed to address these differences, and to codify a common set of criteria for lawful naval bombardment. It formulated a general rule with two exceptions:

> The Convention prohibits naval bombardment of undefended ports, towns, villages, dwellings, or buildings, but (in implicit recognition of the different character of naval warfare) excludes from the prohibition coastal targets which represent a military objective or whose local authorities refuse to comply with legitimate requisitions for supplies necessary for the immediate use of the naval force.[506]

Almost one hundred years later, information operations are the new equivalent of naval bombardment—military operations undertaken less for immediate tactical advantage (although that remains a possibility) or force protection, and more for the accomplishment of larger ends. The criteria that justified naval bombardment at the dawn of the 20th Century may provide a useful analogy for the lawful application of information operations as the century closes.

> "The bombardment by naval forces of undefended ports, towns, villages, dwellings, or buildings is forbidden."[507]

This is the general rule, which comports with the law of land warfare. It suggests that the law of information conflict should prohibit

---

[505] SCHINDLER & TOMAN, ARMED CONFLICTS, *supra* note 5, at 811.

[506] ROBERTS & GUELFF, *supra* note 4, at 93.

[507] 1907 Hague Convention IX, Bombardment, *supra* note 503, at art. I *reprinted in* ROBERTS & GUELFF, *supra* note 4, at 95. *See also* REISMAN & ANTONIOU, *supra* note 432, at 81 *and* SCHINDLER & TOMAN, ARMED CONFLICTS, *supra* note 5, at 812.

attacks on entire towns that have no cybernetic protection for their critical infrastructures.

> "Military works, military or naval establishments, depots of arms or war materiel, workshops or plants which could be utilized for the needs of the hostile fleet or army, and the ships of war in the harbor, are not, however, included in this prohibition"[508]

This is the first exception to the general rule, permitting the destruction of military targets within an undefended town. It applies not only to operational weapons and units, but also to those facilities that provide direct support to them. In the case of information operations, this may include not only a military radar jammer hidden in a populated area, but a civilian television station which broadcasts the locations of enemy units in the local area.

> "The commander of the naval force may destroy them with artillery, after a summons followed by a reasonable time of waiting, if all other means are impossible, and when the local authorities have not themselves destroyed them within the time fixed."[509]

This is a relic from the age of chivalry, when the reduction of a town had to be preceded by a formal request for surrender. In the Middle Ages, a refusal to surrender transferred the moral responsibility for the ensuing action to the defenders, who could be characterized as having brought it upon themselves. It remained over the centuries as a gesture of

---

[508] 1907 Hague Convention IX, Bombardment, *supra* note 503, at art. 2 *reprinted in* ROBERTS & GUELFF, *supra* note 4, at 95. *See also* REISMAN & ANTONIOU, *supra* note 432, at 82 *and* SCHINDLER & TOMAN, ARMED CONFLICTS, *supra* note 5, at 812.

[509] 1907 Hague Convention IX, Bombardment, *supra* note 503, at art. 2 *reprinted in* ROBERTS & GUELFF, *supra* note 4, at 95. *See also* REISMAN & ANTONIOU, *supra* note 432, at 82 *and* SCHINDLER & TOMAN, ARMED CONFLICTS, *supra* note 5, at 812.

fairness, allowable before the 20th Century accelerated the speed and intensity of warfare. Interestingly, psychological operations today frequently make use of this technique, using leaflets or radio broadcasts to warn an enemy of an impending strike against a physical military target. After carrying out several such threats, the force employing PSYOPS effectively gives the information contained on the leaflet or in the broadcast the same military capability as an actual striking force—the target is abandoned and the defenders surrender without any fatalities on either side.

> "He [the commander of the naval force] incurs no responsibility for any unavoidable damage which may be caused by a bombardment under such circumstances."[510]

The key word in this provision is "unavoidable," and is applicable to any militarily necessary operation. This is more difficult to gauge in information operations, where the ripple effect of unintended consequences may well be beyond anything contemplated by the striking force.

> If for military reasons immediate action is necessary, and no delay can be allowed the enemy, it is understood that the prohibition to bombard the undefended town holds good, as in the case given in paragraph I, and that the commander shall take all due measures in order that the town may suffer as little harm as possible.[511]

---

[510] 1907 Hague Convention IX, Bombardment, *supra* note 503, at art. 2 *reprinted in* ROBERTS & GUELFF, *supra* note 4, at 95. *See also* REISMAN & ANTONIOU, *supra* note 432, at 82 *and* SCHINDLER & TOMAN, ARMED CONFLICTS, *supra* note 5, at 812.

[511] 1907 Hague Convention IX, Bombardment, *supra* note 503, at art. 2 *reprinted in* ROBERTS & GUELFF, *supra* note 4, at 95. *See also* REISMAN & ANTONIOU, *supra* note 432, at 82 *and* SCHINDLER & TOMAN, ARMED CONFLICTS, *supra* note 5, at 812.

This special case has, over the course of the century, become the norm. During WWII, maneuver warfare replaced the static model of WWI and the centuries before, with armor, mechanized infantry, and, of course, aircraft accelerating the pace of military operations. Today, information operations can deliver, at almost the speed of light, devastating blows to an enemy who may not even realize he is at war.

> After due notice has been given, the bombardment of undefended ports, towns, villages, dwellings, or buildings may be commenced, if the local authorities, after a formal summons has been made to them, decline to comply with requisitions for provisions or supplies necessary for the immediate use of the naval force before the place in question.
>
> These requisitions shall be in proportion to the resources of the place. They shall only be demanded in the name of the commander of the said naval force, and they shall, as far as possible, be paid for in cash; if not, they shall be evidenced by receipts.[512]

This is the second exception to the general rule, allowing naval bombardment only if necessary supplies are withheld when payment is offered. This is probably an anachronism in modern conflict, and certainly in the age of information operations.

> "Undefended ports, towns, villages, dwellings, or buildings may not be bombarded on account of failure to pay money contributions."[513]

---

[512] 1907 Hague Convention IX, Bombardment, *supra* note 503, at art. 3 *reprinted in* ROBERTS & GUELFF, *supra* note 4, at 95. *See also* REISMAN & ANTONIOU, *supra* note 432, at 82 *and* SCHINDLER & TOMAN, ARMED CONFLICTS, *supra* note 5, at 813.

[513] 1907 Hague Convention IX, Bombardment, *supra* note 503, at art. 4 *reprinted in* ROBERTS & GUELFF, *supra* note 4, at 95. *See also* REISMAN & ANTONIOU, *supra* note 432, at 82 *and* SCHINDLER & TOMAN, ARMED CONFLICTS, *supra* note 5, at 813.

This would render unlawful as a military operation any terrorist extortion of a civilian populace. The groups likely to employ such tactics against such targets are unlikely to concern themselves with the nuanced legality of their actions, but the civilized world may employ such a legal provision against them once they have been identified and apprehended.

> In bombardments by naval forces all the necessary measures must be taken by the commander to spare as far as possible sacred edifices, buildings used for artistic, scientific, or charitable purposes, historic monuments, hospitals, and places where the sick and wounded are collected, on the understanding that they are not used at the same time for military purposes.
>
> It is the duty of the inhabitants to indicate such monuments, edifices, or places by visible signs . . .[514]

These kinds of targets exist in cyberspace, and an adversary must avoid any such systems that are legitimately marked as having a protected function. It will be a challenge for future information operators to discriminate between genuinely marked targets, and those that have been perfidiously marked to protect an otherwise lawful information target.

> "If the military situation permits, the commander of the attacking naval force, before commencing the bombardment, must do his utmost to warn the authorities."[515]

This is an echo of the chivalric duty to warn. While the loss of surprise would seem to act as a deterrent to such warnings, they must only

---

[514] 1907 Hague Convention IX, Bombardment, *supra* note 503, at art. 5 *reprinted in* ROBERTS & GUELFF, *supra* note 4, at 96. *See also* REISMAN & ANTONIOU, *supra* note 432, at 94 *and* SCHINDLER & TOMAN, ARMED CONFLICTS, *supra* note 5, at 813.

[515] 1907 Hague Convention IX, Bombardment, *supra* note 503, at art. 6 *reprinted in* ROBERTS & GUELFF, *supra* note 4, at 96. *See also* REISMAN & ANTONIOU, *supra* note 432, at 82 *and* SCHINDLER & TOMAN, ARMED CONFLICTS, *supra* note 5, at 813.

be made against whole cities that are targeted. Today, weapons are precise enough that such a citywide warning may not be necessary.

> "A town or place, even when taken by storm, may not be pillaged."[516]

This prohibition reinforces the norm that military operations are not to be undertaken for financial gain. An information operation to zero-out a dictator's Swiss bank account would not allow for the balance of that account to be transferred to the information operator's personal account. Whether such sums could be transferred to national treasuries or, more poetically, to the dictator's victims, remain open questions.

---

[516] 1907 Hague Convention IX, Bombardment, *supra* note 503, at art. 7, *reprinted in* ROBERTS & GUELFF, *supra* note 4, at 96. *See also* REISMAN & ANTONIOU, *supra* note 432, at 82 *and* SCHINDLER & TOMAN, ARMED CONFLICTS, *supra* note 5, at 813.

11. 1907 Hague Convention XI Relative to Certain Restrictions with Regard to the Exercise of the Right of Capture in Naval War[517]

At the turn of the century, naval weapons were far less accurate, destructive, or capable of long-distance strikes. For these reasons, naval engagements were not always a sudden, devastating surprise to the disadvantaged ship. Battles could continue for hours, with small amounts of damage done by individual shells, all while the adversaries were in sight of each other. This slower, more incremental form of naval warfare often resulted in the surrender of one vessel to the other. For this reason, rules of naval warfare developed to govern the targeting, capture, processing, and release of certain types of ships and cargoes. Schindler and Toman inform us:

> Some of the restrictions on capture provided for in this Convention were previously applied by virtue of settled practice. This was the case particularly for coastal fishing boats and for vessels engaged in scientific discovery. The Convention extended the immunity to further cases. For military hospital ships, immunity from capture is provided for in Article 22 of Geneva Convention (II) of 1949.[518]

This Convention addresses three areas of such capture: postal correspondence, the exemption from capture of certain vessels, and the treatment of captured merchant crews.[519] The first two of these have direct analogs in the law of information conflict.

---

[517] Convention (XI) Relative to Certain Restrictions with Regard to the Exercise of the Right of Capture in Naval War, October 18, 1907, [hereinafter 1907 Hague Convention XI, Right of Capture] *reprinted in* ROBERTS & GUELFF, *supra* note 4, at 103. *See also* REISMAN & ANTONIOU, *supra* note 432, at 76-77 *and* SCHINDLER & TOMAN, ARMED CONFLICTS, *supra* note 5, at 819-824.

[518] SCHINDLER & TOMAN, ARMED CONFLICTS, *supra* note 5, at 819.

[519] ROBERTS & GUELFF, *supra* note 4, at 101.

> The postal correspondence of neutrals or belligerents, whatever its official or private character may be, found on the high seas on board a neutral or enemy ship, is inviolable. If the ship is detained, the correspondence is forwarded by the captor with the least possible delay.
>
> The provisions of the preceding paragraph do not apply, in case of violation of blockade, to correspondence destined for or proceeding from a blockaded port.[520]
>
> "The inviolability of postal correspondence does not exempt a neutral mail ship from the laws and customs of maritime war as to neutral merchant ships in general. The ship, however, may not be searched except when absolutely necessary, and then only with as much consideration and expedition as possible."[521]

Postal correspondence, the transmission of information, is distinguished from parcel post, the transmission of tangible objects. The latter was never protected, but the former gained protection under this Convention. It was not a codification of existing international customary law (postal correspondence enjoyed no immunity from capture in the 19th Century), but rather instituted a new rule of law.

The protection of postal correspondence was subject to several qualifications in the Convention. Mail in or out of a blockaded port enjoyed no protection. Extending this distinction to the law of information conflict, personal communications, as such, would be protected. The analogy breaks down, however, when discussing

---

[520] 1907 Hague Convention XI, Right of Capture, *supra* note 520, at art. 1 *reprinted in* ROBERTS & GUELFF, *supra* note 4, at 103. *See also* REISMAN & ANTONIOU, *supra* note 432, at 76 *and* SCHINDLER & TOMAN, ARMED CONFLICTS, *supra* note 5, at 820.

[521] 1907 Hague Convention XI, Right of Capture, *supra* note 520, at art. 2 *reprinted in* ROBERTS & GUELFF, *supra* note 4, at 103. *See also* REISMAN & ANTONIOU, *supra* note 432, at 76 *and* SCHINDLER & TOMAN, ARMED CONFLICTS, *supra* note 5, at 820.

blockades. While a city of the last century could be effectively isolated by a besieging force ashore and a blockading force at sea, modern computer systems cannot be so easily quarantined. Information networks are moving from the stovepiped, nationally controlled information pipelines of the 20th Century to the far more decentralized, amorphous networks of the 21st. This process is drastically reducing the effective control of national governments over their citizens' correspondence, and virtually eliminates the possibility of a nationwide or even citywide information blockade.

Still, if traffic analysis of electronic communications allows personal messages to be identified as such (without compromising the actual content of the message itself), it would be at least theoretically possible to provide them some degree of protection in an information conflict. However, the likelihood that such an identifier would be misused (to protect military communications) renders such a distinction far less useful.

Also limiting the usefulness of such protection in the information age is state practice in the 20th Century:

> Although the articles in the Convention relating to postal correspondence are still technically binding, the practice of belligerents during the two world wars has reduced the significance of these provisions. In many cases, enemy merchant vessels were sunk without warning rather than seized, and as a result any correspondence they were carrying was destroyed. In other cases, the use of the mails to forward propaganda, war-related information, and contraband led to postal correspondence being subject to seizure, examination, and either censorship or confiscation. If states regard such practices as compatible with the Convention then the notion of 'inviolability', which is stressed in Articles 1 and 2, has clearly been qualified.[522]

[522] ROBERTS & GUELFF, *supra* note 4, at 101.

> Vessels used exclusively for fishing along the coast or small boats employed in local trade are exempt from capture, as well as their appliances, rigging, tackle, and cargo.
>
> They cease to be exempt as soon as they take any part whatever in hostilities.
>
> The contracting Powers agree not to take advantage of the harmless character of the said vessels in order to use them for military purposes while preserving their peaceful appearance.[523]

The second analogy is to protected vessels. Certain computer systems in the next century could be characterized as performing essentially peaceful, or at least nonbelligerent, functions, similar to coastal fishing and cargo vessels in the 20th Century. Small, local computer systems may receive similar protections for similar reasons.

> "Vessels charged with religious, scientific, or philanthropic missions are likewise exempt from capture."[524]

Information systems operated by religious, scientific, philanthropic, and, possibly, journalistic organizations may be ruled out of bounds for the same reasons their ships were in the last century. However, the state practice of the 20th Century suggests that such theoretical protections may be very limited in practical application:

---

[523] 1907 Hague Convention XI, Right of Capture, *supra* note 520, at art. 3 *reprinted in* ROBERTS & GUELFF, *supra* note 4, at 104 . *See also* REISMAN & ANTONIOU, *supra* note 432, at 76 *and* SCHINDLER & TOMAN, ARMED CONFLICTS, *supra* note 5, at 820.

[524] 1907 Hague Convention XI, Right of Capture, *supra* note 520, at art. 4 *reprinted in* ROBERTS & GUELFF, *supra* note 4, at 104. *See also* REISMAN & ANTONIOU, *supra* note 432, at 76 *and* SCHINDLER & TOMAN, ARMED CONFLICTS, *supra* note 5, at 821.

> In many instances during both world wars, the customary immunity of small coastal fishing and trading boats was not observed in practice. In some cases, action taken against coastal vessels was attributable to the belligerent use of coastal vessels for intelligence purposes: by failing to restrict activity to innocent employment, such vessels thereby forfeited their immunity. The exemption from capture of religious, scientific and philanthropic vessels was interpreted very restrictively by belligerents, and any immunity of particular humanitarian vessels came to rely upon the express agreement of the belligerents.[525]

The greatest difference between the treatment of these ships under the law of armed conflict and the treatment of analogous computer systems under the law of information conflict is the inherent interconnection of modern information systems. The distinctness of well-separated ships on an open sea is precisely the opposite of interlinked computers, networked through the internet. This characteristic of 21st Century information systems suggests that it will be difficult to observe and enforce a regime of protections for personal correspondence or nonbelligerent systems.

---

[525] ROBERTS & GUELFF, *supra* note 4, at 102.

12. 1907 Hague Convention XIII Concerning the Rights and Duties of Neutral Powers in Naval War[526]

By the end of the 19th Century, the broad outlines of a law of neutrality had become customary international law. This Convention was an attempt to codify those principles and provide procedures to implement them. At the most general level, this Convention "prohibits hostile acts by belligerents in neutral ports and waters, and in turn requires a neutral state to use the means at its disposal to prevent such acts."[527]

These provisions have been widely adopted and almost uniformly respected, but state practice has narrowed their application. Often, this was because a belligerent believed (or at least stated) that it was acting against another belligerent in neutral waters solely because the neutral power had itself failed to act against the same belligerent.[528] Even this derogation, however, acknowledges the authority of the standards being upheld.

In the realm of information conflict, the concept of neutrality remains elusive. Distributed systems and networked computers do not respect the 19th or even 20th Century concept of non-porous, sovereign borders. Corporate and national networks often span numerous countries, so the concept of "neutral territory" is more difficult to define and enforce.

> "Belligerents are bound to respect the sovereign right of neutral Powers and to abstain, in neutral territory or neutral

---

[526] Neutral Powers, *supra* note 293, *reprinted in* ROBERTS & GUELFF, *supra* note 4, at 110. *See also* REISMAN & ANTONIOU, *supra* note 432, at137-140 *and* SCHINDLER & TOMAN, ARMED CONFLICTS, *supra* note 5, at 951-960.

[527] ROBERTS & GUELFF, *supra* note 4, at 109.

[528] *Id.*

> waters, from any act which would, if knowingly permitted by any Power, constitute a violation of neutrality."[529]

> "Any act of hostility including capture and the exercise of the right of search, committed by belligerent warships in the territorial waters of a neutral Power, constitutes a violation of neutrality and is strictly forbidden."[530]

These two articles, taken together, describe the essence of belligerents' responsibilities toward neutrals. Reciprocally, neutrals are forbidden to favor either side in an armed conflict, or to act as though they themselves were belligerents.

> "Belligerents are forbidden to use neutral ports and waters as a base of naval operations against their adversaries, and in particular to erect wireless telegraphy stations or any apparatus for the purpose of communicating with belligerent forces on land or sea."[531]

This article is perhaps the most strongly analogous to the law of information conflict, in that it deals with communications during hostilities. The particular emphasis on "wireless telegraphy stations" recognizes the importance of such nodes in wartime. Of course, the technology of the time contemplated the use of radio only for communicating with one's own forces; the idea of jamming the

---

[529] Neutral Powers, *supra* note 293, at art. 1 *reprinted in* ROBERTS & GUELFF, *supra* note 4, at 110. *See also* REISMAN & ANTONIOU, *supra* note 432, at 137 *and* SCHINDLER & TOMAN, ARMED CONFLICTS, *supra* note 5, at 952.

[530] Neutral Powers, *supra* note 293, at art. 2 *reprinted in* ROBERTS & GUELFF, *supra* note 4, at 111. *See also* REISMAN & ANTONIOU, *supra* note 432, at 137 *and* SCHINDLER & TOMAN, ARMED CONFLICTS, *supra* note 5, at 952.

[531] Neutral Powers, *supra* note 293, at art. 5 *reprinted in* ROBERTS & GUELFF, *supra* note 4, at 111. *See also* REISMAN & ANTONIOU, *supra* note 432, at 138 *and* SCHINDLER & TOMAN, ARMED CONFLICTS, *supra* note 5, at 952.

communications of opposing forces, or transmitting false or deceptive information to them, had not yet evolved. What is obvious is the prohibition, under customary and positive international law, was clearly contemplated from the earliest days of this emerging technology.

> "The neutrality of a Power is not affected by the mere passage through its territorial waters of war-ships or prizes belonging to belligerents."[532]

This article may be analogized to the free passage of electronic messages through cyberspace, across national borders. The mere routing of these messages may be the "innocent passage" of the information age, but the active use of routers and servers in neutral territory as part of a military information operation would violate the concept embodied in this article.

> "A neutral power is bound to exercise such surveillance as the means at its disposal allow to prevent any violation of the provisions of the articles occurring in its ports or roadsteads or in its waters."[533]

This article may transfer to the law of information conflict as an affirmative duty on the part of the neutral nation to ensure, to a reasonable standard, that its computer systems are not being co-opted for use by a belligerent in cyberspace. The limits of this duty are far from clear, given the expectation of privacy inherent in any electronic communication.

---

[532] Neutral Powers, *supra* note 293, at art. 10 *reprinted in* ROBERTS & GUELFF, *supra* note 4, at 112. *See also* REISMAN & ANTONIOU, *supra* note 432, at 138 *and* SCHINDLER & TOMAN, ARMED CONFLICTS, *supra* note 5, at 953.

[533] Neutral Powers, *supra* note 293, at art. 25 *reprinted in* ROBERTS & GUELFF, *supra* note 4, at 114. *See also* SCHINDLER & TOMAN, ARMED CONFLICTS, *supra* note 5, at 956.

13. 1923 Hague Rules of Aerial Warfare[534]

Although never formally adopted, these rules were considered at the time of drafting to be an authoritative codification of the existing customary international law on air warfare. Roberts and Guelff point out that, with the possible exception of siege warfare (which is less common in the era of maneuver warfare), air operations against ground targets probably affect more noncombatants than any other type of warfare.[535] Schindler and Toman concur, telling us:

> Although these rules were never adopted in a legally binding form they are of importance 'as an authoritative attempt to clarify and formulate rules of law governing the use of aircraft in war' [footnote omitted]. To a great extent, they correspond to the customary rules and general principles underlying the conventions on the law of war on land and at sea.[536]

The key provision in these rules is the prohibition against aerial bombardment for the purpose of terrorizing civilian populations or destroying private property.[537] This principle was greatly weakened during

---

[534] Rules of Aerial Warfare, 17 *AJIL* (1923) Supplement 245-60, [hereinafter Rules of Aerial Warfare] *reprinted in* ROBERTS & GUELFF, *supra* note 4, at 123. *See also* REISMAN & ANTONIOU, *supra* note 432, at 57, 78-80, 83-84, 95-96, 141-143 *and* SCHINDLER & TOMAN, ARMED CONFLICTS, *supra* note 5, at 207-217.

[535] ROBERTS & GUELFF, *supra* note 4, at 121.

[536] SCHINDLER & TOMAN, ARMED CONFLICTS, *supra* note 5, at 207.

[537] ROBERTS & GUELFF, *supra* note 4, at 122.

> Roberts and Guelff paraphrase British Prime Minister Neville Chamberlain, speaking before the House of Commons on June 21st, 1938, enunciating what were then considered the three fundamental principles of the law of aerial warfare: (1) direct attack against the civilian population is unlawful; (2) targets for air bombardment must be legitimate, identifiable military

the Second World War, when area bombing by both sides was the norm.[538] Since then, however, state practice, at least among civilized countries, appears to be returning to the pre-war norm.

The general principles of the law of armed conflict apply to all military operations, including air warfare. Even so, air warfare remains unique in that, unlike land or sea warfare, there is no binding agreement that addresses it specifically. It is analogous to the law of information conflict in this respect, and in one other: air warfare is the "newest" type of warfare to be assimilated by the law of armed conflict, and may, therefore, provide a model for the adoption of the law of information conflict.

> "Outside the Jurisdiction of any state, belligerent or neutral, all aircraft shall have full freedom of passage through the air and of alighting."[539]

> "In time of war any state, whether belligerent or neutral, may forbid or regulate the entrance, movement or sojourn of aircraft within its jurisdiction."[540]

---

objectives; and (3) reasonable care must be taken in attacking military objectives to avoid bombardment of a civilian population in the neighborhood. These principles were captured in a League of Nations Assembly resolution which was unanimously adopted on September 30th, 1938—less than a year before the outbreak of the Second World War.

Id.

[538] ROBERTS & GUELFF, *supra* note 4, at 122.

[539] Rules of Aerial Warfare, *supra* note 534, at art. 11, *reprinted in* ROBERTS & GUELFF, *supra* note 4, at 123. *See also* SCHINDLER & TOMAN, ARMED CONFLICTS, *supra* note 5, at 209.

[540] Rules of Aerial Warfare, *supra* note 534, at art. 12 *reprinted in* ROBERTS & GUELFF, *supra* note 4, at 123. *See also* SCHINDLER & TOMAN, ARMED CONFLICTS, *supra* note 5, at 209.

> "No aircraft other than a belligerent military aircraft shall engage in hostilities in any form.
>
> The term 'hostilities' includes the transmission during flight of military intelligence for the immediate use of a belligerent."[541]

These three general principles allow uninterrupted flight outside national boundaries in peacetime and controlled access to national territory in wartime. Both of these are common sense measures, but difficult to analogize to the law of information conflict in that digital data packets are routed unpredictably across numerous national boundaries, and do not follow predictable, great-circle arcs as do aircraft. Restricting such a flow in wartime would be difficult, if not impossible.

The third article is the most interesting, specifically because of the standard it sets for equating intelligence collection with hostilities. This is a question directly applicable to the law of information conflict—when does a cyber-probe become a cyber-provocation? The answer given in this article is that tactical intelligence collected for the immediate use of a belligerent is sufficiently threatening to be considered hostile. This might include targeting intelligence, perhaps passed by data link to inbound aircraft or missiles. Presumably, longer-term strategic intelligence collection would not violate this threshold of immediacy.

> "The use of false external markings is forbidden."[542]

---

[541] Rules of Aerial Warfare, *supra* note 534, at art. 16 *reprinted in* ROBERTS & GUELFF, *supra* note 4, at 123. *See also* SCHINDLER & TOMAN, ARMED CONFLICTS, *supra* note 5, at 209.

[542] Rules of Aerial Warfare, *supra* note 534, at art. 19 *reprinted in* ROBERTS & GUELFF, *supra* note 4, at 123. *See also* SCHINDLER & TOMAN, ARMED CONFLICTS, *supra* note 5, at 210.

> "The use of aircraft for the purpose of disseminating propaganda shall not be treated as an illegitimate means of warfare."[543]

These articles, concerning hostilities, are slightly more problematic. Information operations are, almost by definition, conducted in a clandestine manner, so the concept of a national identification within an information weapon may not be an appropriate analogy from these rules. The next article, however, concerning the treatment of aircraft disseminating propaganda, is much more difficult, since information warfare today, as an adjunct to traditional military activities, has as one of its principal goals the curtailing of an enemy leader's propaganda to his people. In Kosovo and elsewhere, transmitting facilities of state-run news and entertainment broadcasters were specifically targeted. The protection these rules accord to aircraft disseminating propaganda appear to be exactly opposed to the emerging standard in the law of information conflict, which considers such facilities among the most dangerous and, therefore, the most appropriate target for a belligerent.

> "Aerial bombardment for the purpose of terrorizing the civilian population, of destroying or damaging private property not of a military character, or of injuring noncombatants is prohibited."[544]

---

[543] Rules of Aerial Warfare, *supra* note 534, at art. 21 *reprinted in* ROBERTS & GUELFF, *supra* note 4, at 123. *See also* SCHINDLER & TOMAN, ARMED CONFLICTS, *supra* note 5, at 210.

[544] Rules of Aerial Warfare, *supra* note 534, at art. 22 *reprinted in* ROBERTS & GUELFF, *supra* note 4, at 126. *See also* REISMAN & ANTONIOU, *supra* note 432, at 83-84 *and* SCHINDLER & TOMAN, ARMED CONFLICTS, *supra* note 5, at 210-211. Article 24's five paragraphs then go into much greater detail about the prohibited activity:

> 1. Aerial bombardment is legitimate only when directed at a military objective, that is to say, and object of which the destruction or injury would constitute a distinct military advantage to the belligerent.

These rules are simply the application of the traditional doctrines of military necessity, proportionality, and chivalry to the field of aerial bombardment. That these rules were widely disregarded during the Second World War was certainly a setback, but the post-war years have seen their return as the standard for air warfare.

Information operations are frequently targeted at civilian audiences, but these operations are more persuasive than destructive. As such, they are not the moral equivalent of bombs being dropped on an undefended populace. Information operations whose end is destructive, however, such as a computer network attack which results in the

---

2. Such bombardment is legitimate only when directed exclusively at the following objectives: military forces; military works; military establishments or depots; factories constituting important and well-known centres engaged in the manufacture of arms, ammunition or distinctively military supplies; lines of communication or transportation used for military purposes.

3. The bombardment of cities, towns, villages, dwellings or buildings not in the immediate neighborhood of the operations of land forces is prohibited. In cases where the objective specified in paragraph 2 are so situated, that they cannot be bombed without the indiscriminate bombardment of the civilian population, the aircraft must abstain from bombardment.

4. In the immediate neighborhood of the operations of land forces, the bombardment of cities, towns, villages, dwellings or buildings is legitimate provided that there exists a reasonable presumption that the military concentration is sufficiently important to justify such bombardment, having regard to the danger thus caused to the civilian population.

5. A belligerent state is liable to pay compensation for injuries to person or to property caused by the violation by any of its officers or forces of the provision of this article.

ROBERTS & GUELFF, *supra* note 4, at 126-127, REISMAN & ANTONIOU, *supra* note 432 at 83-84 and SCHINDLER & TOMAN, ARMED CONFLICTS, *supra* note 5, at 210-211.

immediate shutdown of a civilian telephone system (disabling the national equivalent of a "911" system), would be more analogous to the physical destruction prohibited by this article.

> In bombardment by aircraft, all necessary steps must be taken by the commander to spare as far as possible buildings dedicated to public worship, art, science, or charitable purposes, historic monuments, hospital ships, hospitals and other places where the sick and wounded are collected, provided such buildings, objects or places are not at the time used for military purposes. Such buildings, objects, and places must by day be indicated by marks visible to aircraft. The use of marks to indicate other buildings, objects, or places than those specified above is to be deemed an act of perfidy.[545]

The ability to mark these special protected areas is more challenging in cyberspace, in that targets are more difficult to discriminate in an internetworked world. Under the law of information conflict, the key would be to emphasize warning less and redundant protections more. A hospital might not be able to post a Red Cross banner on every computer screen that could be used to interrupt its power, emergency

---

[545] Rules of Aerial Warfare, *supra* note 534, at art. 25 *reprinted in* ROBERTS & GUELFF, *supra* note 4, at 127. *See also* REISMAN & ANTONIOU, *supra* note 432, at 95 *and* SCHINDLER & TOMAN, ARMED CONFLICTS, *supra* note 5, at 211. Article 25 concludes:

> The marks used as aforesaid shall be in the case of buildings protected under the Geneva Convention the red cross on a white ground, and in the case of other protected buildings a large rectangular panel divided diagonally into two pointed triangular portions, one black and the other white.
>
> A belligerent who desires to secure by night the protection for the hospitals and other privileged buildings above mentioned must take the necessary measures to render the special signs referred to sufficiently visible.

ROBERTS & GUELFF, *supra* note 4, at 127, REISMAN & ANTONIOU, *supra* note 432 at 95 and SCHINDLER & TOMAN, ARMED CONFLICTS, *supra* note 5, at 211.

communications, or distribution of medical supplies. The same hospital, however, could take the reasonable precaution (effective against natural as well as man-made disasters) of having back-up, on-site power generation, redundant communications capability, and enough cushion in its medical supply stock to absorb a temporary disruption in the flow of logistics. The standard could be applied to other protected sites, insulating them from all but physical destruction, which is already covered under the existing rules.

> Any person on board a belligerent or neutral aircraft is to be deemed a spy only if acting clandestinely or on false pretences he obtains or seeks to obtain, while in the air information within belligerent jurisdiction or in the zone of operations of a belligerent with the intention of communicating it to the hostile party.[546]

> "Acts of espionage committed after leaving the aircraft by members of the crew of an aircraft or by passengers transported by it are subject to the provisions of Land Warfare Regulations."[547]

As presumably most information operations would be conducted, or at least launched, from a nation's own territory, this article would apply only to those tactical intelligence collection operations taking place in areas under the enemy's control. The vast majority of information operations would meet this standard of clandestinity, thus leaving the

---

[546] Rules of Aerial Warfare, *supra* note 534, at art. 27 *reprinted in* ROBERTS & GUELFF, *supra* note 4, at 128. *See also* SCHINDLER & TOMAN, ARMED CONFLICTS, *supra* note 5, at 212.

[547] Rules of Aerial Warfare, *supra* note 534, at art. 28 *reprinted in* ROBERTS & GUELFF, *supra* note 4, at 128. *See also* SCHINDLER & TOMAN, ARMED CONFLICTS, *supra* note 5, at 212.

soldiers or intelligence officers carrying them out, whether from aircraft or on the ground, liable to charges of spying. Such in-theater information operations will pose special hazards to those conducting them, not the least of which is exposure to the charges of espionage.

14. 1925 Geneva Protocol for the Prohibition of the Use in War of Asphyxiating, Poisonous or Other Gases, and of Bacteriological Methods of Warfare[548]

On the surface, this Protocol is not directly related to the law of information operations. However, the history surrounding it reveals that absolute prohibitions of any weapon or means of warfare can be far from absolute in practice.

First, state practice can at least threaten, if not completely undermine, such an agreement. With the exception of Japan's use of chemical weapons against China, all belligerents in the Second World War assiduously avoided the use of chemical weapons, probably based less on *opinio juris* and more on a fear of in-kind retaliation. As evidence of this approach, consider that the major powers maintained large stocks of such weapons, to be used in response to another's first-use.[549]

---

[548] 1925 Geneva Protocol, Bacteriological Methods of Warfare, *supra* note 13, *reprinted in* ROBERTS & GUELFF, *supra* note 4, at 139. *See also* REISMAN & ANTONIOU, *supra* note 432, at 57-58 *and* SCHINDLER & TOMAN, ARMED CONFLICTS, *supra* note 5, at 115-127.

[549] ROBERTS & GUELFF, *supra* note 4, at 138. Reisman and Antoniou concur, noting:

> "The exact meaning of this Protocol is contested, and for some states its prohibition is contingent. The United States has interpreted the Protocol of 1925 as not prohibiting the use of riot control agents such as tear gas. A number of Western European allies have interpreted the Protocol as prohibiting all such gasses. Additionally, a significant number of nations, including the United States, issued a reservation stating that first use by a belligerent of chemical and biological weapons authorizes the state subject to the attack to respond in kind, that is, to retaliate with chemical and biological weapons."

REISMAN & ANTONIOU, *supra* note 432, at 58.

Second, the apparently clear language of the prohibition was subject to enough reservations that the Protocol became not a prohibition of the use of such weapons, but a prohibition of *first* use.[550]

Third, the distinction between lethal and nonlethal agents has been very difficult to draw. The problem of definitions, and the variables of employment leave an enormous gray area for interpretation.

Finally, the agreement itself did not guarantee compliance by rogue states. At best, it gave the civilized world a legal basis to protest Iraqi and Russian use of chemical weapons in recent years.[551]

> Whereas the use in war of asphyxiating, poisonous or other gases, and of all analogous liquids, materials or devices, has been justly condemned by the general opinion of the civilised world; and
>
> Whereas the prohibition of such has been declared in Treaties to which the majority of Powers of the world are Parties; and
>
> To the end that this prohibition shall be universally accepted as a part of International Law, binding alike the conscience and the practice of nations;
>
> DECLARE:
>
> That the High Contracting Parties, so far as they are not already Parties to Treaties prohibiting such use, accept this prohibition, agree to extend this prohibition to the use of bacteriological methods of warfare and agree to be bound as

---

[550] ROBERTS & GUELFF, *supra* note 4, at 138.

[551] *Id.*

> between themselves according to the terms of this declaration.[552]

Considering all this, the deceptively simple text of the protocol, reproduced above, was not the conclusion of the legal wrangling, but merely the starting point. Even for a relatively coherent, almost universally reviled set of weapons it took decades to effectively outlaw them, and today rogue nations still have the capability to employ them. The means and methods governed by the law of information conflict could prove far less amenable to such sweeping, generalized control. Controlling the worst effects of information weapons will depend more on the correct application of the principles of the law of armed conflict (discrimination, necessity, proportionality, and chivalry) as they translate into the law of information conflict than on documentary arms control.

---

[552] 1925 Geneva Protocol, Bacteriological Methods of Warfare, *supra* note 13, *reprinted in* ROBERTS & GUELFF, *supra* note 4, at 139-40. *See also* REISMAN & ANTONIOU, *supra* note 432, at 57-58 *and* SCHINDLER & TOMAN, ARMED CONFLICTS, *supra* note 5, at 116.

15. 1936 London Procès-Verbal Relating to the Rules of Submarine Warfare Set Forth in Part IV of the Treaty of London of 22 April 1930[553]

The London Procès-Verbal, also known as the London Protocol, was an attempt to capture and extend a provision of the Treaty of London of 22 April 1930. Article 22 set forth the standards, under international law, for submarine warfare. This treaty, however, had few signatories, and was approaching the end of its time in force (31 December 1936) when the Procès-Verbal was drafted. Consisting of Article 22 from the Treaty and no other substance, the Procès-Verbal was quickly signed by most of the world's naval powers, including almost all of the belligerents in the Second World War.[554]

State practice during the war, however, severely limited the effect of the Procès-Verbal. The Germans conducted unrestricted submarine warfare (attacks against all ships flying a belligerent flag), and gave two legal explanations. First, they claimed that the attacks were reprisals, permitted under international law. Second, they argued that since the British and other belligerents had been arming their transports, it was impossible to discriminate between warships and merchant ships, and therefore impossible to comply with the Procès-Verbal. The Allies soon followed suit, targeting Axis merchant shipping.[555] The lack of any significant state practice after the Second World War leaves little additional action for interpretation.[556]

---

[553] Procès-Verbal Relating to the Rules of Submarine Warfare Set Forth in Part IV of the Treaty of London, 6 November, 1936, [hereinafter Submarine Warfare] *reprinted in* ROBERTS & GUELFF, *supra* note 4, at 149. *See also* REISMAN & ANTONIOU, *supra* note 432, at 77-78 *and* SCHINDLER & TOMAN, ARMED CONFLICTS, *supra* note 5, at 883-885.

[554] ROBERTS & GUELFF, *supra* note 4, at 147.

[555] *Id.* at 147-48.

[556] ROBERTS & GUELFF, *supra* note 4, at 148. Roberts and Guelff describe the awkward legal outcome of the war crimes trials of the leaders of the German Navy:

The Procès-Verbal is an object lesson in the difficulties of fashioning a durable rule of law that is too far out of step with the practical incentives before the parties. Given that any conflict will have at least one aggressor who feels unbound by the inconvenient legal conventions of the day, the defending power is faced with violating (or interpreting very liberally) the existing rule of law to maintain military parity.

A second but related problem is the creation of an unforeseen gray area in military operations—in this case, the light, defensive arming of merchant ships. The Procès-Verbal contemplated a clear distinction between combatants and noncombatants at sea, so the blurring of categories in the real world undercut the object and purpose of the agreement. Furthermore, it provided a ready pretext for the aggressor—here, one sufficient to avoid conviction at trial.

> "(1) In their action with regard to merchant ships, submarines must conform to the rules of International Law to which surface vessels are subject."[557]
>
> (2) In particular, except in cases of persistent refusal to stop on being duly summoned, or of active resistance to visit or search, a warship, whether surface vessel or submarine, may not sink or render incapable of navigation a merchant

---

> In the trials of Admirals Doenitz and Raeder at Nuremberg in 1945-6, the International Military Tribunal found the accused had violated the Procès-Verbal, but because it recognized that the United States and Great Britain had also carried on unrestricted submarine warfare, it was neither prepared to hold them guilty, nor to assess their sentences, on the basis of such violations.

Id.

[557] Submarine Warfare, *supra* note 553, at para 1 *reprinted in* ROBERTS & GUELFF, *supra* note 4, at 149. *See also* REISMAN & ANTONIOU, *supra* note 432, at 77 *and* SCHINDLER & TOMAN, ARMED CONFLICTS, *supra* note 5, at 884.

> vessel without having first placed passengers, crew and ship's papers in a place of safety unless the safety of the passengers and crew is assured, in the existing sea a and weather conditions, by the proximity of land, or the presence of another vessel which is in a position to take them on board.[558]

This text is a remarkably succinct statement of the perceived international legal standard of the day. The drafters, however, appear to have traded clarity and fidelity for the most desirable legal standard of practicability in warfare. As the law of information conflict coalesces, this procedural tension will remain, and can only be solved by structuring incentives so that the parties comply out of a sense, not just of legal obligation, but of practical self-benefit.

Reisman and Antoniou inform us that, "[l]ike many efforts at codifying the law of war, the London Rules did not receive wide acceptance. Because of developments in communications and aviation, they were obsolete with regard to submarine warfare by the time they were drafted."[559] Keeping this in mind, future attempts to codify the law of information operations must take into account the incredible rate of speed at which technology develops as opposed to the comparably slow pace with which international laws and treaties are usually written and accepted.

A substantive analogy to information operations may be found in the concept of first safeguarding the noncombatant contents of a computer also containing military information or applications before destroying it. The difficulty in segregating cybernetic contraband from the digital equivalent of ship's papers (let alone passengers and crew) is more difficult in cyberspace than on the open sea. The small amount of

---

[558] Submarine Warfare, *supra* note 553, at para 2 *reprinted in* ROBERTS & GUELFF, *supra* note 4, at 149-50. *See also* REISMAN & ANTONIOU, *supra* note 432, at 77-78 *and* SCHINDLER & TOMAN, ARMED CONFLICTS, *supra* note 5, at 884.

[559] REISMAN & ANTONIOU, *supra* note 432, at 78.

time allowed by the incredible speed of information operations, and the lack of access to an entire system for a thorough separation of materials, makes such discrimination virtually impossible. In any future information conflict, the incentives seem to favor the electronic equivalent of unrestricted submarine warfare. The law of information conflict will be most effective if it takes this incentive into account before fashioning a workable rule of law.

16. 1946 Judgment of the International Military Tribunal at Nuremberg: Extracts on Crimes Against International Law[560]

Although the Charter of the International Military Tribunal at Nuremberg was drafted to cover a single, albeit immense, proceeding, the "Nuremberg Principles" contained therein are declaratory of customary international law.[561] Specifically, the definitions of offenses contained in Article 6 of the Charter Provisions is the authoritative statement on the subject:

> The following acts, or any of them, are crimes coming within the jurisdiction of the Tribunal for which there shall be individual responsibility:
>
> (a) Crimes against Peace: namely, planning, preparation, initiation, or waging of a war of aggression, or a war in violation of international treaties, agreements, or assurances, or participation in a common plan or conspiracy for the accomplishment of any of the foregoing[;]
>
> (b) War Crimes: namely, violations of the laws or customs of war. Such violations shall include, but not be limited to, murder, ill-treatment, or deportation to slave labor or for any other purpose of civilian population of or in occupied territory, murder or ill-treatment of prisoners of war or persons on the seas, killing of hostages, plunder of public or private property, wanton destruction of cities, towns, or villages, or devastation not justified by military necessity[;]

---

[560] Judgment of the International Military Tribunal at Nuremberg: Extracts on Crimes Against International Law (1946) [hereinafter Crimes Against International Law] *reprinted in* ROBERTS & GUELFF, *supra* note 4, at 155. *See also* REISMAN & ANTONIOU, *supra* note 432, at 318-336 *and* SCHINDLER & TOMAN, ARMED CONFLICTS, *supra* note 5, at 921-924.

[561] ROBERTS & GUELFF, *supra* note 4, at 153.

> (c) Crimes against Humanity: namely, murder, extermination, enslavement, deportation, and other inhumane acts committed against any civilian population, before or during the war, or persecutions on political, racial or religious grounds in execution of or in connection with any crime within the jurisdiction of the Tribunal, whether or not violation of the domestic law of the country where perpetrated.
>
> Leaders, organizers, instigators, and accomplices participating in the formulation or execution of a common plan or conspiracy to commit any of the foregoing crimes are responsible for all acts performed by any persons in execution of such plan.[562]

Whether crimes against peace, war crimes, or crimes against humanity are committed through brute force, or with the sophistication allowed by modern information systems, the offenses are still judged by their effects in the real world. Information operations may have the aura of clean, modern, tightly focused, violence-free raids into computers or command-and-control systems, but the reality is more closely linked with traditional military operations.

---

[562] Crimes Against International Law at art. 6 reprinted *in* ROBERTS & GUELFF, *supra* note 4, at 155. *See also* REISMAN & ANTONIOU, *supra* note 432, at 319-320 *and* SCHINDLER & TOMAN, ARMED CONFLICTS, *supra* note 5, at 924.

The incitement of racial hatred in Rwanda and the Balkans, inflaming citizens to make the short but crucial leap from passive prejudice to active complicity in genocide, was made possible by the clever manipulation of mass media by individual tyrants and small groups of followers. It is not an overstatement to say that such information operations are the difference between sporadic outbursts of the "lynch mob" mentality and the mobilization of entire populations to rid the country of "the other." The law of information conflict may begin by governing the actions of a few technicians surrounded by stacks of state-of-the-art equipment, but it will have its ultimate effect in the preservation or destruction of those terrorized by the world's remaining tyrants.[563]

---

[563] C.S. Lewis was the first to identify that the focus of genocidal evil was not on the traditional battlefield:

> The greatest evil is not now done in those sordid 'dens of crime' that Dickens loved to paint. It is not done even in concentration camps and labour camps. In those we see its final result. But it is conceived and ordered (moved, seconded, carried, and minuted) in clean, carpeted, warmed and well-lighted offices, by quiet men with white collars and cut fingernails and smooth-shaven cheeks who do not need to raise their voice.

C.S. LEWIS, THE SCREWTAPE LETTERS, X (1982). *Cf.*, DANIEL GOLDHAGEN, HITLER'S WILLING EXECUTIONERS: ORDINARY GERMANS AND THE HOLOCAUST (1997).

17. 1948 United Nations Convention on the Prevention and Punishment of the Crime of Genocide[564]

The Genocide Convention translated the Nuremberg Principles into an obligation of states, first as parties to the agreement, and then under customary international law. Although it is admirably clear and direct as a statement of law and principle, it is weak in its enforcement provision. In particular, article VI provides that a person accused of genocide may be tried in the courts of any nation that accepts jurisdiction, including the defendant's own country. This is problematic, in that genocide is almost always domestic, and therefore carried out at the direction of the sovereign. The option in the second half of the same article, allowing for trial by an international tribunal, has become plausible only with the recent creation of the *ad hoc* tribunals for the Balkans and Rwanda, and the permanent International Criminal Court. Even with the administrative machinery in place, the political will to seize and try individuals above the lowest level has yet to be demonstrated.[565]

> "The Contracting Parties confirm that genocide, whether committed in time of peace or in time of war, is a crime under international law which they undertake to prevent and punish."[566]
>
> In the present Convention, genocide means any of the following acts committed with intent to destroy, in

---

[564] Convention on the Prevention and Punishment of the Crime of Genocide, 11 December 1948, [hereinafter Genocide] *reprinted in* ROBERTS & GUELFF, *supra* note 4, at 158. *See also* REISMAN & ANTONIOU, *supra* note 432, at 84-85 *and* SCHINDLER & TOMAN, ARMED CONFLICTS, *supra* note 5, at 231-249.

[565] ROBERTS & GUELFF, *supra* note 4, at 158.

[566] Genocide, *supra* note 564, at art. I *reprinted in* ROBERTS & GUELFF, *supra* note 4, at 158. *See also* REISMAN & ANTONIOU, *supra* note 432, at 84 *and* SCHINDLER & TOMAN, ARMED CONFLICTS, *supra* note 5, at 232.

whole or in part, a national, ethnical, racial or religious group, as such:

(a) Killing members of the group;

(b) Causing serious bodily or mental harm to members of the group;

(c) Deliberately inflicting on the group conditions of life calculated to bring about its physical destruction in whole or in part;

(d) Imposing measures intended to prevent births within the group;

(e) Forcibly transferring children of the group to another group.[567]

"The following acts shall be punishable:

(a) Genocide;

(b) Conspiracy to commit genocide;

(c) Direct and public incitement to commit genocide;

(d) Attempt to commit genocide

(e) Complicity in genocide."[568]

---

[567] Genocide, *supra* note 564, at art. II *reprinted in* ROBERTS & GUELFF, *supra* note 4, at 158. *See also* REISMAN & ANTONIOU, *supra* note 432, at 84-85 *and* SCHINDLER & TOMAN, ARMED CONFLICTS, *supra* note 5, at 232.

[568] Genocide, *supra* note 564, at art. III *reprinted in* ROBERTS & GUELFF, *supra* note 4, at 159. *See also* REISMAN & ANTONIOU, *supra* note 432, at 85 *and* SCHINDLER & TOMAN, ARMED CONFLICTS, *supra* note 5, at 232.

"Persons responsible for committing genocide or any of the other acts enumerated in article III shall be punished, whether they are constitutionally responsible rulers, public officials or private individuals."[569]

"The contracting Parties undertake to enact, in accordance with their respective Constitutions, the necessary legislation to give effect to the provisions of the present Convention and, in particular, to provide effective penalties for persons guilty of genocide or of any of the other acts enumerated in article III."[570]

Persons charged with genocide or any of the other acts enumerated in article III shall be tried by a competent tribunal of the State in the territory of which the act was committed, or by such international penal tribunal as may have jurisdiction with respect to those Contracting Parties which shall have accepted its jurisdiction.[571]

---

[569] Genocide, *supra* note 564, at art. IV *reprinted in* ROBERTS & GUELFF, *supra* note 4, at 159. *See also* REISMAN & ANTONIOU, *supra* note 432, at 85 *and* SCHINDLER & TOMAN, ARMED CONFLICTS, *supra* note 5, at 232.

[570] Genocide, *supra* note 564, at art. V *reprinted in* ROBERTS & GUELFF, *supra* note 4, at 159. *See also* REISMAN & ANTONIOU, *supra* note 432, at 85 *and* SCHINDLER & TOMAN, ARMED CONFLICTS, *supra* note 5, at 232.

[571] Genocide, *supra* note 564, at art. VI *reprinted in* ROBERTS & GUELFF, *supra* note 4, at 159. *See also* REISMAN & ANTONIOU, *supra* note 432, at 85 *and* SCHINDLER & TOMAN, ARMED CONFLICTS, *supra* note 5, at 232-233. Article VII prohibits states from characterizing genocide as a 'political offense' for the purpose of avoiding extradition, and affirmatively calls upon states to grant extradition in such cases. Genocide, *supra* note 564, at art. VII *reprinted in* ROBERTS & GUELFF, *supra* note 4, at 159. *See also* REISMAN & ANTONIOU, *supra* note 432, at 85 *and* SCHINDLER & TOMAN, ARMED CONFLICTS, *supra* note 5, at 233. Article VIII allows states to call upon the UN to assist in the fight against genocide. Genocide, *supra* note 564, at art. VIII *reprinted in* ROBERTS & GUELFF, *supra* note 4, at 159. *See also* REISMAN & ANTONIOU, *supra* note 432, at 85 *and* SCHINDLER & TOMAN, ARMED CONFLICTS, *supra* note 5, at 233. Article IX directs that disputes under the Convention be submitted to the International Court of Justice (ICJ) for adjudication. Genocide, *supra* note 564, at art. IX *reprinted in* ROBERTS & GUELFF, *supra* note 4, at 159-

With the stated exception of article VI, these articles outline a clear, logical approach to the crime of genocide. The law of information conflict will have to grow beyond the stereotypical image of hacker vs. hacker, and accept a crucial, possibly central, role in these largest questions of the law of armed conflict. Specifically, when is an information facility (computer, telephone exchange, radio or TV station) a lawful target? How can article III (c)'s prohibition of "direct and public incitement to commit genocide" be fashioned into a concrete rule of engagement? When this type of information operation by a tyranny is the largest single factor in organizing hundreds of thousands of deaths, what counter-operations would be a lawful response?

The law of information conflict must address these large questions in future military and intelligence operations, remaining consistent with existing international law, but acknowledging the true center-of-gravity in genocidal conflicts.

---

160. *See also* REISMAN & ANTONIOU, *supra* note 432, at 85 *and* SCHINDLER & TOMAN, ARMED CONFLICTS, *supra* note 5, at 233.

18. 1949 Geneva Convention I for the Amelioration of the Condition of the Wounded and Sick in Armed Forces in the Field[572]

Although lengthy and of pivotal importance in the development of the law of armed conflict, the four Geneva Conventions have only limited applicability to the law of information conflict.[573] Beyond codifying and systematizing the pre-existing customary norms relating to soldiers rendered *hors de combat,* the principal contribution of the first of the Geneva Conventions is Article 2, common to all four Conventions:

> In addition to the provisions which shall be implemented in peacetime, the present Convention shall apply to all cases of declared war or of any other armed conflict which may arise between two or more of the High Contracting Parties, even if the state of war is not recognized by one of them.
>
> The Convention shall also apply to all cases of partial or total occupation of the territory of a High Contracting Party, even if the said occupation meets no armed resistance.[574]

This article establishes three conditions to meet the threshold of hostilities required to trigger the protections of the law of armed conflict. These three are declared war, partial or total occupation, or "any other

---

[572] 1949 Geneva Convention I, *supra* note 148, *reprinted in* ROBERTS & GUELFF, *supra* note 4, at 171. *See also* REISMAN & ANTONIOU, *supra* note 432, at 168-179 *and* SCHINDLER & TOMAN, ARMED CONFLICTS, *supra* note 5, at 373-399.

[573] ROBERTS & GUELFF, *supra* note 4, at 169.

[574] 1949 Geneva Convention I, *supra* note 148, *reprinted in* ROBERTS & GUELFF, *supra* note 4, at 171-72. *See also* SCHINDLER & TOMAN, ARMED CONFLICTS, *supra* note 5, at 376.

armed conflict."[575] While the first two are relatively straightforward, the third condition presents an undefined, purely *de facto* standard. This is unsurprising, given that the drafters of the four Conventions were determined to have these protections apply to the greatest number of people at the earliest possible time.

In the absence of a declaration of war (rare in the second half of the 20th Century), or any kind of occupation (virtually impossible through the manipulation of cyberspace alone), the remaining standard of "any other hostilities" is determinative for the application of the law of information conflict.

---

[575] 1949 Geneva Convention I, *supra* note 148, *reprinted in* ROBERTS & GUELFF, *supra* note 4, at 171-2. *See also* SCHINDLER & TOMAN, ARMED CONFLICTS, *supra* note 5, at 376.

19. 1949 Geneva Convention II for the Amelioration of the Condition of the Wounded, Sick, and Shipwrecked Members of Armed Forces at Sea[576]

Like the first Geneva Convention, the second contains articles common to the series of four. Common Article 13 explains to whom the Conventions apply. Three of these categories are of interest under the law of information conflict. First, there is the *de jure* military:[577]

> "(1) Members of the armed forces of a Party to the conflict, as well as members of militias or volunteer corps forming part of such armed forces."[578]

These are the traditional armed forces, and there is little difficulty in determining who is or is not a member. More problematic are separate militias, the *de facto* military:

> (2) Members of other militias and members of other volunteer corps, including those of organized resistance movements, belonging to a Party to the conflict and operating in or outside their own territory, even if this territory is occupied, provided that such militias or volunteer corps, including organized resistance movements, fulfill the following conditions:
>
> (a) that of being commanded by a person responsible for his subordinates;

---

[576] 1949 Geneva Convention II, *supra* note 148, *reprinted in* ROBERTS & GUELFF, *supra* note 4, at 194. *See also* REISMAN & ANTONIOU, *supra* note 432, at 153-168 *and* SCHINDLER & TOMAN, ARMED CONFLICTS, *supra* note 5, at 401-422.

[577] ROBERTS & GUELFF, *supra* note 4, at 191.

[578] 1949 Geneva Convention II, *supra* note 148, at art. 13 *reprinted in* ROBERTS & GUELFF, *supra* note 4, at 199. *See also* REISMAN & ANTONIOU, *supra* note 432, at 157 *and* SCHINDLER & TOMAN, ARMED CONFLICTS, *supra* note 5, at 408.

(b) that of having a fixed distinctive sign recognizable at a distance;

(c) that of carrying arms openly;

(d) that of conducting their own operations in accordance with the laws and customs of war.[579]

This presents problems for a nation's targeting policy. Since Article 13 captures the customary international legal standard for determining the lawfulness of human targets, ascertaining whether the attacker is a mere individual criminal or a soldier in a national army depends on meeting these criteria. Unfortunately, the nature of information operations is such that the identity of the attacker is only discovered weeks or months after the attack, if at all. Since the legal contours of an appropriate reply depend on the attacker's identity, this is an intolerable situation for the defending power. The necessity of instituting a presumption of identity, based on the nature of the *target*, would seem to be a key requirement for the law of information conflict.

Parallel to this concern is the categorization of the attacker once he has been identified. His chain of command may be informal to the point of at-will employment; he may wear no uniform or any insignia; his "arms," such as they are, may not be immediately identifiable, and in any case would not be visible to the victim; and his conduct of operations may well fall outside the accepted boundaries of the law of armed conflict. Given this, it may be difficult to classify an attacker within the Article 13(2) criteria.

The one remaining category for our purposes is somewhat broader:

---

[579] 1949 Geneva Convention II, *supra* note 148, at art. 13 *reprinted in* ROBERTS & GUELFF, *supra* note 4, at 199. *See also* REISMAN & ANTONIOU, *supra* note 432, at 157 *and* SCHINDLER & TOMAN, ARMED CONFLICTS, *supra* note 5, at 408.

> Persons who accompany the armed forces without actually being members thereof, such as civilian members of military aircraft crews, war correspondents, supply contractors, members of labour units or of services responsible for the welfare of the armed forces, provided that they have received authorization from the armed forces which they accompany.[580]

The legal status of contractors is a topic of great interest today, in that numerous high-tech military systems cannot operate without on-scene civilian support personnel from the manufacturer. These contractors are intermingled with military members, and are exposed to the same dangers and threats. This problem will be exacerbated when these contractors become necessary to not only maintain information weapons, but to actually employ them. This may be the category in which civilian information operators find themselves in future conflicts.

---

[580] 1949 Geneva Convention II, *supra* note 148, at art. 13 *reprinted in* ROBERTS & GUELFF, *supra* note 4, at 199. *See also* REISMAN & ANTONIOU, *supra* note 432, at 157 *and* SCHINDLER & TOMAN, ARMED CONFLICTS, *supra* note 5, at 408.

20. 1949 Geneva Convention III Relative to the Treatment of Prisoners of War[581]

The third Geneva Convention has a peripheral, but interesting, contact with the law of information conflict. The restrictions on the flow of information to and from prisoners reflects the tension between security and humanity:

> "Books may not be included in parcels of clothing and foodstuffs."[582]
>
> The censoring of correspondence addressed to prisoners of war or despatched by them shall be done as quickly as possible. Mail shall be censored only by the despatching State and the receiving State, and once only by each.
>
> The examination of consignments intended for prisoners of war shall not be carried out under conditions that will expose the goods contained in them to deterioration; except in the case of written or printed matter, it shall be done in the presence of the addressee, or of a fellow-prisoner duly delegated by him. The delivery to prisoners of individual or collective consignments shall not be delayed under the pretext of difficulties of censorship.
>
> Any prohibition of correspondence ordered by Parties to the conflict, either for military or political reasons, shall be only temporary and its duration shall be as short as possible.[583]

---

[581] 1949 Geneva Convention III, *supra* note 148, *reprinted in* ROBERTS & GUELFF, *supra* note 4, at 244. *See also* REISMAN & ANTONIOU, *supra* note 432, at 39, 179-219 *and* SCHINDLER & TOMAN, ARMED CONFLICTS, *supra* note 5, at 423-493.

[582] 1949 Geneva Convention III, *supra* note 148, at art. 72 *reprinted in* ROBERTS & GUELFF, *supra* note 4, at 215. *See also* REISMAN & ANTONIOU, *supra* note 432, at 196-197 *and* SCHINDLER & TOMAN, ARMED CONFLICTS, *supra* note 5, at 455.

Article 76's concern for the speed of correspondence to and from the prisoner is balanced by Article 72's prohibition of books. Correspondence to and from interned prisoners is obviously necessary for the morale of the prisoners and their families, and is a useful check on the "disappearing" of prisoners.

At some point, however, this benefit is outweighed by the danger of burying secret messages in the sheer volume of text in a book. Since cryptographic analysis of a large number of books would take an exorbitant amount of time, manpower, and computer strength, a flat prohibition is a draconian but effective solution.

As information systems evolve in the future, this problem will become more acute. While real-time communications may still be prohibited, even a short e-mail message could carry reams of undetected data. Prisoners in future conflicts may find their access to 21st Century means of communications severely curtailed. The ease with which simple paper correspondence can be inspected and controlled may keep 19th Century hardcopy letters as the medium for communicating with future prisoners of war.

---

[583] 1949 Geneva Convention III, *supra* note 148, at art. 76 *reprinted in* ROBERTS & GUELFF, *supra* note 4, at 246. *See also* SCHINDLER & TOMAN, ARMED CONFLICTS, *supra* note 5, at 456.

21. 1949 Geneva Convention IV Relative to the Protection of Civilian Persons in Time of War[584]

While this Convention does not address the military use of information as a weapon *per se*, it does outline the "information rights" of those civilians detained or imprisoned in wartime.[585]

> Where in occupied territory an individual protected person is detained as a spy or saboteur, or as a person under definite suspicion of activity hostile to the security of the Occupying Power, such person shall, in those cases where absolute military security so requires, be regarded as having forfeited rights of communication under the present Convention.[586]

The necessity for holding spies and saboteurs incommunicado is obvious and needs no further elaboration. The implicit concern of this article, however, is that such reasoning could be used pretextually to isolate other prisoners or detainees. This accounts for the requirement that the person be detained *as* a spy or saboteur, or that the person be under *definite* suspicion of hostile activity.

> All persons in the territory of a Party to the conflict, or in a territory controlled by it, shall be enabled to give news of a strictly personal nature to members of their families, wherever they may be, and to receive

---

[584] 1949 Geneva Convention IV, *supra* note 148, *reprinted in* ROBERTS & GUELFF, *supra* note 4, at 272. *See also* REISMAN & ANTONIOU, *supra* note 432, at 233-291 *and* SCHINDLER & TOMAN, ARMED CONFLICTS, *supra* note 5, at 495-556.

[585] ROBERTS & GUELFF, *supra* note 4, at 271.

[586] 1949 Geneva Convention IV, *supra* note 148, at art. 5 *reprinted in* ROBERTS & GUELFF, *supra* note 4, at 274. *See also* REISMAN & ANTONIOU, *supra* note 432, at 234 *and* SCHINDLER & TOMAN, ARMED CONFLICTS, *supra* note 5, at 503.

> news from them. This correspondence shall be forwarded speedily and without undue delay.
>
> If the Parties to the conflict deem it necessary to restrict family correspondence, such restrictions shall be confined to the compulsory use of standard forms containing twenty-five freely chosen words, and to the limitation of the number of these forms despatched to one each month.[587]
>
> Each Party to the conflict shall facilitate enquiries made by members of families dispersed owing to war, with the object of renewing contact with one another and of meeting, if possible. It shall encourage, in particular, the work of organizations engaged on this task provided they are acceptable to it and conform to its security regulations.[588]

These regulations govern the general population of occupied territories.[589] They are intended to prevent the restriction of information as an instrument of mass punishment, and as a bargaining chip in wartime negotiations between the Parties to the conflict.

---

[587] 1949 Geneva Convention IV, *supra* note 148, at art. 25 *reprinted in* ROBERTS & GUELFF, *supra* note 4, at 2781-82. *See also* REISMAN & ANTONIOU, *supra* note 432, at 239 *and* SCHINDLER & TOMAN, ARMED CONFLICTS, *supra* note 5, at 509-510.

[588] 1949 Geneva Convention IV, *supra* note 148, at art. 26 *reprinted in* ROBERTS & GUELFF, *supra* note 4, at 282. *See also* REISMAN & ANTONIOU, *supra* note 432, at 239 *and* SCHINDLER & TOMAN, ARMED CONFLICTS, *supra* note 5, at 510.

[589] Once these protected persons are interned as detainees, a parallel information regime is put into place. *See* 1949 Geneva Convention IV, *supra* note 148, at arts. 106-07 *reprinted in* ROBERTS & GUELFF, *supra* note 4, at 309. This regime provides for immediate notification of detention to an international organization, the right to send two letters and four cards each month, the right to send telegrams under pressing conditions, and the right to correspond in one's own language. *Id. See also* REISMAN & ANTONIOU, *supra* note 432, at 265 *and* SCHINDLER & TOMAN, ARMED CONFLICTS, *supra* note 5, at 534-35.

These articles implicitly recognize access to some form of telecommunications, even if not in real time, as a fundamental human need worthy of protection, even under the trying conditions of wartime. To what extent this standard of customary international law extends to the use of e-mail and telephones remains to be seen.

22. 1954 Hague Convention for the Protection of Cultural Property in the Event of Armed Conflict[590]

This Convention was the first to provide a comprehensive, international set of protections for "cultural property."[591] What this term includes is specified in the Convention's first article:

> For the purposes of the present Convention, the term 'cultural property' shall cover, irrespective of origin or ownership:
>
> *(a)* movable or immovable property of great importance to the cultural heritage of every people, such as monuments of architecture, art or history, whether religious or

---

[590] 1954 Hague Convention for the Protection of Cultural Property in the Event of Armed Conflict, 14 May 1954, [hereinafter 1954 Hague Convention, Cultural Property] *reprinted in* ROBERTS & GUELFF, *supra* note 4, at 340. *See also* REISMAN & ANTONIOU, *supra* note 432, at 97-105 *and* SCHINDLER & TOMAN, ARMED CONFLICTS, *supra* note 5, at 745-75.

[591] ROBERTS & GUELFF, *supra* note 4, at 339. Also, Schindler and Toman provide outstanding depth of background on this Convention:

> Following the signature of the Roerich Pact by the American States in 1935 attempts were undertaken to draft a more comprehensive Convention for the protection of monuments and works of art in time of war. In 1939, a draft convention, elaborated under the auspices of the International Museums Office, was presented to governments by the Netherlands. On account of the outbreak of World War II no further steps could be taken. After the war, a new proposal was submitted to UNESCO by the Netherlands in 1948. The General Conference of UNESCO in 1951 decided to convene a committee of government experts to draft a convention. This committee met in 1952 and thereafter submitted its drafts to the General Conference. The revised drafts were then transmitted to governments for advice. The Intergovernmental Conference which drew up and adopted the Convention and the further Acts took place at The Hague from 21 April to 14 May 1954. 56 States were represented. The Final Act has no force of law.

SCHINDLER & TOMAN, ARMED CONFLICTS, *supra* note 5, at 741.

> secular; archaeological sites; groups of buildings which, as a whole, are of historic or artistic interest; works of art; manuscripts, books, and other objects of artistic, historical or archaeological interest; as well as scientific collections and important collections of books or archives or of reproductions of the property defined above;
>
> *(b)* buildings whose main and effective purpose is to preserve or exhibit the movable cultural property defined in subparagraph *(a)* such as museums, large libraries and depositories of archives, and refuges intended to shelter, in the event of armed conflict, the movable cultural property defined in subparagraph *(a)*;
>
> *(c)* centres containing a large amount of cultural property as defined in subparagraphs *(a)* and *(b)*, to be known as 'centres containing monuments'.[592]

Roberts and Guelff explain that the protection of such cultural property is not absolute:

> If cultural property is used for military purposes, an opposing belligerent is released from the obligation to ensure immunity so long as the particular violation persists. In addition, exceptional cases of unavoidable military necessity may remove the special protection of cultural property only for such time as that necessity continues. In such circumstances, advance notice of the withdrawal of immunity is required.[593]

---

[592] 1954 Hague Convention, Cultural Property, *supra* note 590, at art. 1 *reprinted in* ROBERTS & GUELFF, *supra* note 4, at 341. *See also* REISMAN & ANTONIOU, *supra* note 432, at 97 *and* SCHINDLER & TOMAN, ARMED CONFLICTS, *supra* note 5, at 747-48.

[593] ROBERTS & GUELFF, *supra* note 4, at 340.

The precise circumstances under which cultural property would lose its protection are obviously fact-driven, but the Convention presents the broad criteria:

> A centre containing monuments shall be deemed to be used for military purposes whenever it is used for the movement of military personnel or material, even in transit. The same shall apply whenever activities directly connected with military operations, the stationing of military personnel, or the production of war material are carried out within the centre.[594]

For the purposes of the law of information conflict, it is clear that use of a facility for command and control of enemy forces ("directly connected with military operations") renders it a lawful target, despite the presence of cultural property.

Information attacks that targeted the military presence near cultural property would not appear to run afoul of the Convention, in that there would be no physical threat to the protected items.

---

[594] 1954 Hague Convention, Cultural Property, *supra* note 590, at art. 8 *reprinted in* ROBERTS & GUELFF, *supra* note 4, at 343. *See also* REISMAN & ANTONIOU, *supra* note 432, at 100 *and* SCHINDLER & TOMAN, ARMED CONFLICTS, *supra* note 5, at 750.

23. 1954 Hague Protocol for the Protection of Cultural Property in the Event of Armed Conflict[595]

This Protocol was signed the same day as the Convention, and made clear the responsibilities of the parties in wartime. Specifically, the Protocol has five requirements. First, to prevent the export of cultural property from occupied territory.[596] Second, to seize cultural property improperly imported from occupied territory.[597] Third, to return improperly exported cultural property at the end of hostilities.[598] Fourth, to reimburse holders in good faith from whom cultural property was

---

[595] 1954 Hague Protocol for the Protection of Cultural Property in the Event of Armed Conflict, 14 May 1954, [hereinafter 1954 Hague Protocol, Protection of Cultural Property] *reprinted in* ROBERTS & GUELFF, *supra* note 4, at 363. *See also* SCHINDLER & TOMAN, ARMED CONFLICTS, *supra* note 5, at 777-82. Schindler and Toman again provide some history:

> The purpose of the present Protocol is to prevent the exportation of cultural property and to provide for the restitution of illegally exported objects. In view of the difficulties of several governments in adopting provisions on the restitution of property (the original draft contained more far-reaching provisions) it was decided to separate them from the Convention and to adopt them in the form of a separate Protocol.

SCHINDLER & TOMAN, ARMED CONFLICTS, *supra* note 5, at 777.

[596] 1954 Hague Protocol, Protection of Cultural Property, *supra* note 595, at art. 1 *reprinted in* ROBERTS & GUELFF, *supra* note 4, at 363. *See also* SCHINDLER & TOMAN, ARMED CONFLICTS, *supra* note 5, at 777.

[597] 1954 Hague Protocol, Protection of Cultural Property, *supra* note 595, at art. 2 *reprinted in* ROBERTS & GUELFF, *supra* note 4, at 363. *See also* SCHINDLER & TOMAN, ARMED CONFLICTS, *supra* note 5, at 777.

[598] 1954 Hague Protocol, Protection of Cultural Property, *supra* note 595, at art. 3 *reprinted in* ROBERTS & GUELFF, *supra* note 4, at 364. *See also* SCHINDLER & TOMAN, ARMED CONFLICTS, *supra* note 5, at 777.

seized in order to effect its return.[599] Fifth and finally, to return cultural property which has been stored outside the territory of a belligerent for safekeeping once hostilities are concluded.[600]

This Protocol is related to the law of information conflict in that the states parties have an obligation, under international law, to track the records of ownership, transport, and storage of all missing cultural property in an occupied area. Such a tracking effort could be lengthy and intrusive, and would require access to private business and personal records that would be otherwise safe from government intrusion.

---

[599] 1954 Hague Protocol, Protection of Cultural Property, *supra* note 595, at art. 4 *reprinted in* ROBERTS & GUELFF, *supra* note 4, at 364 *See also* SCHINDLER & TOMAN, ARMED CONFLICTS, *supra* note 5, at 777.

[600] 1954 Hague Protocol, Protection of Cultural Property, *supra* note 595, at art. 5 *reprinted in* ROBERTS & GUELFF, *supra* note 4, at 364. *See also* SCHINDLER & TOMAN, ARMED CONFLICTS, *supra* note 5, at 778.

24. 1971 Zagreb Resolution of the Institute of International Law on Conditions of Application of Humanitarian Rules of Armed Conflict to Hostilities in Which United Nations Forces May be Engaged[601]

The Zagreb Resolution is unusual in several respects, not the least of which is that it was issued by an unofficial body.[602] More substantively, it addressed the application of the law of war to a non-state actor that is a party to none of the controlling international agreements.[603]

---

[601] Conditions of Application of Humanitarian Rules of Armed Conflict to Hostilities in which United Nations Forces May Be Engaged, 4 September 1971, [hereinafter Conditions of Application of Humanitarian Rules] *reprinted in* ROBERTS & GUELFF, *supra* note 4, at 372. *See also* SCHINDLER & TOMAN, ARMED CONFLICTS, *supra* note 5, at 903-905.

[602] Roberts and Guelff explain:

> Founded in 1873, the Institute of International Law is an unofficial body consisting of sixty members and approximately sixty associates, all of whom are distinguished international lawyers from various states throughout the world. According to its Statute, the objective of the Institute is to assist in the progressive development of international law by, *inter alia*, contributing to the maintenance of peace and the observance of the laws of war. Since its founding, the Institute has adopted many resolutions relating to the laws of war.

ROBERTS & GUELFF, *supra* note 4, at 371.

[603] Schindler and Toman concur:

> No international treaty expressly provides for the application of the law of armed conflicts to hostilities in which United Nations Forces are engaged. There are, however, several resolutions, adopted by governmental and non-governmental bodies which take up this question. The resolution of the Institute of International Law is the first which sets out more detailed rules for the problems involved. It was adopted by a vote of 42 in favour, none against and 1 abstention.

SCHINDLER & TOMAN, ARMED CONFLICTS, *supra* note 5, at 903.

The unique nature of the United Nations requires a unique legal means of implementing customary international law. As Roberts and Guelff point out:

> While all national contingents of a UN force are equally bound by customary law, they may not be equally bound by such other parts of the laws of war as do not have the status of customary law. Also, despite the measure of international personality possessed by the UN, it does not have the same attributes as a state, and this may make the application of certain specific provisions of the international agreements on the laws of war (as opposed to their 'principles and spirit') less practicable.[604]

The requirement for a single standard among the national contingents in a United Nations force animated the international community to accept the Resolution.

> "For the purposes of the present Articles, the term 'United Nations Forces' shall apply to all armed units under the control of the United Nations."[605]
>
> The humanitarian rules of the law of armed conflict apply to the United Nations as of right, and they must be complied with in all circumstances by United Nations forces which are engaged in hostilities.
>
> The rules referred to in the preceding paragraph include in particular:

---

[604] ROBERTS & GUELFF, *supra* note 4, at 372. For example, the application of the concept of "protecting powers" in article 6 of the resolution. *See infra* note 6.

[605] Conditions of Application of Humanitarian Rules, *supra* note 601, at art. I *reprinted in* ROBERTS & GUELFF, *supra* note 4, at 372. *See also* SCHINDLER & TOMAN, ARMED CONFLICTS, *supra* note 5, at 904.

> *(a)* the rules pertaining to the conduct of hostilities in general and especially those prohibiting the use or some uses of certain weapons, those concerning the means of injuring the other party, and those relating to the distinction between military and non-military objectives;
>
> *(b)* the rules contained in the Geneva Conventions of August 12, 1949;
>
> *(c)* the rules which aim at protecting civilian persons and property.[606]

Although not a Security Council Resolution, this passage makes the clearest single statement of the extent to which the law of armed conflict applies to UN forces. It is a logical extension of a long line of *ad hoc* statements on the topic by the UN itself.[607]

---

[606] Conditions of Application of Humanitarian Rules, *supra* note 601, at art. 2 *reprinted in* ROBERTS & GUELFF, *supra* note 4, at 372. *See also* SCHINDLER & TOMAN, ARMED CONFLICTS, *supra* note 5, at 904.

[607] Roberts and Guelff review theses statements:

> In late 1950, during the Korean War, the UN Command in Korea specifically instructed all the forces under its command to observe all provisions of 1949 Geneva Convention III on prisoners of war. The UN Emergency Force in Egypt was subject to Regulations (drawn up by the Secretary-General and dated 20 February 1957) of which Article 44 stated: 'The Force shall observe the principles and spirit of the general international Conventions applicable to the conduct of military personnel.' The UN Operation in the Congo was subject to an identical Article in its Regulations dated 15 July 1963. In a letter to the ICRC, published in *International Review of the Red Cross* in January 1962, UN Secretary-General U Thant confirmed that the 'UNO insists on its armed forces in the field applying the principles of these 1949 Geneva Conventions as scrupulously as possible.'

ROBERTS & GUELFF, *supra* note 4, at 371-72.

> If United Nations Forces are formed through individual recruitment, the United Nations shall issue regulations defining the rights and duties of the members of such forces...
>
> If United Nations Forces are composed of national contingents with regard to which the United Nations has not issued any regulations such as those mentioned in the preceding paragraph, effective compliance with the humanitarian rules of armed conflict must be secured through agreements concluded between the Organization and the several States which contribute contingents.[608]

Here, the UN becomes responsible for providing sufficient rules of engagement (ROE) and law of war training so that its forces can comply with the standards of customary international law. This responsibility exists independently of the same responsibility borne by each state contributing a national contingent to any given military operation. Roberts and Guelff are quite clear on this point: "National contingents in the service of the UN remain bound, to the same extent and to the same degree, by the laws of war which would apply if the same forces were engaged in international armed conflict for their own states; states retain a responsibility for their contingents."[609]

> Without prejudice to the individual or collective responsibility which derives from the very fact that the party opposing the United Nations Forces has committed

---

[608] Conditions of Application of Humanitarian Rules, *supra* note 601, at art. 3 *reprinted in* ROBERTS & GUELFF, *supra* note 4, at 372. Article 4 calls for instruction of armed forces on the law of armed conflict. Article 5 provides for health services sufficient to meet the requirements of humanitarian law. Article 6 allows the United Nations to designate or create a protecting power, if one is necessary to comply with the law of armed conflict. Zagreb Resolution, *supra* note 601. *See also* SCHINDLER & TOMAN, ARMED CONFLICTS, *supra* note 5, at 904-905.

[609] ROBERTS & GUELFF, *supra* note 4, at 372.

> aggression, that party shall make reparation for injuries caused in violation of the humanitarian rules of armed conflict. The United Nations is entitled to demand compliance with these rules for the benefit of its Forces and to claim damages for injuries suffered by its Forces in violation of these rules.[610]
>
> "The United Nations is liable for damage which may be caused by its Forces in violation of the humanitarian rules of armed conflict, without prejudice to any possible recourse against the State whose contingent has caused the damage."[611]

Here, the Resolution makes a clear statement on liability: that the UN will press its claims against those who violate the laws of war in action against its forces, and that the UN will accept the liability attendant under customary international law for the activities of forces under its direction. Parenthetically, the above passages make a surprisingly clear statement on the emerging doctrine that military activity against UN forces is aggression *per se*. This departure from the Grotian doctrine of equal application suggests that the international community, although eager to have humanitarian protection attach to both sides of a conflict at the lowest possible level of hostilities, is beginning to take note of the lopsided nature of military conflict—that is, that hostilities are rarely conducted by two parties of equal moral culpability, but rather, are conducted by an aggressor and a defender.

---

[610] Conditions of Application of Humanitarian Rules, *supra* note 601, at art. 7 *reprinted in* ROBERTS & GUELFF, *supra* note 4, at 372. *See also* SCHINDLER & TOMAN, ARMED CONFLICTS, *supra* note 5, at 905.

[611] Conditions of Application of Humanitarian Rules, *supra* note 601, at art. 8 *reprinted in* ROBERTS & GUELFF, *supra* note 4, at 372. *See also* SCHINDLER & TOMAN, ARMED CONFLICTS, *supra* note 5, at 905.

25. 1977 United Nations Convention on the Prohibition of Military or Any Other Hostile Use of Environmental Modification Techniques[612]

While the 1977 Geneva Protocol I prohibits any means or method of warfare that results in damage *to* the environment,[613] the "EnMod" Convention prohibits the use of the environment *as* a weapon.[614] This latter convention is implicated by information operations only to the extent that such operations could be construed as manipulating the forces of nature.

> "1. Each State Party to this Convention undertakes not to engage in military or any other hostile use of environmental modification techniques having widespread, long-lasting or severe effects as the means of destruction, damage or injury to any other State Party."[615]

---

[612] Convention on the Prohibition of Military or Any Other Hostile Use of Environmental Modification Techniques, 18 May 1977, [hereinafter Environmental Modification Techniques] *reprinted in* ROBERTS & GUELFF, *supra* note 4, at 379. *See also* REISMAN & ANTONIOU, *supra* note 432, at 68 *and* SCHINDLER & TOMAN, ARMED CONFLICTS, *supra* note 5, at 163-175.

[613] 1977 Geneva Protocol I, *supra* note 355, at art. 35(3), *reprinted in* ROBERTS & GUELFF, *supra* note 4, at 389. Protocol I prohibits attacks causing "widespread, long-term *and* severe damage to the natural environment." [emphasis added] *Id. See also* SCHINDLER & TOMAN, ARMED CONFLICTS, *supra* note 5, at 645.

The EnMod Convention prohibits attacks which modify the environment to the point of causing "widespread, long-lasting *or* severe effects." [emphasis added] Environmental Modification Techniques, *supra* note 612, at Art. 1 *reprinted in* ROBERTS & GUELFF, *supra* note 4, at 379. *See also* REISMAN & ANTONIOU, *supra* note 432, at 68 *and* SCHINDLER & TOMAN, ARMED CONFLICTS, *supra* note 5, at 164. The EnMod Convention's disjunctive test sets a lower threshold of damage than that of Protocol I's conjunctive test.

[614] ROBERTS & GUELFF, *supra* note 4, at 378.

[615] Environmental Modification Techniques, *supra* note 612, at Art. 1 *reprinted in* ROBERTS & GUELFF, *supra* note 4, at 379.

> "As used in article I, the term 'environmental modification techniques' refers to any technique for changing—through the deliberate manipulation of natural processes—the dynamics, composition or structure of the Earth, including its biota, lithosphere, hydrosphere and atmosphere, or of outer space."[616]

A computer network attack on an oil refinery's control system, for instance, or on a nuclear power plant's control system, could have catastrophic effects in the real world. Such attacks may meet the high standard set by Protocol I, and could almost always be construed qualitatively as an attempt to modify the environment (by creating a choking atmosphere of petrochemicals, or of releasing radioactive material), thus violating the EnMod Convention.[617]

There are, however, two strict limitations on the application of this treaty to the law of information conflict. First, the damage that has been described here has, up to this point, only been accomplished through

---

[616] Environmental Modification Techniques, *supra* note 612, at Art. II *reprinted in* ROBERTS & GUELFF, *supra* note 4, at 380.

[617] The precise meanings of the terms "widespread," "long-lasting," and "severe" were transmitted on 2 September 1976 by the Conference of the Committee on Disarmament to the UN General Assembly. The first Understanding included these definitions:

> (a) "widespread:" encompassing an area on a scale of several hundred square kilometres;
>
> (b) "long-lasting:" lasting for a period of months, or approximately a season;
>
> (c) "severe:" involving serious or significant disruption or harm to human life, natural and economic resources or other assets

ROBERTS & GUELFF, *supra* note 4, at 377.

the use of traditional kinetic weapons. The multiple layers of redundant safety features at such facilities pose a serious challenge to the information operation planner. Second, the customary legal doctrines of proportionality and discrimination provide an independent bar to information operations which would affect so many civilians so adversely. It is difficult to imagine a widespread, long-lasting, and severe effect on the environment that would not be grossly beyond any reasonable balancing of military necessity and collateral damage to civilian life and property.

26. 1977 Geneva Protocol I Additional to the Geneva Conventions of 12 August 1949, and Relating to the Protection of Victims of International Armed Conflicts[618]

Protocol I was adopted to update and expand upon the four Geneva Conventions of 1949. Three trends drove this requirement: First, most of the wars in the latter half of the 20th Century were predominantly if not exclusively civil wars, and the existing legal regime for non-international conflicts needed strengthening. Second, guerrillas played a

---

[618] 1977 Geneva Protocol I, *supra* note 355, *reprinted in* ROBERTS & GUELFF, *supra* note 4, at 389. *See also* REISMAN & ANTONIOU, *supra* note 432, at 22, 26, 43-45, 69, 87-93, 105 *and* SCHINDLER & TOMAN, ARMED CONFLICTS, *supra* note 5, at 621-688. Schindler and Toman provided the historical context of the Protocol:

> The present Protocol brings mainly the following innovations: Article I(4) provides that armed conflicts in which peoples are fighting against colonial domination, alien occupation or racist regimes are to be considered international conflicts.
>
> Part II (Articles 8-34) develops the rules of the First and the Second Geneva Conventions on wounded, sick and shipwrecked. It extends the protection of the Conventions to civilian medical personnel, equipment and supplies and to civilian units and transports and contains detailed provisions on medical transportation.
>
> Part III and several chapters of Part IV (Articles 35-60) deal with the conduct of hostilities, *i.e.* questions which hitherto were regulated by the Hague Conventions of 1899 and 1907 and by customary international law. Their reaffirmation and development is important in view of the age of the Hague Conventions and of the new States which had no part in their elaboration. Article 43 and 44 give a new definition of armed forces and combatants. Among the most important Articles are those on the protection of the civilian population against the effects of hostilities. They contain a definition of military objectives and prohibitions of attack on civilian persons and objects. Further Articles (61-79) deal with the protection of civil defense organizations, relief actions, and the treatment of persons in the power of a party to a conflict.

REISMAN & ANTONIOU, *supra* note 432, at 621.

major role in these new conflicts, and their activities challenged the traditional definitions of 'combatants.' Finally, civilians became a primary target in these conflicts, and so their protection in wartime acquired a new urgency.[619]

Several aspects of Protocol I have proven controversial.[620] First, Protocol I broadens the application of the Geneva Conventions to "armed conflicts in which people are fighting against colonial domination and alien occupation and against racist regimes in the exercise of their right of self-determination."[621] In addition, the Protocol permits the application of the Geneva Conventions to both sides by means of a unilateral declaration by the head of such a revolutionary movement.[622]

Second, the Protocol broadens the definition of "combatants" to include guerrillas wearing distinctive clothing and carrying their weapons openly *only* at the actual moment of attack, permitting them combatant status even when approaching or leaving the target in civilian clothing and with weapons hidden.[623]

---

[619] REISMAN & ANTONIOU, *supra* note 432, at 387.

[620] *Id.* at 388.

[621] 1977 Geneva Protocol I, *supra* note 355, at art. 1(4) *reprinted in* ROBERTS & GUELFF, *supra* note 4, at 390. *See also* REISMAN & ANTONIOU, *supra* note 432, at 22 *and* SCHINDLER & TOMAN, ARMED CONFLICTS, *supra* note 5, at 628.

[622] 1977 Geneva Protocol I, *supra* note 355, at art. 96(3) *reprinted in* ROBERTS & GUELFF, *supra* note 4, at 389. *See also* SCHINDLER & TOMAN, ARMED CONFLICTS, *supra* note 5, at 676.

[623] 1977 Geneva Protocol I, *supra* note 355, at art. 44 *reprinted in* D ROBERTS & GUELFF, *supra* note 4, at 411-12. *See also* REISMAN & ANTONIOU, *supra* note 432, at 43 *and* SCHINDLER & TOMAN, ARMED CONFLICTS, *supra* note 5, at 647-648.

Finally, Protocol I denies combatant and prisoner of war status to mercenaries, although it does permit belligerent nations to voluntarily grant these protections on an *ad hoc* basis.[624]

Reisman and Antoniou provide further illumination on these issues:

> When combatants are clearly marked as such, noncombatants also are apparent, and the capacity of fighters to distinguish them and to avoid firing on them in error is enhanced. Hence the emphasis in many of the earlier Conventions on uniforms and on carrying arms openly. A majority of the states that participated in the international conferences that adopted the Additional Protocols of 1977 were sympathetic to the insurgent sides in the so-called "wars of national liberation." Such insurgents usually did not wear uniforms or carry their arms openly, but concealed themselves among the population at large.
>
> The concealed combatant certainly has an advantage over the uniformed soldier, but the advantage comes with a price that others must pay. It inevitably leads to increased casualties among the civilian population, as the uniformed adversary can no longer clearly distinguish between combatant and noncombatant. In some wars of national liberation, this side effect may have been intentionally sought by the guerrilla force as a way of forcing the government to act in ways that would alienate the population and drive it in to the guerrilla camp. These provisions have been of particular concern to states that have refused to become party to Additional Protocol I.

---

[624] 1977 Geneva Protocol I, *supra* note 355, at art. 47 *reprinted in* ROBERTS & GUELFF, *supra* note 4, at 414. *See also* REISMAN & ANTONIOU, *supra* note 432, at 44-45 *and* SCHINDLER & TOMAN, ARMED CONFLICTS, *supra* note 5, at 649.

> Protocol I attempts to expand the category of individuals considered combatants to include guerrilla forces. It also attempts to exclude a class of individuals that always had been considered combatants—mercenaries. The effect of Protocol I's downgrading of the status of mercenaries, however, is unclear.[625]

These changes to the Geneva law reflect the strong influence of Third World nations in the drafting of the Protocol. Its terms favor guerrillas, irregulars, and terrorists, but limit protections for mercenaries, who were almost always in the employ of the more wealthy belligerent. The language of Article 1(4) is particularly reminiscent of the communist jargon of the era.

These areas of controversy affect the law of information conflict. Specifically, the increased likelihood that the law of war would be applied to any given conflict could render criminals and terrorists, at least in international eyes, as lawful combatants. This would circumscribe a nation's choices in dealing with such malefactors. In addition, the expertise required for high-end information operations is often found in the private sector, and these "cyber-mercenaries," once again in the employ of the wealthier side, would not enjoy the immunities of lawful combatants. To take just one example, such a contracted hacker could become personally, criminally liable for the damage he does as a "mercenary," and not be considered of or treated as a lawful combatant—personally immune for all actions not constituting war crimes.

Article 37—Prohibition of Perfidy

> 1. It is prohibited to kill, injure or capture an adversary by resort to perfidy. Acts inviting the confidence of an adversary to lead him to believe that he is entitled to, or is obliged to accord, protection under the rules of international law applicable in armed conflict, with the intent to betray that

---

[625] REISMAN & ANTONIOU, *supra* note 432, at 43-44.

confidence, shall constitute perfidy. The following acts are examples of perfidy:

(a) the feigning of an intent to negotiate under a flag of truce or of surrender;

(b) the feigning of an incapacitation by wounds or sickness;

(c) the feigning of civilian, noncombatant status; and

(d) the feigning of protected status by the use of signs, emblems or uniforms of the United Nations or of neutral or other States not Parties to the conflict.

2. Ruses of war are not prohibited. Such ruses are acts which are intended to mislead an adversary or to induce him to act recklessly but which infringe no rule of international law applicable in armed conflict and which are not perfidious because they do not invite the confidence of an adversary with respect to protection under the law. The following are examples of such ruses: the use of camouflage, decoys, mock operations and misinformation.[626]

Article 37 of Protocol I is perhaps the single most important passage in any treaty applicable to the law of information conflict. Here, the crucial distinction between perfidious acts and ruses of war is explained clearly and authoritatively. This distinction is the razor's edge along which information operations would be conducted.

The goal of any information operation is, ultimately, to shape the perceptions of the enemy leader to direct his decision-making process toward a desired end. The means and methods available to do this are

---

[626] 1977 Geneva Protocol I, *supra* note 355, at art. 37 *reprinted in* ROBERTS & GUELFF, *supra* note 4, at 409. *See also* SCHINDLER & TOMAN, ARMED CONFLICTS, *supra* note 5, at 645.

virtually unlimited, but they do not include any techniques or targets that would compromise the integrity of international law. The legal adviser to the warfighter must keep clear in his own mind, and make clear to his client, that this distinction cannot be disregarded in the rush of war but, rather, that it is the difference between being an honorable and criminal warfighting.

The policy behind this distinction was born out of the law of chivalry, which evolved over a thousand years to capture the intuitive reaction of warriors, clerics, nobles, and common folk to actions that were intrinsically dishonorable or despicable. Once enshrined in international law, the policy took on a new dimension, that of protecting law itself from malefactors who would undermine the foundations of civilization for their own short-term gain. Today, perfidious acts are viewed as those that threaten faith in and compliance with the fundamental protections accorded noncombatants.

Article 47—Mercenaries

> 1. A mercenary shall not have the right to be a combatant or a prisoner of war.
>
> 2. A mercenary is any person who . . .
>
> (e) is neither a national of a Party to the conflict nor a resident of territory controlled by a Party to the conflict . . .[627]

Although information operations rely to a great extent upon civilian contractors (whether computer science professors or hackers) outside of a nation's traditional military structure, such a contractor would not meet the Protocol's definition of a "mercenary" of Article

---

[627] 1977 Geneva Protocol I, *supra* note 355, at art. 47 *reprinted in* ROBERTS & GUELFF, *supra* note 4, at 414. *See also* REISMAN & ANTONIOU, *supra* note 432, at 44-45 *and* SCHINDLER & TOMAN, ARMED CONFLICTS, *supra* note 5, at 649.

47(2)(d), excludes nationals of belligerent powers. A foreign national, contracted for similar services, could conceivably meet the Protocol's definition of a "mercenary."

Article 48—Basic Rule

> In order to ensure respect for and protection of the civilian population and civilian objects, the Parties to the conflict shall at all times distinguish between the civilian population and combatants and between civilian objects and military objectives and accordingly shall direct their operations only against military objectives.[628]

"Article 51—protection of the civilian population

> 2. The civilian population as such, as well as individual civilians, shall not be the object of attack. Acts or threats of violence the primary purpose of which is to spread terror among the civilian population are prohibited."[629]

The Protocol captures the customary international norm of discrimination. Although incontrovertibly the law, this norm does present two distinct problems for the planner of an information operation. First, while a primary target may be easily identifiable, "ripple" or "cascade" effects beyond the targeted system may wreak catastrophic damage against the civilian population. The classic example is the targeting of an electrical distribution grid to cut power to a county's integrated air defense system (IADS) immediately before an airstrike. Such an attack would be lawful in and of itself, but if that same power grid supported a

---

[628] 1977 Geneva Protocol I, *supra* note 355, at art. 48 *reprinted in* ROBERTS & GUELFF, *supra* note 4, at 414. *See also* REISMAN & ANTONIOU, *supra* note 432, at 87 *and* SCHINDLER & TOMAN, ARMED CONFLICTS, *supra* note 5, at 650.

[629] 1977 Geneva Protocol I, *supra* note 355, at art. 51(2) *reprinted in* ROBERTS & GUELFF, *supra* note 4, at 415. *See also* REISMAN & ANTONIOU, *supra* note 432, at 88 *and* SCHINDLER & TOMAN, ARMED CONFLICTS, *supra* note 5, at 651.

hospital, the calculus may change. This becomes even more true of very sophisticated information attacks, where subtly nuanced messages, intended for regime elites, could have almost unlimited unintended consequences. The key for these operations is accurately predicting the reasonably foreseeable damage at the time the attack is planned, and the use of means and methods of warfare that can be focused on a single intended target.

The second area of concern is the concept of punishing a civilian population for its support of an aggressive tyrant—such as the decision to "inconvenience" the Serbian population with sporadic power outages during the most recent war in the Balkans. The civilian population as a whole was at least intentionally affected, if not targeted. The gray area between this sort of societal punishment and the unlawful targeting of civilians is dangerously thin, if it exists at all, and modern information technology renders traditionally civilian targets, such as critical infrastructures, vulnerable to such attacks. More than kinetic weapons, which are point-directed, information weapons, affecting large areas of enemy countries, have the potential to violate this international norm.

> *Article 56—Protection of works and installations containing dangerous forces*
>
> 1. Works or installations containing dangerous forces, namely dams, dikes and nuclear electrical generating stations, shall not be made the object of attack, even where these objects are military objectives, if such attack may cause the release of dangerous forces and consequent severe losses among the civilian population.[630]

Article 56 provides a specific prohibition against the worst potential violations of the general principle of discrimination. This

---

[630] 1977 Geneva Protocol I, *supra* note 355, at art. 56, *reprinted in* ROBERTS & GUELFF, *supra* note 4, at 418 . *See also* REISMAN & ANTONIOU, *supra* note 432, at 91 *and* SCHINDLER & TOMAN, ARMED CONFLICTS, *supra* note 5, at 653.

prohibition is particularly applicable to information operations which are targeted against a country's critical infrastructure and may have unintended secondary or tertiary consequences. Reisman and Antoniou agree, noting that: "The primary innovations of Protocol I concern the prohibition of targeting objects indispensable to the survival of the civilian population such as foodstuffs and water (Article 54) and objects that contain dangerous forces, such as dams, dikes, and nuclear power stations (Article 56)."[631]

The incredible efficiencies afforded by modern information systems come with a concomitantly greater vulnerability when these central control systems are compromised by an adversary in wartime. The effects of shutting down a nation's systems for power distribution, telephone communications, air traffic control, or financial management would be nothing short of catastrophic, and clearly within the ambit of the Article 56 prohibition. While more surgically targeted information attacks against the regime elites of an adversary country are well within these broad limitations, the electronic "carpet bombing" of an enemy nation by the release of dangerous forces through information attacks would clearly violate international law.

---

[631] REISMAN & ANTONIOU, *supra* note 432, at 93.

27. 1977 Geneva Protocol II Additional to the Geneva Conventions of 12 August 1949, and Relating to the Protection of Victims of Non-International Armed Conflicts[632]

Protocol II is the civil war complement to the international war focus of Protocol I. Before World War II, the customary law of war applied only to inter-, and not intra-, state conflicts.[633]

The drive to develop a law of war for civil wars began early in the 20th Century, but did not bear fruit until 1949, when common Article 3 of the four Geneva Conventions addressed "conflict not of an international character."[634] Common Article 3 provides for only the most basic humanitarian principles, and as such was ineffective as a clear guide for warfighters or jurists.[635]

---

[632] 1977 Geneva Protocol II, *supra* note 353, *reprinted in* ROBERTS & GUELFF, *supra* note 4, at 449. *See also* REISMAN & ANTONIOU, *supra* note 432, at 385-386 *and* SCHINDLER & TOMAN, ARMED CONFLICTS, *supra* note 5, at 689-700.

[633] ROBERTS & GUELFF, *supra* note 4, at 447. As Roberts and Guelff point out:

> "The laws of war, as embodied in customary international law, were only regarded as applicable to civil war if the government of the state in which the insurrection existed, or a third state, chose to recognize the belligerent status of the insurgent group and thereby acknowledge the law's application."

Id.

[634] 1949 Geneva Convention I, *supra* note 148, at art. 3, *reprinted in* REISMAN & ANTONIOU, *supra* note 432, at 171. *See also* SCHINDLER & TOMAN, ARMED CONFLICTS, *supra* note 5, at 376-77. 1949 Geneva Convention II, *supra* note 148, at art. 3, *reprinted in* ROBERTS & GUELFF, *supra* note 4, at 194. *See also* REISMAN & ANTONIOU, *supra* note 432, at 153-54, *and* SCHINDLER & TOMAN, ARMED CONFLICTS, *supra* note 5, at 404-05. 1949 Geneva Convention III, *supra* note 148, at art. 3, *reprinted in* ROBERTS & GUELFF, *supra* note 4, at 244. *See also* SCHINDLER & TOMAN, ARMED CONFLICTS, *supra* note 5, at 430. 1949 Geneva Convention IV, *supra* note 148, at art. 3 *reprinted in* ROBERTS & GUELFF, *supra* note 4, at 272. *See also* SCHINDLER & TOMAN, ARMED CONFLICTS, *supra* note 5, at 501-02.

[635] Schindler and Toman concur:

Protocol II was an attempt to advance the protections outlined in common Article 3, and provide a more specific framework for their application and enforcement.

Article 1—Material field of application

1. This Protocol, which develops and supplements Article 3 common to the Geneva Conventions of 12 August 1949 without modifying its existing conditions of application, shall apply to all armed conflicts which are not covered by Article 1 of the Protocol Additional to the Geneva Conventions of 12 August 1949, and relating to the Protection of Victims of International Armed Conflicts (Protocol I) and which take place in the territory of a High Contracting Party between its armed forces and dissident armed forces or other organized armed groups which, under responsible command, exercise such control over a part of its territory as to enable them to carry out sustained and concerted military operations and to implement this protocol.

2. This Protocol shall not apply to situations of internal disturbances and tensions, such as riots, isolated and sporadic acts of violence and other acts of a similar nature, as not being armed conflicts.[636]

The only provision applicable to non-international armed conflicts before the adoption of the present Protocol was Article 3 common to all four Geneva Conventions of 1949. This Article proved to be inadequate in view of the fact that about 80% of the victims of armed conflicts since 1945 have been victims of non-international conflicts and that non-international conflicts are often fought with more cruelty that international conflicts. The aim of the present Protocol is to extend the essential rules of law of armed conflicts to internal wars.

SCHINDLER & TOMAN, ARMED CONFLICTS, *supra* note 5, at 689.

[636] 1977 Geneva Protocol II, *supra* note 353, *reprinted in* ROBERTS & GUELFF, *supra* note 4, at 449-50. *See also* SCHINDLER & TOMAN, ARMED CONFLICTS, *supra* note 5, at 691.

This first article describes the scope of the Protocol, emphasizing that there is no analytical gap between armed conflicts covered by Protocol I and those covered by Protocol II. If there is a gap, it appears to be below the Article I threshold, where a recalcitrant state could easily mischaracterize the nature of an internal disturbance to prevent the application of Protocol II, so as to avoid its attendant legal protections and international attention for the rebels. In addition, the Protocol contemplates more or less "traditional" civil wars, and not the "failed state" conflicts that have appeared since the end of the Cold War.

> *"Article 13—Protection of the civilian population*
>
> 2. The civilian population as such, as well as individual civilians, shall not be the object of attack. Acts or threats of violence the primary purpose of which is to spread terror among the civilian population are prohibited."[637]
>
> *Article 15—Protection of works and installations containing dangerous forces*
>
> Works or installations containing dangerous forces, namely dams, dikes and nuclear electrical generating stations, shall not be made the object of attack, even where these objects are military objectives, if such an attack may cause the release of dangerous forces and consequent severe losses among the civilian population.[638]

These passages are identical to the ones in Protocol I, which clearly restate the customary international law prohibition of directly

[637] 1977 Geneva Protocol II, *supra* note 353, at art. 13(2) *reprinted in* ROBERTS & GUELFF, *supra* note 4, at 455. *See also* SCHINDLER & TOMAN, ARMED CONFLICTS, *supra* note 5, at 696.

[638] 1977 Geneva Protocol II, *supra* note 353, at art. 15 *reprinted in* ROBERTS & GUELFF, *supra* note 4, at 456. *See also* SCHINDLER & TOMAN, ARMED CONFLICTS, *supra* note 5, at 697.

targeting civilians. Protocol II extends precisely the same standard to civil wars. In practice, however, the application of this standard is extremely difficult, due to the far greater involvement of civil populations in domestic conflicts.

28. 1978 Red Cross Fundamental Rules of International Humanitarian Law Applicable in Armed Conflicts[639]

Although only a single page in length, the Red Cross Fundamental Rules capture the basic principles of the law of armed conflict, distilled for those unable or unwilling to digest the longer restatements.[640] It has no independent force under international law, but the concepts embodied in its seven articles sketch out the corpus of the customary international law of armed conflict.[641] While its intrinsic legal value is marginal, it is useful as a baseline for the minimum widely published standards of the laws of war, which is commonly considered the least common denominator in the law of armed conflict.

---

[639] Fundamental Rules of International Humanitarian Law Applicable in Armed Conflicts, September-October 1978, [hereinafter Humanitarian Law] *reprinted in* ROBERTS & GUELFF, *supra* note 4, at 470. *See also* SCHINDLER & TOMAN, ARMED CONFLICTS, *supra* note 5, at 733-734.

[640] Schindler and Toman explain:

> By the adoption of the 1977 Protocols Additional to the Geneva Conventions the law of armed conflicts had become so voluminous and complex that it seemed desirable to establish a short text containing the most essential rules of this law. This text was drafted by experts of the ICRC, the League of Red Cross Societies and of national Red Cross Societies. It has no official character and no force of law.

SCHINDLER & TOMAN, ARMED CONFLICTS, *supra* note 5, at 733.

[641] ROBERTS & GUELFF, *supra* note 4, at 469. Roberts and Guelff quote the ICRC statement which accompanied the release of the Fundamental Rules:

> "This test does not posses the authority of an international legal instrument, nor is it in any way intended to take the place of existing treaties. Its only purpose is to facilitate the dissemination of knowledge of international humanitarian law."

Id.

> "6. Parties to the conflict and members of the armed forces do not have an unlimited choice of methods and means of warfare. It is prohibited to employ weapons or methods of warfare of a nature to cause unnecessary losses or excessive suffering."[642]
>
> "7. Parties to a conflict shall at all times distinguish between the civilian population and combatants in order to spare civilian population and property. Neither the civilian population as such nor civilian persons shall be the object of attack. Attacks shall be directed solely against military objectives."[643]

These two articles succinctly restate the core principles of humanity and discrimination. Far from being buried in the fine print of international law, these concepts are the foundation of the law of information conflict.

---

[642] Humanitarian Law at art. 6 *reprinted in* ROBERTS & GUELFF, *supra* note 4, at 470. *See also* SCHINDLER & TOMAN, ARMED CONFLICTS, *supra* note 5, at 734.

[643] Humanitarian Law at art. 7 *reprinted in* ROBERTS & GUELFF, *supra* note 4, at 470. *See also* SCHINDLER & TOMAN, ARMED CONFLICTS, *supra* note 5, at 734.

29. 1981 United Nations Convention on Prohibitions or Restrictions on the Use of Certain Conventional Weapons Which May be Deemed to be Excessively Injurious or to have Indiscriminate Effects[644]

The 1981 UN Convention grew out of the conferences that produced Protocols I and II, motivated by the desire on the part of the international community to clarify particular applications of the existing customary law of armed conflict. Three specific areas were chosen: weapons that produced fragments not visible in x-ray photography, land mines and booby traps, and incendiary weapons.[645]

These limitations were derived from two interrelated principles of the law of war: that the right of belligerents to adopt means and methods of warfare is not unlimited, and that the use of means and methods of warfare to produce unnecessary suffering is prohibited.[646] The Convention itself is purely preambular and procedural; the actual substance is found in three "plug-in" protocols, each relating to a specific means or method.[647]

This Convention is instructive for the development of the law of information conflict in two respects. First, it shows that the modern

---

[644] Indiscriminate Effects, *supra* note 384, *reprinted in* ROBERTS & GUELFF, *supra* note 4, at 473. *See also* REISMAN & ANTONIOU, *supra* note 432, at 49-55 *and* SCHINDLER & TOMAN, ARMED CONFLICTS, *supra* note 5, at 179-196.

[645] ROBERTS & GUELFF *supra* note 4, at 471.

[646] *Id.* at 472.

[647] Schindler and Toman describe the Convention's ambit:

> The Convention is a pure umbrella containing only general provisions but no prohibitions of weapons. Prohibitions of weapons are the object of the three Protocols annexed to the Convention. The Convention can be adhered to only by States which express their consent to be bound by at least two of the Protocols (Art. 4, par. 3, of the Convention).

SCHINDLER & TOMAN, ARMED CONFLICTS, *supra* note 5, at 179.

trend toward outright prohibition of means or method of warfare begins with a reaffirmation that it may not be employed against civilians. Second, the definitions provided in the Convention are very helpful in capturing the current meaning of several key terms of art.

> "Protocol on Non-Detectable Fragments
> (Protocol I)
>
> It is prohibited to use any weapon the primary effect of which is to injure by fragments which in the human body escape detection by X-rays."[648]

This is the entire text of the Convention's first protocol. While new nonlethal weapons may be capable of causing damage to the human body that is not easily diagnosed or treated, no true information weapon is *per se* able to inflict this kind of damage on human beings.

> Protocol on Prohibitions or Restrictions on the Use of Mines, Booby-Traps and Other Devices
> (Protocol II)
>
> *Article 3—General restrictions of the use of mines, booby-traps and other devices*
>
> 2. It is prohibited in all circumstances to direct weapons to which this Article applies, either in offence, defence or by way of reprisals, against the civilian population as such or against individual civilians.
>
> 3. The indiscriminate use of weapons to which this Article applies is prohibited.[649]

---

[648] Indiscriminate Effects, *supra* note 384, *reprinted in* ROBERTS & GUELFF, *supra* note 4, at 479. *See also* REISMAN & ANTONIOU, *supra* note 432, at 50 *and* SCHINDLER & TOMAN, ARMED CONFLICTS, *supra* note 5, at 185.

Although superseded by the Land Mine Convention of 1997, which prohibited any use of land mines, whether discriminate or not, this Protocol remains of interest. First, restating the existing customary international norm against targeting civilians, it was the greatest single step toward the eventual criminalization of all uses of land mines. Perhaps more useful in the development of the law of information conflict, Article 3 goes on to define two key terms: "indiscriminate use,"[650] and "feasible precautions."[651] These are directly applicable to the law of information conflict.

---

[649] Indiscriminate Effects, *supra* note 384, at art. 3 *reprinted in* ROBERTS & GUELFF, *supra* note 4, at 480. *See also* REISMAN & ANTONIOU, *supra* note 432, at 51 *and* SCHINDLER & TOMAN, ARMED CONFLICTS, *supra* note 5, at 186.

[650] Article 3 states:

> Indiscriminate use is any placement of such weapons:
>
> (a) which is not on, or directed against, a military objective; or
>
> (b) which employs a method or means of delivery which cannot be directed at a specific military objective; or
>
> (c) which may be expected to cause incidental loss of civilian life, injury to civilians, damage to civilian objects, or a combination thereof, which would be excessive in relation to the concrete and direct military advantage anticipated.

Indiscriminate Effects, *supra* note 384, at art. 3(3) *reprinted in* ROBERTS & GUELFF, *supra* note 4, at 480-81. *See also* REISMAN & ANTONIOU, *supra* note 432, at 51 *and* SCHINDLER & TOMAN, ARMED CONFLICTS, *supra* note 5, at 186.

[651] Article 3 states that "[f]easible precautions are those precautions which are practicable or practically possible taking into account all circumstances ruling at the time, including humanitarian and military considerations." Indiscriminate Effects at art. 3(4), 10 April 1981, *reprinted in* ROBERTS & GUELFF, *supra* note 4, at 481. *See also* REISMAN & ANTONIOU, *supra* note 432, at 51 *and* SCHINDLER & TOMAN, ARMED CONFLICTS, *supra* note 5, at 186.

Article 6 of the same Protocol lists ten specific situations in which the land mines or booby traps would be "constructively treacherous"—that is, prohibited by the Convention, but "[w]ithout prejudice to the rules of international law applicable in armed conflict relating to treachery and perfidy."[652] This signals an interesting trend toward expanding the ambit of "treachery" and "perfidy" to cover examples not strictly forbidden by the traditional application of the definitions of each under customary international law. Although such a similar attempt may be made in the law of information conflict, the mind-numbing complexity of the information means and methods render them far less amenable to comprehensive prohibition by definition.

> Protocol on Prohibitions or Restrictions on the Use of Incendiary Weapons (Protocol III)
>
> *Article 1—Definitions*
>
> For the purpose of this Protocol:
>
> 1. "Incendiary weapon" means any weapon or munition which is primarily designed to set fire to objects or to cause burn injury to persons through the action of flame, heat, or a combination thereof, produced by a chemical reaction of a substance delivered on the target.[653]

This Protocol is helpful for the same reasons as Protocol II: it highlights the difficulties in defining a means or method to be outlawed,

---

[652] Indiscriminate Effects, *supra* note 384, at art. 6(1) *reprinted in* ROBERTS & GUELFF, *supra* note 4, at 482. *See also* REISMAN & ANTONIOU, *supra* note 432, at 52 *and* SCHINDLER & TOMAN, ARMED CONFLICTS, *supra* note 5, at 187.

[653] Indiscriminate Effects, *supra* note 384, at art. 1(1) *reprinted in* ROBERTS & GUELFF, *supra* note 4, at 484. *See also* REISMAN & ANTONIOU, *supra* note 432, at 54 *and* SCHINDLER & TOMAN, ARMED CONFLICTS, *supra* note 5, at 190.

and it defines several terms of art. As for the definition of "incendiary weapons," the Protocol frames it in terms of its effects, which were more limited than the broad range of weapons capable of producing them. In addition, it goes on to give examples not only of weapons that were incendiary, but also weapons that were *not*, making the task of applying the Protocol far simpler. Both of these techniques, effects-based definition and definition by example and counter-example, will be useful in framing discussions of information weapons.

As for the helpful definitions of legal terms, Protocol III defines "concentration of civilians,"[654] "military objective,"[655] "civilian objects,"[656] and repeats the Protocol II definition of "feasible

---

[654] Article I states that "'[c]oncentration of civilians' means any concentration of civilians, be it permanent or temporary, such as in camps or columns of refugees or evacuees, or groups of nomads." Indiscriminate Effects, *supra* note 384, at art. I(2) *reprinted in* ROBERTS & GUELFF, *supra* note 4, at 485. *See also* REISMAN & ANTONIOU, *supra* note 432, at 55 *and* SCHINDLER & TOMAN, ARMED CONFLICTS, *supra* note 5, at 190.

[655] Article I states:

> "'Military objective' means, so far as objects are concerned, any object which by its nature, location, purpose or use makes an effective contribution to military action and whose total or partial destruction, capture or neutralization, in the circumstances ruling at the time, offers a definite military advantage."

Indiscriminate Effects, *supra* note 384, at art. I(3) *reprinted in* ROBERTS & GUELFF, *supra* note 4, at 485. *See also* REISMAN & ANTONIOU, *supra* note 432, at 55 *and* SCHINDLER & TOMAN, ARMED CONFLICTS, *supra* note 5, at 190.

[656] Article I states:

> "'Civilian objects' are all objects which are not military objectives as defined in paragraph 3."

Indiscriminate Effects, *supra* note 384, at art. I(4) *reprinted in* ROBERTS & GUELFF, *supra* note 4, at 485. *See also* REISMAN & ANTONIOU, *supra* note 432, at 55, *and* SCHINDLER & TOMAN, ARMED CONFLICTS, *supra* note 5, at 190.

precautions."[657] These terms will be as important in evaluating the lawfulness of information attacks in the 21st Century as they were in evaluating kinetic attacks in the 20th Century.

[657] Indiscriminate Effects, *supra* note 384, at art. 1(5) *reprinted in* ROBERTS & GUELFF, *supra* note 4, at 485. *See also* REISMAN & ANTONIOU, *supra* note 432, at 55 *and* SCHINDLER & TOMAN, ARMED CONFLICTS, *supra* note 5, at 190.

CHAPTER 9: EFFECT OF HOSTILITIES ON THE PEACETIME REGIME OF SEA, AIR, SPACE, AND FOREIGN DOMESTIC LAW†

These four geographic regimes carry with them additional legal requirements and concerns. The peacetime regime for each is altered, to a greater or lesser extent, by the existence of hostilities. Determining the precise contours of these changes requires an understanding of the state of the law in peacetime, and a further understanding of the standards by which the legal obligations of belligerents are altered by the *jus in bello*.

Two other disciplines, telecommunications law and intellectual property law, are implicated by the emerging law of information conflict, but to varying degrees. Telecommunications law suffuses each of the geographic regimes, and is most properly reviewed in the context of the law of sea, air, space, and foreign nations, rather than as a separate category.[658] Intellectual property law, although crucial at the national level, is less relevant to the international picture. Intelligence collection operations through cyberspace would violate only the domestic intellectual property laws of the target state (just as they violate the domestic espionage laws of the target state), and, in any event, are intended to remain secret, without even the target becoming aware of the loss. Those violated by information operations in wartime would almost

† The Law of the Sea and the Foreign Domestic Law sections of this chapter are drawn substantially from work by Prof. Walter Gary Sharp, Sr., for Aegis Research Corporation. The Space Law portion of this chapter is drawn substantially from the author's article, *Legal Aspects of Offensive Information Operations in Space*, appearing in the Fall 1999 issue of the UNITED STATES AIR FORCE ACADEMY JOURNAL OF LEGAL STUDIES.

[658] For example, the UNCLOS contains provisions prohibiting unauthorized broadcasting at sea, the Chicago and Nairobi Conventions address interference with aircraft communications, and the international agreements governing the use of the INTELSAT and INMARSAT communications satellite constellations prohibit certain military activities through their platforms.

certainly not survive in the *jus in bello* as enforceable legal norms for the duration of the hostilities.[659]

This chapter will analyze the law of the sea, air law, space law, and international standards for the application of any given foreign domestic law in turn.

The Law of the Sea

Mare liberum—mare clausum

The principles inherent in the law of the sea continue to be a driving force in shaping international law—and the law of the sea

---

[659] The evanescent nature of information in cyberspace also makes it difficult to smoothly analogize from intellectual property law. Kanuck points out:

> Information represents a special commodity because information is easily reproducible yet destructible. Unlike the objects of an intellectual property law regime, however, it has no tangible manifestation that can be regulated under public international law. For example, a patent infringement consists of producing a similar product or utilizing a similar method, but simply possessing the knowledge itself is no violation. Similarly, while one can apply the international law paradigm to the platforms for C4I technology, one cannot apply it to the information itself. A satellite can be owned, identified, and located. These three traits make it possible to evaluate actions conducted by or against it. Satellite emissions, cellular telephone calls, and e-mail messages, on the other hand, may be "owned" in some sense, but their identification or location can easily be prevented, thereby rendering attribution and regulation ineffective. At the same time, it would be equally difficult for any state or individual to assert ownership of a specific datum residing in a transnational database—or the electrons that temporarily comprise it—under current theories. The uncertainties surrounding the actual ownership of information have led to the development of cryptographic techniques that protect the very ideas, or information itself, for which the law offers no aegis.

Sean P. Kanuck, *Information Warfare: New Challenges for Public International Law*, 37 HARV. INT'L L. J. 272 (1996) at 287-88.

continues to reflect the culture of contemporary international society.[660] Modern international law of the sea can best be understood in the context of the continual conflict of two opposing, yet complementary, principles—those of territorial sovereignty and the freedom of the high seas.[661] In the 15th century, Spain and Portugal attempted to partition the world's oceans between themselves.[662] Later, in the 17th century, the same countries continued to strongly defend the principle of *mare clausum, i.e.,* the exercise of sovereignty over the high seas, in their monopolistic efforts to deny the United Kingdom access to the lucrative trade of the East Indies.[663] Until the end of the 18th century, the world's oceans and all of the seas surrounding Europe were subject to proprietary claims by sovereigns.[664]

Hugo Grotius challenged these overreaching claims by the Iberian states in defense of his client, the Dutch East India Company.[665] His powerful argument for the principle of *mare liberum*—the freedom of the high seas—gained the support of those states, such as the United Kingdom, which would gain by breaking the Iberian monopoly.[666] As the balance of power shifted from Spain and Portugal to the Netherlands in the 17th century and later to the United Kingdom at the beginning of the

---

[660] 1 E.D. BROWN, THE INTERNATIONAL LAW OF THE SEA 6 (1994). [hereinafter BROWN, LAW OF THE SEA].

[661] *Id.* at 6.

[662] *Id.* at 7.

[663] *Id.*

[664] *Id.*

[665] *Id.*

[666] *Id.*

19th century, *mare liberum* became firmly established as a fundamental principle of international law.[667]

The First United Nations Conference on the Law of the Sea (UNCLOS I) that met in Geneva in the late 1950s reinforced the principle of *mare liberum*.[668] UNCLOS I concluded in 1958 with the signing of four Geneva Conventions on the Law of the Sea.[669] Although these conventions reflected a degree of progressive development, they were predominantly a codification of most of the fundamental rules of the law of the sea.[670] Since the work of UNCLOS I, however, the law of the sea has been one of the most dynamic areas of international law.[671]

The United Nations Convention on the Law of the Sea

The Third United Nations Convention on the Law of the Sea (UNCLOS III) involved 157 states in 93 weeks of negotiations between 1973 and 1982.[672] These negotiations resulted in the conclusion of the United Nations Convention on the Law of the Sea,[673] which is considered one of the most complex treaties in the history of international relations.[674]

---

[667] *Id.* at 7-8.

[668] *Id.* at 1.

[669] *Id.*

[670] *Id.* at 1, 3.

[671] *Id.* at 1.

[672] *Id.* at 3.

[673] United Nations Convention on the Law of the Sea, Dec. 10, 1982, U.N. Doc. A/CONF.62/122, 21 I.L.M. 1261 [hereinafter 1982 LOS Convention], *reprinted in* UNITED NATIONS, THE LAW OF THE SEA 1 (1983), U.N. Sales No. E.83.V.5.

[674] BROWN, LAW OF THE SEA, *supra* note 660, at 5.

The purpose of the 1982 LOS Convention was to establish a comprehensive regime "dealing with all matters relating to the law of the sea, . . . bearing in mind that the problems of ocean space are closely interrelated and need to be considered as a whole."[675] The convention entered into force on 16 November 1994, one year after Guyana deposited the sixtieth ratification.[676]

The conflicting state interests between *mare liberum* and *mare clausum* are also balanced in the 1982 LOS Convention.[677] For example, while the freedom of the high seas was expanded by the recognition of unrestricted passage through the territorial sea, straits, and archipelagic waters, the declaration of a new 200-mile Exclusive Economic Zone by some nations evidenced the countervailing proprietary interests of *mare clausum*.[678] The 1982 LOS Convention, with the exception of the provisions concerning deep sea-bed mining and particular arrangements for settling disputes, is considered by most states as declaratory of customary international law.[679] These provisions were particularly troubling to the major maritime powers, but the New York Agreement adopted on 28 July 1994 ameliorated the objectionable provisions and opened the way for almost universal acceptance.[680]

---

[675] BERNARDO ZULETA, *Introduction* to UNITED NATIONS, THE LAW OF THE SEA, xix (1983).

[676] BROWN, LAW OF THE SEA, *supra* note 660, at 1. Article 308 of the 1982 LOS Convention requires that sixty instruments of ratification or accession be deposited before the convention enters into force.

[677] BROWN, LAW OF THE SEA, *supra* note 660, at 10.

[678] *Id.*

[679] COMMANDER'S HANDBOOK, *supra* note 109, at § 1.1 fn.5.

[680] BROWN, LAW OF THE SEA, *supra* note 660, at 1, 3, 11.

Although some provisions of the 1982 LOS Convention permit a temporary suspension of the rights of other states if essential for national security,[681] no provisions of the 1982 LOS Convention discuss the effect of armed conflict on the Convention or address information warfare or information operations. Several articles, however, do restrict a state's ability to conduct offensive and defensive information warfare.

Innocent Passage

The concept of "innocent passage" is a key component of the law of the sea:

> International law provides that ships (but not aircraft) of all nations enjoy the right of innocent passage for the purpose of continuous and expeditious traversing of the territorial sea or for proceeding to or from internal waters. Innocent passage includes stopping and anchoring, but only insofar as incidental to ordinary navigation, or as rendered necessary by *force majeure* or by distress. [footnote omitted] Passage is *innocent* so long as it is not prejudicial to the peace, good order, or security of the coastal nation. [footnote omitted][682]

---

[681] *See, e.g.*, 1982 LOS Convention, *supra* note 673, art. 25(3) that provides for the suspension of innocent passage when in the national security interests of the coastal State.

[682] COMMANDER'S HANDBOOK, *supra* note 109, at § 2.3.2.1 (1995). The Commander's Handbook goes on to explain:

> Military activities considered to be *prejudicial* to the peace, good order, and security of the coastal nation, and therefore inconsistent with innocent passage, are:
>
> 1. Any threat or use of force against the sovereignty, territorial integrity, or political independence of the coastal nation
>
> 2. Any exercise or practice with weapons of any kind

If these prejudicial acts are undertaken, the offended nation may take action:

> The coastal nation may take affirmative actions in its territorial sea to prevent passage that is not innocent, including, where necessary, the use of force. If a foreign ship enters the territorial sea and engages in non-innocent activities, the appropriate remedy, consistent with customary international law, is first to inform the vessel of the reasons why the coastal nation questions the innocence of the passage,

---

3. The launching, landing, or taking on board of any aircraft or of any military device

4. Intelligence collection activities detrimental to the security of that coastal nation

5. The carrying out of research or survey activities

6. Any act aimed at interfering with any system of communication of the coastal nation

7. Any act of propaganda aimed at affecting the defense or security of the coastal nation

8. The loading or unloading of any commodity, currency or person contrary to the customs, fiscal, immigration or sanitary laws and regulations of the coastal nation

9. Any act of willful and serious pollution contrary to the 1982 LOS Convention

10. Any fishing activities

11. Any other activity not having a direct bearing on passage. [footnote omitted]

Id. See also Convention on the Territorial Sea and Contiguous Zone, Geneva, 29 April 1958, art. 18, 15 U.S.T. 1601, 516 U.N.T.S. 205.

> and to provide the vessel a reasonable opportunity to clarify its intentions or to correct its conduct in a reasonably short period of time. [footnote omitted][683]

Unauthorized Broadcasting

Article 109 provides that all "States shall co-operate in the suppression of unauthorized broadcasting from the high seas" and defines unauthorized broadcasting, for the purposes of this Convention, as "the transmission of sound radio or television broadcasts from a ship or installation on the high seas intended for reception by the general public contrary to international regulations." The international regulations referred to in article 109 primarily include the International Telecommunication Conventions of 1973 (Malaga-Torremolinos Convention[684]) and 1982 (Nairobi Convention[685]), and their Administrative Regulations.

Liability for Damage

There are three articles in the 1982 LOS Convention that specifically provide for State accountability for damage. Articles 31 and 42(5) impose international responsibility for any damage to a State caused by a government ship or aircraft under its control in a non-commercial setting.[686] A much more comprehensive expression of liability is found in article 235. This article provides, in pertinent part, that

---

[683] COMMANDER'S HANDBOOK. at § 2.3.2.1

[684] International Telecommunications Convention, Malaga Torremolinos, 25 October 1973, *TIAS* 8572.

[685] International Telecommunications Convention, with Final Protocol, Additional Protocols I to VII and Optional Additional Protocol, Nairobi, 6 November 1982; 1985 UKTS No. 33, Cmnd. 9557.

[686] *See id.* at arts. 31, 42(5).

> 1. States are responsible for the fulfillment of their international obligations concerning the protection and preservation of the marine environment. They shall be liable in accordance with international law...
>
> 3. With the objective of assuring prompt and adequate compensation in respect of all damage caused by pollution of the marine environment, States shall co-operate . . .[687]

These provisions of the 1982 LOS Convention are unequivocal. They make it clear that during peacetime States have an international obligation to protect and preserve the marine environment, and have international responsibility for any damage they cause. Although some provisions of the 1982 LOS Convention may be inconsistent with a state of hostilities,[688] a continuing duty to protect the environment under a multilateral convention or customary international law is not.

Under this treaty, liability could be imposed for damages indirectly caused by information operations. All of Part XII of the 1982 LOS Convention concerns the "Protection and Preservation of the Marine Environment."[689] Article 192 imposes a general duty to protect and preserve the marine environment.[690] Article 194 is a lengthy provision that imposes an imprecise and subjective obligation to take necessary measures to "prevent, reduce and control pollution of the marine environment from any source."[691] The more specific provisions that applied to the environmental

---

[687] *See id.* at art. 235.

[688] For example, an enemy warship passing through the territorial waters of a warring State can no longer claim, by definition, a right of innocent passage. *See id.* at arts. 17-32.

[689] *Id.* at arts. 192-237.

[690] *Id.* at art. 192.

[691] *Id.* at art. 194.

damage occurring in the Persian Gulf are found in article 207 for pollution from land-based sources,[692] article 210 for pollution by dumping,[693] article 211 for pollution from vessels,[694] and article 212 for pollution from or through the atmosphere.[695]

In summary, the law of information conflict is most clearly implicated by two facets of the law of the sea: the ban on unauthorized broadcasting, and the limitations on and remedies for direct or indirect environmental damage caused by information operations. Information operations verging on these areas should be examined closely in light of the relevant treaties for a fact-driven determination of the treaties' application in wartime.

Air Law

The Concept of "Airspace"

Those spaces superjacent to land and water can be defined as national airspace, international airspace, and outer space. This delineation

---

[692] *See id.* at art. 207 ("States shall take other measures as may be necessary . . . to minimize, to the fullest extent possible, the release of toxic, harmful or noxious substances, especially those which are persistent, into the marine environment.").

[693] *See id.* at art. 210 ("[State] . . . laws, regulations and measures shall ensure that dumping is not carried out without the permission of the competent authorities of [coastal and affected] States.").

[694] *See id.* at art. 211 (imposes a general duty to adopt laws and regulations for the prevention, reduction and control of pollution of the marine environment from vessels under their control).

[695] *See id.* at art. 212 (imposes a general duty to adopt laws and regulations to prevent, reduce and control pollution of the marine environment from or through the airspace under its control).

first began in 1919 with the Paris Convention on Aerial Navigation that declared the "High Contracting Parties recognise that every power has complete and exclusive sovereignty over the air space above its territory."[696] Article I of the Paris Convention defined territory as including all national land territory and adjacent territorial waters. The sovereignty over national airspace, *i.e.*, that which is superjacent to national territory and territorial waters, was reaffirmed by the international community in 1944 with the Chicago Convention on International Civil Aviation,[697] in 1958 with the Geneva Convention on the Territorial Sea and the Contiguous Zone,[698] and in 1982 with the United Nations Convention on the Law of the Sea.[699]

International airspace was first defined, albeit indirectly, by the 1958 Geneva Convention on the High Seas.[700] High seas were defined by article I of the convention as all parts of the sea that are not included in either the territorial sea or internal waters of a state. Article 2(4) of the convention declared that the freedom of the high seas comprises the freedom to fly over the high seas. Accordingly, international airspace is that airspace superjacent to the high seas. This principle was also reaffirmed by the 1982 LOS Convention.

---

[696] Convention on Aerial Navigation, Paris, 13 October 1919, article I (League of Nations Treaty Series, No. 297, Vol. XI, p. 174.

[697] Convention on International Civil Aviation, Chicago, 7 December 1944, 59 Stat. 1693, 84 U.N.T.S. 389.

[698] Convention on the Territorial Sea and Contiguous Zone, Geneva, 29 April 1958, 15 U.S.T. 1606, 516 U.N.T.S. 205.

[699] 1982 LOS Convention, *supra* note 673, *reprinted in* UNITED NATIONS, THE LAW OF THE SEA I (1983), U.N. Sales No. E.83.V.5.

[700] Convention on the Territorial Sea and Contiguous Zone, Geneva, 29 April 1958, 15 U.S.T. 1606, 516 U.N.T.S. 205.

The most difficult issues with the establishment of an international law of the air was the tension between a state's sovereignty over its own territory and the international community's interest in uniform rules and procedures for international civil aviation. The 1944 Chicago Convention came near the end of the Second World War, at the high point of internationalism.

The delegates at the Chicago Convention responded to the need for international cooperation. Two major themes of the Convention were freedom of traffic and freedom of transit. Unlike the law of the sea, air law provides no right of innocent passage. Aircraft of any country may fly over the high seas, but must have authorization to fly over another country's territory. Once the Convention had settled on the broad principles of air navigation, more detailed regulations were developed. The Convention created the International Civil Aviation Organization (ICAO) to implement and oversee the Convention's regulatory scheme for international air navigation.[701]

The next major convention, the Tokyo Convention of 1963,[702] addressed the problems created by lack of jurisdiction. Criminal acts perpetrated above the high seas, where no country could claim national jurisdiction, were difficult to prosecute. The Convention allowed all signatory nations to exercise jurisdiction over those committing acts of aerial piracy over the high seas.

In 1976, the Hague Convention[703] was held in response to the increasing number of hijackings, adding sectors of concurring and multiple

---

[701] Convention on International Civil Aviation, Chicago, 7 December 1944, 59 Stat. 1693, 84 U.N.T.S. 389.

[702] Convention on Offenses and Certain Other Acts Committed on Board Aircraft, 14 September 1963, Tokyo, 20 U.S.T. 2941, 704 U.N.T.S. 219.

[703] Convention for the Suppression of Unlawful Seizure of Aircraft, The Hague, 16 December 1970, 22 U.S.T. 1641.

jurisdiction. The Hague Convention left a lacuna, which was the impetus for the Montreal Convention of 1971.[704] The Hague Convention applied only to aircraft in flight; acts committed against aircraft on the ground and airport installations were not covered. Two prohibitions were directly relevant to technology which would be developed later: the prohibition of acts which endanger an aircraft in flight and a prohibition on placing devices which may result in the aircraft's destruction.[705] The Chicago, Tokyo, Hague, and Montreal Conventions apply to civil, and not military, aircraft.

The principal issues of air law affecting the law of information conflict are the prohibitions against interfering with the safety of flight of civil aircraft, specifically navigational and communications signals.

Interference with Communications

The need for absolute reliability and clarity in communications with aircraft is obvious; a clear channel for weather reports, navigational information, separation warnings, and distress calls is a prerequisite to safe flight. For that reason, "pirate broadcasting"—the "illegal transmission of radio and television programmes from ships, installations and aircraft on or above international waters,"[706] has been prohibited since 1938.[707]

---

[704] Convention for the Suppression of Unlawful Acts against the Safety of Civil Aviation, Montreal, 23 September 1971, 24 U.S.T. 564.

[705] *Id.* at 546.

[706] NICHOLAS GRIEF, PUBLIC INTERNATIONAL LAW IN THE AIRSPACE OF THE HIGH SEAS, 119

[707] *Id.* at 119.

The Geneva Convention on the High Seas applies not only to international waters, but also to the superadjacent airspace.[708] Art. 2 reads, in pertinent part: "These freedoms, and others which are recognized by the general principles of international law, shall be exercised by all States with reasonable regard to the interests of other States in their exercise of freedom of the High Seas."[709]

As Grief points out, "unauthorized broadcasting from the airspace of the High Seas would be likely to interfere with essential radio communications and so violate the 'reasonableness' requirement."[710] Art. 2 also requires that activities on the High Seas be undertaken "under the conditions laid down by...other rules of international law."[711]

One such set of rules is the 1990 Radio Regulations.[712] Under the 1992 Convention of the International Telecommunications Union (ITU), they are binding on all members.[713] Radio Regulation (RR) 2665 prohibits "the establishment and use of broadcasting stations on board ships, aircraft, or any other floating or airborne objects outside national territory."[714] RR 3603 prohibits "the operation of a broadcasting service over the sea."[715]

---

[708] Geneva Convention of the High Seas 1958, UNTS, vol. 450, p. 82, in GRIEF, at 119.

[709] GRIEF at 119.

[710] *Id.* at 120.

[711] Geneva Convention on the Law of the High Seas 1958, Art. 2, in GRIEF, 120.

[712] Radio Regulations 1990, vol. 1, ITU, General Secretariat (Geneva 1990).

[713] Article 4(3), ITU Constitution.

[714] GRIEF at 121.

[715] *Id.* at p. 121.

The UN Convention on the Law of the Sea contains similar restrictions on unauthorized broadcasting,[716] but it is unclear whether the framer intended to include the superadjacent airspace in the prohibition.[717]

The Chicago Convention addresses the issue only obliquely. Art. 29(e) requires civil aircraft engaged in international navigation to carry a radio and the license for its use.[718] Art. 30(b) limits the use of that radio to trained and licensed crewmembers.[719]

The most vital communications from aircraft, distress signals, receive the greatest legal protection. Article 40 of the ITU Constitution provides:

> Radio stations shall be obliged to accept, with absolute priority, distress calls and messages regardless of their origin, to reply in the same manner to such messages, and immediately take such action in regard thereto as may be required.[720]

RR 3086 echoes this:

> The distress call shall have absolute priority over all other transmissions. All stations which shall hear it shall immediately cease any transmission capable of interfering with

---

[716] "All states shall co-operate in the suppression of unauthorized broadcasting from the high seas," UN Convention on the Law of the Sea 1982, ILM, Article 109(1).

[717] GRIEF at 126.

[718] Convention on International Civil Aviation, Chicago, 7 December 1944, 59 Stat. 1693, 84 U.N.T.S. 389, Article 29(e).

[719] *Id.* at Article 30(b).

[720] Article 46, ITU Constitution.

> the distress traffic and shall continue to listen on the frequency used for the emission of the distress call.[721]

To reinforce the priority of such transmissions, false or deceptive distress signals are prohibited. Article 47 of the ITU Constitution provides:

> Members agree to take the steps required to prevent the transmission or circulation of false or deceptive distress, urgency, safety or identification signals, and to collaborate in locating and identifying stations under their jurisdiction transmitting such signals.[722]

In addition, Annex 10 to the Chicago Convention prohibits transmitting false distress signals.[723]

Article 48(1) of the ITU Constitution grants Member States almost complete freedom in the operation of military radio installations, but, according to Article 48(2), those stations must:
"...so far as possible, observe statutory provisions relative to giving assistance in case of distress and to the measures taken to prevent harmful interference...."[724]

The prohibition against harmful interference is laid down in Article 45(1) of the ITU Constitution:

> All stations, whatever their purpose, must be established and operated in such a manner as not to cause

---

[721] RR 3086.

[722] Art. 47, ITU Constitution.

[723] Convention on International Civil Aviation, Chicago, 7 December 1944, 59 Stat. 1693, 84 U.N.T.S. 389, Annex 10.

[724] Art. 48(2), ITU Constitution.

> harmful interference to the radio services or communications of other Members or recognised private operating agencies, or of other duly authorised operating agencies which carry on radio service, and which operate in accordance with the provisions of the Radio Regulations.[725]

"Harmful interference" is defined in the Annex to the ITU Constitution as:

> ...interference which endangers the functioning of a radio navigation service or of other safety services or which degrades, obstructs or repeatedly interrupts a radio communications service operating in accordance with the Radio Regulations.[726]

Since such activities are precisely the goal of some information operations, employing these techniques against aircraft in peacetime may be problematic under the ITU Constitution and the Radio Regulations.

Physical Attacks on Aircraft

The Convention on International Civil Aviation, better known as the Chicago Convention, contains an additional limitation. As a result of KAL 007 being shot down by the Soviets in 1983, the delegates to the International Civil Aviation Organization adopted article 3*b*, which provides that "every State must refrain from resorting to the use of weapons against civil aircraft in flight and that, in case of interception, the lives of the persons on board and the safety of the aircraft must not be endangered." While article 3*b* is not in force for the United States, the Legal Counsel for the Department of Justice opined in a formal

---

[725] *Id.* at Art. 45(1).

[726] *Id.* at Annex.

memorandum of law dated 14 July 1994 that article 3*b* is declaratory of customary international law.[727]

While it is difficult to imagine a scenario in which a civil airliner would be targeted in an offensive information operation, the possibility exists that damage to such an airliner could be sustained as a result of an attack on another target. The destruction of an adversary's Global Positioning System (GPS) satellite may result in degraded navigational information to military and civil aircraft alike. A more subtle attack could occur if the attacker were able to degrade the accuracy of the navigational information below the level useful for military targeting, but maintain a level necessary for civilian airliners' safety of flight. To the extent that such an effect is foreseeable, it must be taken into account in planning such an operation.

Space Law

Delimiting Air and Space

There is no internationally recognized line between air and space. The absence of a clear delimitation is due to several factors. First, the environment itself is not defined by a sharp transition; the atmosphere gradually thins with altitude. Second, technology has not created a legal need for such a line. Aircraft have for decades operated well within any reasonable definition of airspace, and spacecraft have orbited well above any plausible minimum limit of outer space. Third, no insoluble legal problem has arisen as a result of the absence of a sharp defining line. In those cases where an arbitrary line has been required, an arbitrary line has been negotiated.

---

[727] Walter Gary Sharp, Sr., Cyberspace and the Use of Force: Information Warfare, the Law of War, and the United States Standing Rules of Engagement, presented Nov. 3, 1995 to the Legal Aspects of Information Warfare Symposium, Maxwell AFB, Alabama.

Even so, there may be a value in at least an international default standard that operates in the absence of a context-specific limit. The principal reason for this, beyond the desire for academic clarity in the law, is that transatmospheric flight over national borders is becoming technologically feasible. In the near future, aerospace craft will operate in the "gray area" between air and space, not yet in orbit, but far above conventional notions of national airspace. A theoretical framework for evaluating potential dividing lines will then become useful in determining the limits of national jurisdiction.

The first possible line is the limit of the atmosphere. The earliest air law treaties[728] provided that each state has sovereignty over the airspace above its territory. Since this limit is a part of customary international law, this is an attractive place to begin. Unfortunately, it begs the question of where the atmosphere ends. There is also a legal objection, noted by van Bogaert:

> ...when the principal conventions on air law were drafted, governments were not aware that there would be future spaceflights. The clauses of the treaties concerning the right of sovereignty were never drafted to impose a limit to airspace and it is not inconformity with their *ratio juris* to use these stipulations for the delimitation of airspace and outer space.[729]

---

[728] Art. I, Convention on the Regulation of Aerial Navigation, 1919; Art. I, Spanish-American Convention on Aerial Navigation, 1926; Art. 1, Convention on Commercial Aviation (Havana Convention), 1928; and Art. I, Convention on International Civil Aviation (Chicago Convention), 1944.

[729] Van Bogaert, Aspects of Space Law, 1986, p. 12.

The aerodynamical doctrine is the second theory. Under this theory, the atmosphere would end and space begin where aircraft were no longer capable of controlled, sustained flight. This operational definition has the benefit of producing a verifiable, acknowledged limit coincident with the world aircraft altitude record. The flaw in this definition is that such a limit is in flux. The greater an aircraft's speed, the less atmospheric density is required to provide oxygen to its air-breathing engines and atmospheric pressure for lift on its control surfaces. As technology advances with the development of hypersonic aircraft technology, this line will extend upward until it merges with the lower limits of orbital space.

The biological theory sets a limit at that altitude at which human life is possible. This is an extremely low limit, in that pilots of unpressurized aircraft go on oxygen at 3 km, and that human life cannot be sustained at altitudes above 10 to 16 km.[730] This limit would remove from the jurisdiction of air law much of today's civil and military flight.

The orbital theory sets the lower limit of space at the lowest altitude a satellite can maintain orbit. This theory has the benefit of widespread acceptance no nation asserts that satellites in orbit violate its airspace. This limit is not absolute; satellites may orbit the earth at extremely low altitudes before deorbiting. The lower the orbit, the sooner a satellite will burn up. Still, many satellites orbit as low as 100-110 km for at least part of their orbit (the "perigee'" or closest approach to earth, as opposed to the "apogee," or greatest distance from earth).[731] At the very least, the orbital theory sets the highest practical limit on the lowest extent of space: any object orbiting at a lower altitude would be governed by air law.

More technical is the line advocated by scientist Theodor von Karman. He demonstrated mathematically that at an altitude of 83 km,

---

[730] *Id.* at 13.

[731] *Id.* at 14.

the density of the atmosphere has fallen to the point where centrifugal or Keplerian force (7 km/sec) is required to maintain altitude and forward motion. In layman's term, this is the point at which an aircraft stops flying and starts orbiting. The problem with this line is that is in reality a broad range, within which a low altitude may be offset by a high orbital speed, and vice versa. Technological advances would increase the aerodynamic efficiency of aerospace craft, extending this limit upward.[732]

The last major theory is that of effective control, under which a state could lay claim to "territorial airspace" as high as its military capabilities reached. A variation of this theory would append an additional buffer zone for the spatial equivalent of innocent passage.[733] The problem with this theory is that setting different limits for different nations, based on their air and space defense capabilities, would create an unworkable patchwork. It would also strongly favor nations with advanced high altitude capabilities. Another problem is that there is no effective upper limit to atmospheric defense now that several nations have space defense capabilities. For these nations, "territorial airspace" would encompass every object in earth orbit.

None of these theories is wholly satisfactory, but together they demonstrate the factors that any workable theory would have to address. Nations would have sovereignty over their airspace, however high that limit was set. The practical limits of air and spacecraft, determining the aerodynamic, orbital, and Karman limits, narrow the range for the placement of the line separating air from space. Finally, the attractiveness of an arbitrary limit, well within these broad criteria, may allow a default of 100 or 110 km as the line of separation. This last suggestion, advanced by what was then the Soviet Union in 1979,[734] appears

---

[732] *Id.* at 14.

[733] *Id.* at 15.

[734] *Id.* at 15-16.

reasonable on its face. In the absence of any of these alternative limits, however, the legal gray area between air and space remains as broad as the altitudes it affects.

The Treaty Regime

Outer Space Treaty

Four space treaties define the positive law environment for evaluating the legality of offensive information operations in space. The first of these is the Outer Space Treaty. In its preamble, it declares that space is "the province of all mankind," and, therefore, not subject to sovereign claims by any nation. It goes on to describe "the common interest of all mankind in the progress of the exploration and use of outer space for peaceful purposes."[735] From this has evolved the concept of "peaceful purposes," aspirational here and mandatory in other treaties. Nowhere in the Outer Space Treaty is the term defined, and two opposing views have developed. The majority opinion, certainly among spacefaring nations, is that "peaceful" means "nonaggressive," a relatively high standard allowing for considerable military operations in space. The minority view, more common among the less advanced, nonspacefaring nations, is that "peaceful" means "non-military," setting such a low threshold that even routine, peacetime military business, such as communications and weather observation, would be prohibited.

Article III of the treaty extends the application of international law, including the UN Charter articles on use of force, to outer space.[736] Article IV details specific prohibitions: weapons of mass destruction

---

[735] Treaty on Principles Governing the Activities of States in the Exploration and Use of Outer Space, including the Moon and other Celestial Bodies, Jan. 27, 1967, 18 U.S.T. 2410 T.I.A.S. No. 6347, 610 U.N.T.S.205, pmbl.

[736] *Id.* at art. III.

anywhere in space, and military bases, weapons testing, and maneuvers on the surface of any celestial body.[737]

State responsibility is governed by Articles VI and VII. Article VI makes states responsible for the actions of all of their nationals in space, and not merely government representatives.[738] Article VII assigns liability for damage caused by activities in space to the launching state[739]

Article IX carries a notice requirement, instructing states to notify the office of the UN Secretary General of each space activity "to the greatest extent feasible and practical."[740] This language creates a broad gray area within which notice of most if not all military activities in space are not reported in any detail. Finally, Article XII grants states parties the right of inspection of any facility located on the surface of another celestial body, but no access to orbital platforms.[741]

Liability Convention

The second major space treaty is the Liability Convention, which sets a liability standard for activities in outer space, provides an exception to that standard, and details the procedures for pursuing a claim. The liability standard is bifurcated: on the surface of the earth or in its atmosphere, the launching state has absolute liability for damage caused

---

[737] *Id.* at art. IV.

[738] *Id.* at art. VI.

[739] *Id.* at art. VII.

[740] *Id.* at art. IX.

[741] *Id.* at art. XII.

by its platforms.[742] In space, the standard is simple negligence.[743] This standard is limited in one circumstance: if the damaged state acts with gross negligence or willful intent, and the launching state acts in accordance with international law, there is no absolute liability for the launching state.[744]

To pursue a claim under the treaty, Articles VIII and IX require that the claim be submitted by a state through diplomatic channels.[745] Article X adds the requirement that the claim be submitted within one year of the event or the discovery of the responsible state's identity.[746] Finally, Article IX states that local remedies need not be exhausted before a claim may be made under the treaty.[747]

Under the law of war, a belligerent has no duty to pay for damage done to lawful targets. Where the Liability Convention has the greatest effect, though, is in peacetime information operations. A nation launching an offensive information operation in space would be liable to the launching state for the damage done to the satellite, and by the satellite as a result of the operation against it. For example, an operation that simply destroyed a space platform would leave the nation that conducted the operation liable for the value of the satellite. An operation

---

[742] Convention on International Liability for Damage Caused by Space Objects, June 21, 1971, 24 U.S.T. 2389, T.I.A.S. No. 7762, art. II.

[743] *Id.* at art. III.

[744] *Id.* at art. VI.

[745] *Id.* at art. VIII and IX.

[746] *Id.* at art. X.

[747] *Id.* at art. IX.

that deorbited a satellite, sending it crashing into a terrestrial building, would be liable for that damage as well.

INTELSAT Agreement

The third major treaty is the INTELSAT Agreement. It governs the use of INTELSAT communications satellites, which link fixed ground stations. The Agreement divides satellite services into two categories: "public telecommunications services" and "specialized telecommunications services."[748] Public telecommunications services include:

> Fixed or mobile telecommunications services which can be provided by satellite and which are available for use by the public, such as telephony, telegraphy, telex, facsimile data transmission, transmission of radio and television programs between approved earth stations having access to the INTELSAT space segment for further transmission to the public, and leased circuits for any of these purposes; but excluding those mobile services...which are provided through mobile stations operating directly to a satellite which is designed , in whole or in part, to provide services relating to the safety or flight control of aircraft or to aviation or maritime radio navigation...[749]

Specialized telecommunications services, on the other hand, include:

> ...telecommunications services which can be provided by satellite, other than those defined in paragraph (k) of this Article [*supra*], including, but not limited to, radio navigation services, broadcasting satellite services for reception by the

---

[748] Agreement Relating to the International Telecommunications Satellite Organization, June 29, 1971, 23 U.S.T. 3813, T.I.A.S. No. 7532, art. I.

[749] *Id.* at art. I, para. (k).

> general public, space research activities, meteorological services, and earth resource services...[750]

Article III, on the scope of INTELSAT activities, contains two restrictions on "military purposes," not defined in Article I. Paragraph (d) allows the INTELSAT space segment "to be utilized for the purpose of specialized telecommunications services, either international or domestic, [for] other than military purposes..."[751] Paragraph (e) states that "INTELSAT may provide satellites or associated facilities separate from the INTELSAT space segment for specialized telecommunications services, other than for military purposes..."[752] Taken together, it appears that the INTELSAT organization may offer "public telecommunications services" to the military of a nation, but not "specialized telecommunications services." Offensive information operations, routed through INTELSAT's public telecommunications services appear to be nonviolative of the Agreement

INMARSAT Convention

The fourth and final major treaty is the INMARSAT Convention. This treaty covers the use of INMARSAT communications satellites, which service mobile "ground" stations, such as ships at sea. Article 3(3) of the Convention states that "[t]he Organization shall act exclusively for peaceful purposes,"[753] but does not specifically prohibit military use in accordance with international law and the UN Charter.

---

[750] *Id.* at art. I, para. (l).

[751] *Id.* at art. III, para. (d).

[752] *Id.* at art. III, para. (e).

[753] Convention at the International Maritime Satellite Organization, Sept. 3, 1976, 31 U.S.T. 1, T.I.A.S. No. 7532, art. 3(3).

Again, a nonaggressive operation, legal under the Charter, would not violate the Convention.

Other Treaties

Beyond these four major treaties, several other international agreements tangentially affect offensive information operations in space.

The International Telecommunications Convention of 1982, better known as the Nairobi Convention, contains several provisions relating to interference with transmissions. Article 19, paragraph 132, allows signatories to "stop the transmission of any private telegram" that is threatening to state security,[754] while paragraph 133 of the same article allows Member States to do the same to any private telecommunication appearing to pose such a threat.[755]

Article 20, paragraph 134, permits Members "to suspend all international telecommunication service" to their own country, provided they notify the UN Secretary General of the action.[756] Article 35, paragraph 158, requires that all "stations, whatever their purpose, must be established and operated in such a manner as not to cause harmful interference to the radio services or communications of other Members. . . ."[757] Article 38, paragraph 164, extends this provision to military radio installations, with a qualification: they must observe measures to prevent harmful interference, "so far as possible."[758] Annex 2

---

[754] International Telecommunication Convention, with Annexes and Protocols, Nov. 6, 1982, U.S.T.__, T.I.A.S. No. ___, Senate Treaty Document 99-6, art. 19.

[755] *Id.* at art. 19.

[756] *Id.* at art. 20.

[757] *Id.* at art. 35.

[758] *Id.* at art. 38.

to the Convention defines harmful interference as that "which endangers the functioning of a radio navigation service or of other safety services or seriously degrades, obstructs, or repeatedly interrupts a radio communication service...."[759]

The principal question under the Nairobi Convention is determining which portions remain in effect at which levels of hostility. Articles 19 and 20 deal only with a country's control over its *own* communications, and so are always in effect. Articles 35 and 38, with their prohibition of "harmful interference," are more problematic in international conflict. Clearly, essential humanitarian communications should continue to operate in wartime. Equally clearly, military and intelligence traffic would become legitimately vulnerable to attack. What is less clear is the gray area in-between, in which some civilian communications would be interrupted in the process of affecting military transmissions. In a traditional military attack, it is relatively simple to drop bombs on a known military communications site and not drop bombs on a separate civilian radio station. With the advent of subtle information technology, however, it may be possible to discriminate in the same way by insinuating signals into mixed military-civilian facilities, targeting only those circuits or programs that are lawful targets. This technology, like the technology of precision-guided munitions, may raise expectations and later legal standards for discrimination and reduction of collateral damage.

Several other types of treaties affect the legal environment of information operations. Arms control treaties, such as the ABM Treaty,[760] frequently contain prohibitions on "interference and concealment," in evading national technical means (NTM) of verification. NTM include national-level strategic reconnaissance platforms, primarily specialized

---

[759] *Id.* at Annex 2.

[760] Treaty on the Limitation of Anti-Ballistic Missile Systems, May 26, 1972, art. XII (2).

aircraft and satellites. These prohibitions are designed to provide a confidence-building transparency and deter cheating. An offensive information operation, at least in peacetime, against a platform performing this sort of verification work would run afoul of these treaties. In wartime, these reconnaissance platforms would become lawful targets.

Although discussed more fully in the final section of this chapter, stationing arrangements, such as the Status of Forces Agreements (SOFAs) the United States has with numerous allies, have provisions on the coordination required to launch offensive (distinct from "aggressive") operations from the host country's soil. A legal factor to be considered is whether or not the contemplated operation is governed by such an agreement. Perhaps more important, the political ramifications of launching such an operation could be immense, particularly if the government of the host nation were unwitting of the capability resident within its borders.

The last such agreements are the contracts governing the use of the new cellular constellations. These voice and data networks, such as GlobalStar and Teledesic, each consisting of dozens of satellites, permit instantaneous communications from any cellular phone user to any other cellular phone user, anywhere on the planet. If owners of these networks are concerned with potential military interference with their a network and the prospect of their satellites becoming lawful targets in time of war, they could place contractual obligations on governments using their systems—obligations that may limit offensive information operations through them.

Operations

The first step in analyzing the legality of offensive information operations in space to develop a taxonomy for describing them. For ease of analysis, this field may be divided into three broad areas: intelligence collection, offensive information operations *through* satellites, and offensive information operations *against* satellites.

Intelligence Collection

The two principal types of intelligence collection involving space platforms are those collection operations *against* a given platform, and those *through* a given platform.

There are at least four levels of intrusiveness for platform reconnaissance against a satellite: passive collection, active or intrusive collection, installing trapdoors, and planting Trojan Horses.

Passive collection operations involve observing a satellite (usually photographically) or monitoring its radio or other electronic emissions. This type of collection, requiring the least technical sophistication, is analogous to a police officer staking out a suspect's home from across a street, observing only those details the suspect reveals to the public.

Active or *intrusive* collection became possible with recent advances in computer technology, enabling an adversary to "hack" into an opponent's satellite in order to collect intelligence from the inside. In a police analogy, this would be the equivalent of the police entering the suspect's home while he is away.

Installing trapdoors is the next level of intrusiveness. Here, the adversary leaves a hidden code in the space platform's software that only he can activate, enabling him to gain fast, simple, and stealthy access to the enemy system in the future. These operations are similar to the police obtaining a copy of the suspect's house key, permitting rapid reentry at any time.

Planting Trojan Horses is the most intrusive sort of intelligence collection. This method involves leaving behind a code that does not merely grant repeat access, but that will at some time in the future do actual harm to the system. This code, dormant and hidden in peacetime, may be activated in wartime if certain programmed criteria are met or, if the adversary has a trapdoor, by an affirmative command. These operations are similar to the officers staking out a violent suspect's home from the inside, ready to surprise him when he returns home and use force to subdue him. To be effective, the actual *planting* of such a Trojan Horse must be done without the adversary's knowledge and without affecting his information operations. It is only in time of crisis or during armed conflict, when the wartime criteria are met or the command is given, that this potential capability becomes an information operation under the DoD definition. This is the point at which intelligence collection transitions to offensive operations.

The four previously described types of intelligence collection are all aimed at the platform *per se*. The other broad category of collection is *"piggybacking,"* which is altogether different. Here, the intelligence collection target is not the satellite, but the satellite's own target. By piggybacking, the adversary receives the same information as the satellite's unwitting owner—the same video images, the same signals intercepts. In this way, the adversary not only has increased his knowledge base, but also knows precisely what intelligence information the satellite's owner is receiving. Undetected piggybacking would require careful placement of intercept equipment in the downlink footprint, and sophisticated cryptographic algorithms, operating in near real time. Completing the police analogy, the police have planted a wiretap in the office of a crime boss meeting with his informants and are able to listen in on what that boss knows about the police investigation.

Offensive Operations Through Satellites

Offensive information operations that simply use a satellite as a means to reach another, terrestrial target—also known as taking a *"bank shot,"*—require two levels of analysis. The primary analysis is an application of the law of war to the attack on the terrestrial target. The secondary analysis is a review of space treaties to determine the legality of using a satellite in the operation. The more attenuated the role of the satellite in question, the greater the chance that international law will not be violated. To give two examples, a cyber weapon (perhaps a digital sequence intended to disrupt the operations of a target computer) routed, *intact*, through a satellite whose governing treaty forbids such use may be problematic. However, a cyber weapon that is merely *activated* by a simple, satellite-routed phone call, having reached its target through terrestrial communications links, may cause less difficulty. In any case, these two lines of analysis, for the target and for any satellite platform used in reaching it, are distinct.

Offensive Operations Against Satellites

A different analysis is required when the target *is* the satellite. There are five distinct types of offensive operations against satellites, and all involve analyses of their legality as targets under the law of war and under the special treaty regime of space law. The five are blinding, shutdown, movement, destruction, and appropriation or impressment.

Blinding operations include anything from a temporary "dazzling" with a laser to a permanent burnout of optical or other receivers with an intense energy burst. This type of attack would be directed toward intelligence collection satellites, and is intended to prevent them from gathering or relaying information to an enemy.

In *shutdown* operations, the adversary gains access to a satellite's control program and directs that it cease functioning for some length of time. Again, this may be for only the first few critical moments of a terrestrial attack, or a permanent command to never resume operations. While not physically damaging the satellite, it does deprive its owner of its use during precisely the period when it is most needed. Clearly, a permanent shutdown command is tantamount to loss of the platform, especially for any owner not able to re-access the platform and remove the overriding code.

In space, all *movement* is described in two elements, attitude and translation. Attitude refers to the direction a satellite points; translation is its actual movement from one position to another. An attitude movement attack, then, may be mounted by accessing its control program and ordering a satellite to rotate on its axis, pointing it in another direction. Such an attack would be effective against a photographic intelligence satellite whose effectiveness depends on pointing at precisely the right place for imaging a target.

More threatening is a translation movement attack, which involves activating a satellite's thrusters and sending it into a new orbit. In many respects, a satellite's orbit determines its usefulness—geostationary for whole-earth communications, very low for pinpoint intelligence collection, and, of course, a precise ground track to bring it over a high-priority collection target. Moving a satellite's orbit may put it on a collision course with another object in an intersecting orbit. Finally, the altitude of a satellite determines its longevity: at several hundred miles, it may stay aloft for centuries; under a hundred miles, it may last only days or even hours before reentering the atmosphere. For a small satellite, this would involve its complete destruction. For a large one, it would raise the additional legal issue of discrimination, as the residue would slam into a largely undeterminable earth target.

Such a movement, in either attitude or translation, can also be permanent or temporary. A satellite may be momentarily turned away from a terrestrial attack in progress, or permanently lofted into a distant, useless orbit.

Destruction may be an extension of movement, such as a command to translate to such a low orbit that the satellite burns up or crashes into the earth. It could be accomplished by the movement of the target or another satellite so that they collide in orbit. The platform could be attacked with anything from a high-tech laser to a closer but lower-tech cannon. The destruction could even be accomplished by an appropriate command to the satellite's control program, mismanaging its fuel temperature controls to the point of detonation.

Destruction is the closest fit with traditional terrestrial military uses of force, and is perhaps the simplest in a purely analytical light.

Although less physically damaging than destruction, *appropriation* or *impressment* is perhaps more harmful to an enemy. This technique is the offensive operations parallel to piggybacking, but involves a transfer of control to the adversary. The satellite's control program is accessed and altered, denying the launching state use of its own platform. Worse than mere destruction, however, the satellite's capabilities are then at the disposal of the attacking state. This is roughly comparable to a ship being captured at sea. Again, this technique could be temporary or permanent.

Foreign Domestic Law

The sovereignty of the receiving state (*i.e.*, the state upon which another nation has consensually stationed its forces, usually for mutual defense), which includes a jurisdiction over its territory that is *prima facie* exclusive, is the most fundamental doctrine of state relations under international law,[761] and is the cardinal consideration in determining the status of military forces operating outside their flag state.

Nonbelligerent Consensual Forces

Nonbelligerent consensual forces that deploy in the territory of a sovereign state are fully exposed to the domestic law and regulation of the receiving state, subject only to protections provided by a stationing agreement, and those privileges, immunities, and protections afforded by international law. This exposure, if not altered by agreement or international law, includes subjecting members of the force to full criminal and civil jurisdiction of the receiving state, as well as exposing the force itself to receiving state regulation such as import fees, customs, and taxes. Consequently, it is preferable during a consensual deployment to have an agreement with the receiving state to determine the status of the members of the force, to ensure that members of the force are protected by receiving state law, and to exempt the force itself from unnecessary regulation and expense—notwithstanding that some of these privileges and immunities are already granted by international law. Unless additional authorization is granted by the receiving state, military personnel deploying outside their own state may only use armed force for self-defense. Nonbelligerent consensual forces are not lawful targets, even though they may be deployed into areas of ongoing hostilities.

---

[761] *See* BROWNLIE, INTERNATIONAL LAW, *supra* note 91, at 287.

Receiving State Law

The peacetime stationing of military forces on foreign territory did not become general state practice until the 20th century,[762] perhaps as a consequence of the evolving emphasis on collective security. Indeed, prior to 1914, military forces were generally present on foreign territory only as forces of hostile or peaceful occupation.[763] During a hostile occupation, an occupied state retained its sovereignty as a matter of law, but extensive portions of its governmental authority were displaced by the occupying military force.[764] In contrast, during a peaceful occupation, which historically included only small contingents present for a short time under exceptional circumstances, the occupied state retained its full governmental authority.[765]

Consent of the receiving state was required for a peaceful occupation, and the rights and obligations of the occupying force were normally defined by mutual agreement.[766] In the absence of an agreement, an occupying force retained its internal disciplinary power, but, with the exception of jurisdiction, was otherwise considered fully subject to the sovereignty of the receiving state.[767] Proponents of the doctrine of *the law*

---

[762] JOHN WOODLIFFE, THE PEACETIME USE OF FOREIGN MILITARY INSTALLATIONS UNDER MODERN INTERNATIONAL LAW 11 (1992) [hereinafter WOODLIFFE, PEACETIME USE].

[763] Serge Lazareff, STATUS OF MILITARY FORCES UNDER CURRENT NTERNATIONAL LAW 7 (1971). [hereinafter LAZAREFF, CURRENT INTERNATIONAL LAW].

[764] *See id.* at 7-8; 1907 Hague Convention IV, Customs of War on Land, *supra* note 355, Annex at arts. 42-43, *reprinted in* ROBERTS & GUELFF, *supra* note 4, at 48.

[765] LAZAREFF, CURRENT INTERNATIONAL LAW, *supra* note 763, at 8.

[766] *Id.*

[767] *Id.* at 9.

of the flag asserted that a military force operating on foreign territory had exclusive jurisdiction over its personnel; in practice, however, military personnel participating in a peaceful occupation were prosecuted and sentenced for offenses committed outside the performance of their official duty by receiving states when no jurisdictional agreement existed.[768]

It was this provisional custom of peaceful occupation that developed into the practice of a more permanent stationing of forces.[769] As this practice developed, consent of the receiving state remained a requirement for the peacetime stationing of foreign forces, and the conditions of their admission continued to be defined by mutual agreement.[770] The status of forces *stationed* in a foreign territory with a receiving state's consent but without an agreement on the conditions of their admission, however, was a very contentious area of international law for a number of years that involved the exclusive jurisdictional doctrine of *the law of the flag* and a restrictive jurisdictional approach that balanced competing state interests.[771] The law and state practice is now well-established—absent an express agreement to the contrary, foreign forces are fully subject to receiving state sovereignty and members of the force are fully subject to the jurisdiction of a receiving state for offenses committed outside the performance of their official duty.[772]

---

[768] Lazareff presents an excellent summary of this contentious issue of jurisdiction over military forces in the absence of an agreement. *Id.* at 11-18.

[769] *Id.* at 8.

[770] WOODLIFFE, PEACETIME USE, *supra* note 762, at 35-44 (1992).

[771] LAZAREFF, CURRENT INTERNATIONAL LAW, *supra* note 763, at 11-18 (1971); DEP'T OF THE ARMY, PAM. 27-161-1, THE LAW OF PEACE, VOLUME I, §§ 5-2 to 5-4, 10-3 (1979) [hereinafter DA PAM. 27-161-1].

[772] LAZAREFF, CURRENT INTERNATIONAL LAW, *supra* note 763, at 57-58; DA PAM., *supra* note 771, at 27-161-1 §§ 10-3; DEP'T OF THE AIR FORCE, PAM. 110-20, SELECTED INTERNATIONAL AGREEMENTS, p. 2-1 (1981) [hereinafter AFP 110-20]; BROWNLIE, INTERNATIONAL LAW, *supra* note 91, at 369.

This application of receiving state law can be very problematic for foreign forces because it includes subjecting the force to receiving state laws and regulations concerning entry and exit, customs, laws, and taxation, as well as criminal and civil jurisdiction.[773] This exposure to receiving state laws and regulations can adversely effect foreign forces, because they govern the daily activities of the force, such as the requirements for drivers licenses and vehicle insurance, the wearing of uniforms, the display of foreign flags and other national markings, the possession and carrying of weapons, the use of force, the exercise of military discipline, the juridical status of the force and its ability to contract for services and goods, and claims (both contractual and tort) against the foreign force.[774] Even if not misused, the pervasive nature of receiving state law and regulations can interfere with the basic functioning of foreign forces and significantly increase their operational costs. Receiving state sovereignty does *not* preempt the inherent right of foreign forces to defend themselves[775] and does *not* preempt the sending state from using additional forces to protect its own nationals from an imminent threat of injury or death when the receiving state is unwilling or unable to protect them.[776]

---

[773] These are the most important areas of concern when addressing the status of foreign forces and are the most commonly addressed issues in stationing agreements. *See* AFP 110-20, *supra* note 772, at p. 2-1.

[774] The North Atlantic Treaty Organization (NATO) Status of Forces Agreement (SOFA) provides a well-established example of how these and other interests of the receiving and sending state are balanced. Agreement between the Parties to the North Atlantic Treaty Regarding the Status of Forces, June 19, 1951, 4 U.S.T. 1792, 199 U.N.T.S. 67.

[775] *See* U.N. CHARTER art. 51; *Supplement to An Agenda for Peace: Position paper of the Secretary-General*, §§ 33-34, 36, U.N. Doc. A/50/60-S/1995/1 (Jan. 3, 1995), U.N. Sales No. E.95.I.15 (1995); THE CHARTER OF THE UNITED NATIONS: A COMMENTARY, *supra* note 70, at 589 (Bruno Simma ed., 1994).

[776] *See* NATIONAL SECURITY LAW 188-190 (John Norton Moore *et al.* eds., 1990). *But see* BROWNLIE, INTERNATIONAL LAW, *supra* note 91, at 264-280, 431-434 (1963); THE CHARTER OF THE UNITED NATIONS: A COMMENTARY, *supra* note 70, at 124-126, 666.

While the application of receiving state law can interfere with or hinder the purpose underlying the stationing of foreign forces, it can also serve to protect foreign forces. Receiving state criminal law, for example, is intended to prevent harm to society by protecting individuals and property within its territory. To the extent that a receiving state's system of criminal law is effective and enforced, members of foreign forces are protected from injury and the property of foreign forces is protected from damage or theft.

Stationing Agreements

Inherent in the authorization to station forces within a receiving state is the implied authorization that the foreign forces may exercise those rights and powers that are necessary for its effective operation; any further rights and powers, however, must be expressly granted by the receiving state.[777] The uncertainty of the breadth of these implied powers and the exclusive nature of state sovereignty make detailed agreements a fundamental requirement to clarify the application of receiving state law and regulation to foreign forces. These agreements can take many forms, such as a collective defense treaty, a base agreement or lease, a security assistance sales contract, an international headquarters agreement, an exchange of diplomatic notes, or an informal arrangement.

When the rights, privileges, powers, duties, and obligations of a foreign force are addressed in a separate agreement, it is normally referred to as a *status of forces agreement* (SOFA). Key issues that should be addressed in any SOFA include the exercise of criminal jurisdiction over offenses committed by members of the foreign force; the exercise of civil jurisdiction over acts of the members of the foreign force; the settlement of civil claims arising out of the activities of the force; the allocation of responsibility for security over installations; passport and visa requirements; freedom of movement for official duties; on-site rights and

---

[777] WOODLIFFE, PEACETIME USE, *supra* note 762, at 146 (1992).

powers; off-site rights, powers, and authority; rights of overflight; the right to conduct maneuvers and exercises; and, dispute resolution.[778]

Status of forces agreements may be either bilateral or multilateral, and may be utilized to detail the status of United Nations forces. Since a chapter IV or a chapter VI force operates in purely a consensual environment, SOFAs are as important to a United Nations force as to any national force stationed in a receiving state. Accordingly, the United Nations has recognized the importance of SOFAs since its first consensual peacekeeping force. In 1957, the United Nations entered into an Exchange of Letters with Egypt to specify the status of the United Nations Emergency Force (UNEF) in Egypt, which was established under the authority of chapter IV of the Charter.[779] This agreement on the status of UNEF was the first document of its kind to be used by the United Nations.[780] Similarly, in 1989, the United Nations entered into an agreement with South Africa to detail the status of the United Nations Transition Assistance Group (UNTAG) in Namibia (South West Africa), which was established under the authority of chapter VI of the Charter.[781]

---

[778] *Id.* at 135-150.

[779] Exchange of letters constituting an agreement between the United Nations and the Government of Egypt concerning the status of the United Nations Emergency Force in Egypt, Feb. 8, 1957, U.N.-Egypt, 260 U.N.T.S. 61, *reprinted in* BASIC DOCUMENTS ON UNITED NATIONS AND RELATED PEACE-KEEPING FORCES 7 (Robert C.R. Siekmann ed., 2d ed. 1989). [hereinafter BASIC DOCUMENTS].

[780] U.N. DEPARTMENT OF PUBLIC INFORMATION, THE BLUE HELMETS – A REVIEW OF UNITED NATIONS PEACE-KEEPING 54 (2nd ed. 1990), U.N. Sales No. E.90.I.18 [hereinafter THE BLUE HELMETS].

[781] Agreement between the United Nations and the Republic of South Africa concerning the status of the United Nations Transition Assistance Group in Namibia (South West Africa), U.N. Doc. S/20412/Add. 1, Mar. 16, 1989, *reprinted in* BASIC DOCUMENTS, *supra* note 781, at 247.

It is the established preference of the United Nations to attempt to work out an agreement on the status of its forces to be deployed within a receiving state.[782] A review of the practice of the United Nations, however, reveals the lack of a specific written agreement for recent operations; for intrastate disputes, the lack of an agreement is likely caused by the dispute within the receiving state over who has the authority to conclude such an agreement.[783] At the request of the General Assembly,[784] the Secretary-General prepared a model status of forces agreement to be used by the United Nations and receiving states.[785] In his report, the Secretary-General stated that the model status of forces agreement was based upon established practice,[786] and the Office of the Legal Adviser to the United Nations has concluded in the context of a specific operation that:

> in the absence of a signed agreement, the status of the United Nations forces and operations in Croatia are governed by the customary practices and principles applicable to UN peacekeeping or similar operations as codified in the Model Status-of-forces agreement issued as a General Assembly document dated 9 October 1990 (A/45/594).[787]

---

[782] *See* BLUE HELMETS, *supra* note 780, at 87.

[783] THE CHARTER OF THE UNITED NATIONS: A COMMENTARY, *supra* note 70, at 598.

[784] U.N.G.A. Res. 44/49 (Dec. 8, 1989).

[785] *Model status-of-forces agreement for peace-keeping operations: Report of the Secretary-General*, U.N. Doc. A/45/594 (Oct. 9, 1990).

[786] *Id.* at § I.

[787] Letter from Ralph Zacklin, Director and Deputy to the Under-Secretary-General for Legal Affairs, the United Nations, to Robert B. Rosenstock, Minister Counsellor, United States Mission to the United Nations (Apr. 25, 1995) (on file with author).

In 1991, the Secretary-General restated this position in his *Model agreement between the United Nations and Member States contributing personnel and equipment to United Nations peace-keeping operations*, which provides in paragraph 6 that, pending the conclusion of a status of forces agreement, the United Nations "shall apply the customary principles and practices which are embodied in the model status-of-forces agreement."[788]

This precise language from the model agreement recognizing the customary international law status of the United Nations model SOFA was utilized in 1994 by the United Nations in making arrangements with Canada to provide personnel and equipment to the United Nations Mission in Haiti.[789] Although considerably different in format, the SOFA arrangements concluded as a part of the General Framework Agreement for Peace in Bosnia and Herzegovina[790] are substantively consistent with the United Nations model SOFA. The terms of the model SOFA,[791] and thus customary international law, are limited to operations established under the authority of the United Nations *and* conducted under United Nations authority and control.[792]

---

[788] *Model agreement between the United Nations and Member States contributing personnel and equipment to United Nations peace-keeping operations*, U.N. Doc. A/46/185 (May 23, 1991).

[789] Agreement between the United Nations and Canada regarding the provision of Military personnel and equipment to the United Nations Mission in Haiti (UNMIH) (Nov. 30, 1994) (on file with author).

[790] General Framework Agreement for Peace in Bosnia and Herzegovina, Dec. 14, 1995, Republic of Bosnia and Herzegovina-Republic of Croatia-Federal Republic of Yugo., U.N.S.C. Doc. S/1995/999, annex.

[791] *Model status-of-forces agreement* at §§ II-III.

[792] This formula of application is parallel to the definition of *United Nations operations* in art. I of the Convention on the Safety of United Nations and Associated Personnel, U.N.G.A. Res. 49/59 (Dec. 9, 1994).

Section 3 of the United Nations model SOFA provides that the Convention on the Privileges and Immunities of the United Nations[793] [hereinafter Convention (P&I)] applies to the United Nations peacekeeping operation, subject to the provisions of the model agreement. For example, the Convention (P&I) provides for varying levels of privileges and immunities for the representatives of Members, officials of the United Nations, the Secretary-General and Assistant Secretaries-General, and experts on mission.[794] The United Nations model SOFA grants military observers "experts on mission" status, the Commander of the force the protections of a diplomatic envoy accorded the Secretary-General, and members of the force the more limited privileges and immunities as detailed in the model agreement. Members of the force are immune from criminal and civil jurisdiction for all acts performed in their official capacity; exempt from all income taxes except on income received from sources inside the receiving State; exempt from all other direct taxes, registration fees or charges; and, exempt from those laws and regulations governing customs and foreign exchange for personal property required by reason of their presence in the receiving State. Additionally, the model SOFA contains detailed provisions commonly found in a status agreement such as the use of the U.N. flag, wearing of uniforms, carrying of weapons, use of force, arrest and detention powers of the force, vehicle markings, use of U.N. issued drivers license, communications capabilities, postal services, freedom of movement, designation and protection of facilities, cost of utility services, local contracting, hiring of local personnel, currency exchange rates, entry and departure requirements, required identity documents, and dispute settlement provisions.

---

[793] Convention on the Privileges and Immunities of the United Nations, Feb. 13, 1946, 21 U.S.T. 1418, 1 U.N.T.S. 15.

[794] The Convention (P&I) will be discussed in more detail *infra*, this section.

Information Operations and Host Nations

Many information operations may be prosecuted from the safety of the acting state's own territory. However, there are times when third-country law (other than the nations launching or receiving the attack) comes into play. First, when an information operation is launched from a ship, plane, or facility located in a host nation. In this case, the contours of the stationing agreement will strongly influence which platforms may be employed, and the degree to which they may become involved in operations.

Second, an attack launched from, say, the United States may be routed through one or more terrestrial computers located in third countries. Here, the two key variables are the likelihood of detection and the damage done en route. An attack that is electronically indistinguishable from any other stream of data packets may very well pass through a third nation's computers undetected. Even if the data stream were identified, its essentially peaceful nature (until assembled into a weapon within the target nation) would argue for allowing its passage. However, if the attack were launched so as to make the third country appear to be the aggressor, or if it did significant damage (perhaps to discourage hacking back to the origin) of computers en route, then the rights of the third country would be implicated in a serious way. At the very least, such actions could be interpreted as violations of a stationing agreement; at worst, they could be construed as independent uses of force by the attacking nation against the third country.

PART IV. CONCLUSION

◆

CHAPTER 10: DISTINGUISHING COMPUTER NETWORK ESPIONAGE FROM COMPUTER NETWORK ATTACK†

Computer network espionage and computer network attack are two of the most urgent issues facing warfighters, decision-makers, and international lawyers today. Computer network espionage, or the use of computers to gather intelligence against an adversary from his own systems, is lawful under international law. The amount of damage done in pure collection activities is usually negligible, both because the intent is to gather intelligence and not inflict harm, and because any damage done, like a vase broken by a clumsy burglar, would be inadvertent and would serve to alert the target nation that it was being ransacked. Computer network attack, on the other hand, is an intentional use of force by definition. Any detectable damage done in the real world would serve, at the very least, as the basis for a colorable claim of use of force, and, possibly, an armed attack. The intermediate cases between these extremes are the most problematic. Installing a trapdoor to facilitate later intelligence gathering penetrations would probably be at the high end of CNE, while installing a logic bomb to be detonated in the future would probably be at the low end of CNA. The crucial question in differentiating them is the reasonably expected result if each were activated: the former would simply lead to another penetration, while the latter would result in damage done to the target state. Chapter 6 examined use of force norms in cyberspace within the legal framework of the Charter paradigm of conflict management, and this chapter will examine use of force norms in cyberspace as they apply to these distinct but interrelated issues.

---

† This chapter is drawn substantially from Chapter 7 of WALTER GARY SHARP, SR., CYBERSPACE AND THE USE OF FORCE (1999).

Computer Network Espionage

The right to conduct espionage is an essential part of a state's inherent right of self-defense.[795] The 1907 Hague Convention IV explicitly recognizes the lawfulness of espionage during armed conflict.[796] Similarly, the 1961 Vienna Convention on Diplomatic Relations explicitly recognizes the well-established right of nations to engage in espionage during peacetime, and the practice of states has specifically recognized a right to engage in such clandestine intelligence collection activities as an inherent part of foreign relations and policy.[797] Espionage is, however, unlawful under the domestic law of most states during peacetime and armed conflict.[798]

What we now define as information in cyberspace and cyberspace infrastructures have historically been the subjects of lawful espionage during peacetime and armed conflict. Customary international law, for example, has long recognized the lawfulness of collecting information in cyberspace, such as signals intelligence, communications intelligence, electronics intelligence, foreign instrumentation signals intelligence, and imagery intelligence.[799] These terms are currently defined as follows:

Signals intelligence (SIGINT): intelligence information comprising either individually or in combination all communications

---

[795] NATIONAL SECURITY LAW 433 (John Norton Moore *et al.* eds., 1990).

[796] *See* 1907 Hague Convention IV, Customs of War on Land, *supra* note 355, Annex at arts. 24, 29-31, *reprinted in* ROBERTS & GUELFF, *supra* note 4 at 48, 53-54, *Id.* at 435.

[797] *See* NATIONAL SECURITY LAW 435-36 (John Norton Moore *et al.* eds., 1990).

[798] *See id.* at 435.

[799] *See* NATIONAL SECURITY LAW 436-37 (John Norton Moore *et al.* eds., 1990).

intelligence, electronics intelligence, and foreign instrumentation signals intelligence, however transmitted.[800]

Communications intelligence (COMINT): technical and intelligence information derived from foreign communications by other than the intended recipients.[801]

Electronics intelligence (ELINT): technical and geolocation intelligence derived from foreign non-communications electromagnetic radiations emanating from other than nuclear detonation or radioactive sources.[802]

Foreign instrumentation signals intelligence (FISINT): technical and intelligence information derived from the intercept of foreign electromagnetic emissions associated with the testing and operational deployment of non-US aerospace, surface, and subsurface systems such as telemetry, beaconry, electronic interrogators, and video data links.[803]

Imagery intelligence (IMINT): intelligence derived from the exploitation of collection by visual photography, infrared sensors, lasers, electro-optics, and radar sensors such as synthetic aperture radar wherein images of objects are reproduced optically or electronically on film, electronic display devices, or other media.[804]

Similarly, cyberspace infrastructures such as computers, telecommunication systems, and satellites have been used for intelligence

---

[800] JOINT PUB 1-02, *supra* note 292.

[801] *Id.* at 109.

[802] *Id.* at 168.

[803] *Id.* at 195.

[804] *Id.* at 227.

collection since their invention.[805] The tactical and strategic use of this information in cyberspace is embraced in the operational concept of information operations.[806]

As a matter of domestic criminal law during peacetime, any person who is caught engaging in espionage within the criminal jurisdiction of a state is subject to prosecution and can be exchanged for a spy of the apprehending state who has been detained by the sending state on similar charges.[807] Should a spy have diplomatic immunity under the terms of the 1961 Vienna Convention on Diplomatic Relations, then he or she is declared *persona non grata* and sent home.[808]

As a matter of international conflict management during peacetime, the conduct of espionage within foreign territory has caused the territorial state to use force on a number of occasions. During the period April 1950 through April 1969, for example, there were at least 22 incidents where aircraft were suspected of espionage and were attacked, mostly by the Soviet Union.[809] During one of these incidents, a 1960 U-2 flight by the United States over Soviet territory, the Soviet government asserted that military intelligence flights by reconnaissance aircraft over Soviet territory and territorial waters are *per se* acts of aggression.[810]

---

[805] *See* NATIONAL SECURITY LAW at 438-42 (John Norton Moore *et al.* eds., 1990).

[806] *See* JOINT PUB 3-13, *supra* note 47, at I-9 and I-10.

[807] NATIONAL SECURITY LAW at 435-36 (John Norton Moore *et al.* eds., 1990).

[808] *Id.*

[809] *Id.* at 440-41.

[810] BROWNLIE, INTERNATIONAL LAW, *supra* note 91, at 363.

The United Nations Security Council disagreed, and concluded that the U-2 flight was a violation of Soviet airspace but not a use of force within the meaning of Article 2(4).[811] Consistent with the Security Council's view, more than one scholar has concluded that military intelligence flights may be deliberate and unlawful intrusions, but are not necessarily uses of force.[812] The consensus view by these and other scholars is that all nations have complete and exclusive sovereignty over their territory, but, customary international law requires a reasonable self-defense response to aircraft that violate another state's airspace that is proportionate to the danger posed by the presence of the trespassing aircraft.[813]

The essence of espionage is the collection of information, and with the capabilities of today's inter-linked computers, there is perhaps no richer source of information than what can potentially be accessed via the internet. Some of the information is publicly available, and some must be accessed surreptitiously without the consent of the owner of the computer system. When a state accesses *publicly* available databases in a foreign state, no unlawful act occurs under international or domestic law. Publicly available information that has potential intelligence value is called open-source intelligence (OSINT).[814]

If the unlawful physical penetration of a state's airspace by the military aircraft of another state is not a use of force within the meaning

---

[811] NATIONAL SECURITY LAW at 439 (John Norton Moore *et al.* eds., 1990).

[812] BROWNLIE, INTERNATIONAL LAW, *supra* note 91, at 364. *See also* NATIONAL SECURITY LAW at 439 (John Norton Moore *et al.* eds., 1990).

[813] *See* NATIONAL SECURITY LAW at 439 (John Norton Moore *et al.* eds., 1990); COMMANDER'S HANDBOOK at § 4.4; BROWNLIE, INTERNATIONAL LAW, *supra* note 91, at 363-64, 373-74 (1963). For a more detailed discussion of what force a territorial state may take in self-defense, see the previous section concerning violations of international law.

[814] JOINT PUB 1-02, *supra* note 292, at 340.

of Article 2(4), it is certain as well that a virtual penetration of a state's cyberspace is not a use of force. When a state hacks into another state's computer systems it is very likely engaging in acts that are unlawful under the domestic law of the territorial state, but under international law, the hacking state may be engaging in a lawful act of espionage that may not be considered a use of force. This type of computer espionage poses a number of very serious practical problems. It may be virtually impossible, for example, for the territorial state to prosecute anyone under its domestic law for espionage since hacking can be undertaken anonymously from any place in the world. A territorial state's inability to prosecute and the almost assured anonymity of the hacking state reduces the deterrent value of domestic espionage statutes and the public embarrassment of a state getting caught spying.

Another practical problem is that once a state has penetrated another state's information infrastructure to conduct espionage, it is only one keystroke away from the capability of engaging in hostile and potentially destructive activities that are unlawful under international law. A state's penetration of another state's information infrastructure may be espionage, a pre-attack exploration, or an actual attack in progress that has not yet manifested itself. Although some indicia of intent can be monitored, the intentions of a state are very difficult to discern. If a trespassing state is simply looking around and copying files, for example, it may likely be engaging in nothing more than espionage. If those files, however, contained an order of battle and rules of engagement, then the trespassing state may be engaging in a pre-attack exploration of the battlefield. Similarly, if the trespassing state were installing trapdoors to facilitate future penetrations, then it may be engaged in pre-attack preparations. Finally, the planting of offensive cyber-tools may be indicative of an attack that has not yet manifested itself. Short of an actual destructive attack, however, it is very difficult for a state to be sure of the intent of a trespassing state—albeit at the minimum such a trespassing state is engaged in espionage.

The technology of computers and the internet allows a lawful act of espionage to morph into an unlawful use of force at the speed of light. Despite these practical problems, espionage conducted by the nonconsensual penetration of another state's computer systems is lawful under existing international law. Regardless of whether a state committing computer espionage is about to attack immediately or if it has installed the latent hostile capability to attack with a single keystroke at some unknown point in the future, a legal regime that fails to take this drastically truncated warning time into account risks irrelevancy and circumvention.

Existing law, however, does offer at least a partial solution to the threat presented by cyberspace technology. Recall that anticipatory self-defense is permissible when "the necessity of that self-defence is instant, overwhelming and leaving no choice of means, and no moment for deliberation."[815] There seems to be no better circumstance for anticipatory self-defense to apply than when technology allows an unlawful use of force, and potentially an armed attack, to occur literally at the speed of light. The right to respond in anticipatory self-defense does not apply to the penetration of *all* government computer systems during peacetime, but it should apply presumptively to those sensitive systems that are critical to a state's vital national interests. As previously discussed, any use of force in anticipatory self-defense must be necessary and proportional under international law.

Of course, anticipatory self-defense is problematic in that the identity of the attacker may not be known until the conclusion of a domestic investigation by the FBI. This time lag suggests that anticipatory self-defense may be less of an issue than traditional self-defense once the attackers have been identified.

The key to this presumptive solution is a predetermination of the potential threat posed by the penetration of a specific computer system or infrastructure that is dependent upon its value to the vital national interests of a state. Remedies will, of course, vary depending upon the sensitivity and importance of the penetrated computer system or

---

[815] THE CHARTER OF THE UNITED NATIONS: A COMMENTARY, *supra* note 70, at 675.

infrastructure. Penetration in government computers that store personnel, logistical, or administrative data, for example, would not pose an instant and overwhelming need of anticipatory self-defense. Early detection, preventive measures, and public disclosure that results in the embarrassment of the offending state are appropriate remedies under these circumstances absent any other indications of hostile intent.

In contrast, the mere penetration by a state into sensitive computer systems such as early warning or command and control systems, missile defense computer systems, and other computers that maintain the safety and reliability of a nuclear stockpile, should by their very nature be presumed a demonstration of hostile intent. Individually, these computer systems are so important to a state's ability to defend itself that espionage into anyone of them should be presumed to demonstrate hostile intent. In these and similar cases, the necessity of self-defense may be instant, overwhelming and leaving no choice of means, and no moment for deliberation. If the identity of the intruder is known, then a victim state can evaluate the potential threat to its vital national interests and may lawfully respond in anticipatory self-defense with a necessary and proportional use of force, either in kind through cyberspace or by more traditional uses of force. In short, states should adopt a rule of engagement that allows them to use force in anticipatory self-defense against any identified state that demonstrates hostile intent by penetrating a computer system that is critical to their respective vital national interests. The challenge of adopting such a rule of engagement is for all states to strike the right balance between the equities of the law of conflict management and their definition of what information infrastructures are critical to their respective vital national interests.[816]

The concept of vital national interests in today's environment of information technology must include non-government commercial and

---

[816] *See* William Gravell, *Some Observations Along the Road to "National Information Power,"* 9 DUKE J. COMP. & INT'L L. 401 (1999) for an excellent discussion of how the United States should define its vital national interests in the Information Age.

other economic interests of a state.[817] Almost all legal commentators would agree that a state has an unequivocal right to use force to defend the commercial infrastructure within its territory from an armed attack, and most would agree that the inherent right of self-defense includes the right of anticipatory self-defense. An attack on a state's vital commercial infrastructure is an attack upon the state as a whole. Individually, these computer systems may *not* be so important to a state's ability to defend itself that damage to any one of them should be presumed to demonstrate hostile intent. A threat to these infrastructures collectively, however, is important to a state's ability to defend itself. Accordingly, the proposed presumption should still apply to the protection of a state's critical non-governmental infrastructures, although the utmost care must be applied in defining when a state has the right to use force to protect them from computer intrusion.

Recall from Chapter 2, for example, that the United States declared in Executive Order 13010 that eight categories of "national [commercial] infrastructures are so vital that their incapacity or destruction [by either physical or cyber attack] would have a debilitating impact on the defense or economic security of the United States." The key in the commercial sector is to define vital portions of non-governmental systems and infrastructures as vital national interests, not individual computers or smaller systems. Accordingly, a single bank in Kansas would not meet the definition of a vital national interest, and the United States would probably not have the right under international law to use force to protect that bank from a single cyber attack. If, however, there were a pattern of computer attacks that threatened the economic vitality of the national financial infrastructure of the United States, it would have the right to use necessary and proportional force in self-defense to prevent future computer espionage.

Computer espionage is lawful under international law and is not a use of force within the meaning of Article 2(4). However, acts beyond

---

[817] *See id.*

mere intelligence collection may demonstrate the requisite hostile intent of the intruding state necessary to invoke the victim state's right of anticipatory self-defense. The difficulty is to be able to determine the identity of the intruder to evaluate the seriousness of the threat presented by the penetration of these sensitive computer systems. Without the capability of identifying the intruder, the legal authority to use force offers little help. Recognition by states that the penetration of some systems will be considered either an indication of hostile intent or a hostile act which authorizes a use of force in response will restore, at least in part, some deterrence. There is no better solution under the rule of law until technology can assure a rapid identification of the intruding state.

Computer Network Attack

Computer network attack (CNA) is defined as operations that "disrupt, deny, degrade, or destroy information resident in computers and computer networks or the computers and networks themselves."[818] By definition, a computer network attack against another state will cause a destructive effect in that state. Accordingly, these operations pose a much more straightforward issue than computer espionage, and need only be addressed very briefly.

In a previous discussion, it was recognized that unarmed, non-military physical force by any agency of a state could affect another state just as severely as the use of armed military force. It was concluded in this discussion that almost any state activities in cyberspace that intentionally cause almost any destructive effect within the sovereign territory of another state is probably an unlawful use of force within the meaning of Article 2(4) that may produce the effects of an armed attack prompting the right of self-defense.

---

[818] JOINT PUB 3-13, *supra* note 47, at I-9.

It was concluded that the scope, duration, and intensity of the force must be analyzed to determine if an armed attack has occurred and what may constitute a necessary and proportional response. Should a computer network attack cause a single train to crash, it would be an unlawful use of force but likely not an armed attack. In contrast, should a computer network attack cause dozens of such trains and aircraft to crash, it would very likely be considered an armed attack which invokes the victim state's right to use force in self-defense.

Just as with computer espionage, the difficulty of responding to a computer network attack in self-defense is to be able to determine the identity of the attacking state. Without the capability of identifying the attacker, the legal authority to use force offers little help.

CHAPTER 11: CORE PRINCIPLES AND FINAL REFLECTIONS[†]

Previous chapters examined in great detail what peacetime state activities in cyberspace constitute a threat or use of force under the Charter paradigm of conflict management, and when a use of force constitutes an armed attack within the meaning of Article 51 of the Charter. These chapters moored together a number of concepts—such as 'threat of force,' 'use of force,' 'armed force,' 'armed attack,' 'Common Article 2 threshold,' 'hostile intent,' 'hostile act,' 'anticipatory self-defense,' and 'self-defense'—that are frequently found floating about each other, but are not conceptually linked with an internally consistent set of principles.

This final chapter will extract from these previous discussions a review of those guiding principles that form the basis for analyzing *all* use of force issues under international law. These guiding principles are a composite of the peacetime regime of international law, the law of conflict management, and the law of armed conflict. Application of these principles should serve as a useful starting point for an analysis of all use of force issues, to include the use of force by states in cyberspace. Table 1 captures these principles without discussion for ease of reference.

Maintaining a credible ability to use force, in cyberspace and elsewhere, is lawful and a fundamentally important aspect of deterrence and international peace and security. Even clearly defined norms are not adequate to control the behavior of states that continue to use aggressive force. Accordingly, peace-loving states must maintain a strong and credible ability to use force when necessary in self-defense. This ability to threaten and use force is not inherently unlawful or evil—it is essential to international peace and security. Indeed, the collective use of force by the international community is the core principle upon which the Charter of

---

[†] This chapter is drawn substantially from Chapter 8 of WALTER GARY SHARP, SR., CYBERSPACE AND THE USE OF FORCE (1999).

the United Nations is built. International law outlaws only the aggressive use of force, and it specifically acknowledges the lawfulness of the use of force in self-defense. Maintaining the credible ability to use force is an important deterrent element that is required to ensure that the rule of law is enforced. If peace-loving states do not maintain the ability to defend themselves, individually and collectively, then international peace and security is an unachievable goal.

International and national law governs, but does not prohibit *per se*, state activities that involve coercion and the use of force in cyberspace. At the end of every cyber-connection is a computer and a person at a particular point on earth, and that person's behavior—whether acting individually or as a state agent—is governed by the applicable national law of where he or she resides, international law, and the national law of a foreign state where his or her behavior may have an effect. Although stirred up by cyberspace into a complex melt of jurisdiction, conflict of law, and judgment enforcement issues, these laws are all familiar to us. They include civil law that involves tort liability and intellectual property, criminal law that proscribes trespass and theft, international law that regulates the use of force and armed conflict, and foreign domestic law that prohibits espionage or the introduction of certain political or social ideas. All state activities in cyberspace are governed by existing international and national law.

States have an international obligation to act in good faith in settling their disputes by peaceful means. Any state activity in cyberspace that

- is an act of bad faith as it pertains to that state's dispute settlement negotiations;
- otherwise threatens a peaceful settlement of that state's own dispute;
- interferes with a dispute of other states in such a manner that would delay or prevent their dispute settlement; or,
- creates a dispute between other states that threatens international peace and security,

violates the obligations of all Member States under Article 2(3) of the Charter of the United Nations to "settle their international disputes by peaceful means in such a manner that international peace and security, and justice, are not endangered."

Articles 2(4), 39, and 51 of the Charter of the United Nations now redefine and codify the contemporary *jus ad bellum* in its entirety. For four millennia, war has been accepted as a legitimate form of violence. Not until the establishment of the Charter of the United Nations was the aggressive use of force outlawed and punishable as a crime against peace.

If a state activity is a use of force within the meaning of Article 2(4), it is unlawful unless it is an exercise of that state's inherent right of self-defense or unless it is authorized by the Security Council under its coercive Chapter VII authority. While the aggressive use of force was specifically outlawed, states reserved their inherent right to use force in individual and collective self-defense. The international community has conferred upon the Security Council the primary responsibility for the maintenance of international peace, and the power to authorize the use of military force to eliminate breaches of the peace, acts of aggression, and threats to international peace and security.

What constitutes a prohibited 'threat or use of force,' in cyberspace and elsewhere, is a question of fact that must be *subjectively* analyzed in each and every case in the context of all relevant law and circumstances. The terms 'use of force,' 'armed attack,' and 'self-defense' will never be clearly defined by objective rules of law. The international community has struggled for decades to define these terms, but there are simply too many factual variables involved to capture them in a few simple rules. Almost all attempts by statesmen and scholars to craft generally accepted, objective definitions, and certainly all of those in the mainstream of international legal thought, are qualified by exceptions consistent with the principles discussed in this book and summarized in this chapter.

The best way to accurately predict what may be considered an armed attack within the meaning of Article 51 is by studying state

practice. In addition to traditional, universally accepted examples such as an armed cross-border invasion, an armed attack may occur when a use of force or an activity not traditionally considered an armed attack is used in such a way that it becomes tantamount in effect to an armed attack. There is no clear definition of armed attack, but Chapter 6 provided a number of examples of state activities in cyberspace that define what is and what is not an armed attack within the meaning of Article 51 of the Charter.

An international armed conflict exists upon the declaration of war, the occurrence of *de facto* hostilities between two or more contracting parties even if the state of war is not recognized by one of them, and in all cases of partial or total occupation even if met with no armed resistance. This restatement of the Common Article 2 threshold is universally accepted as the test for when an international armed conflict exists.

***De facto* hostilities exist between two states when the scope, duration, and intensity of force between them reaches the level of an armed attack within the meaning of Article 51 of the Charter.** What constitutes a use of force of a scope, duration, and intensity that constitutes an armed attack and triggers the *jus in bello* is a question of fact that must be subjectively analyzed in each and every case in the context of all relevant law and circumstances.

A use of force within the meaning of Article 2(4) may not always rise to the scope, duration, and intensity threshold of an armed attack that invokes a *de facto* international armed conflict and the *jus in bello*. This is based upon the universally accepted view that the thresholds of Articles 2(4) and 51 are different in scope, *i.e.*, the Article 2(4) concept of a threat or use of force is much broader than the more restrictive Article 51 threshold of armed attack.

A state never loses its right to use force in self-defense in response to a use of force within the meaning of Article 2(4), however, the right of self-defense under customary international law may not always justify an armed response. There are no circumstances under which a state does not have the right to self-defense when provoked by a use of force within the meaning of Article 2(4). The right of self-defense, however, is not unlimited. A claim to self-defense must be reasonable under the circumstances, but what constitutes a threat or use of force that invokes the right of self-defense must be evaluated from the subjective perspective of the state claiming self-defense. All lawful uses of force must be necessary and proportional to the threat, which may mean that an armed response is not appropriate under the customary international law of self-defense. Until such time that the Security Council is able to take effective measures to remove such a threat, all states retain the right to use all necessary force in individual and collective self-defense.

State activities in cyberspace that constitute a use of force within the meaning of Article 2(4) may be conducted by any state agent—not just the military. Unarmed, non-military physical force can affect another state just as severely as the use of armed military force. State activities must be analyzed in light of their effect on another state, and not what agency of the state is responsible for the state activity.

Computer espionage is lawful under international law and is not a use of force within the meaning of Article 2(4), however, it may demonstrate the requisite hostile intent of the intruding state necessary to invoke the victim state's right of anticipatory self-defense. The difficulty is to be able to determine the identity of the intruder to evaluate the seriousness of the threat presented by the penetration of sensitive computer systems. Without the capability of identifying the intruder, the legal authority to use force offers little help.

Almost *any* computer network attack that intentionally causes almost *any* destructive effect within the sovereign territory of another state is probably an unlawful use of force within the meaning of Article 2(4) that may produce the effects of an armed attack prompting the right of self-defense. This is a relatively straightforward analysis for determining when a computer network attack constitutes an unlawful use of force. Just as with computer espionage, the difficulty of responding to a computer network attack in self-defense is to be able to determine the identity of the attacking state.

The principle of discrimination requires that, prior to any use of force, the attacker distinguish between combatants and noncombatants, between military objectives and civilian objects. Noncombatants include civilians and "protected persons," such as medics and chaplains. While it is unlawful to attack civilian objects as such, it is not unlawful to collaterally damage civilian objects located so close to lawful military objectives that no reasonable precautions could have prevented the damage. In cyberspace, such distinctions will require the collection and analysis of information to produce intelligence to enable such determinations to be made. The military commander must consider foreseeable secondary effects in his target selection and mission planning.

The principle of necessity permits the use of all force required for mission accomplishment and force protection, but no superfluous force or unnecessary suffering. Superfluous force is that which is inflicted solely for the sake of causing damage. Unnecessary suffering has two components: quantitatively, it is simply the complement of military necessity (necessary force); qualitatively, it includes those means and methods of war that are by nature inhumane (chemical weapons, biological weapons, exploding bullets, transparent bullets that cannot be detected in the body by x-ray equipment).

The principle of proportionality requires the military commander to balance the collateral damage (against civilians and their property) of a planned attack against the concrete and direct military advantage expected to be gained. Proportionality is *not* the requirement to respond with only as much force as was used in the precipitating attack, and is *not* the requirement to employ the same means and methods as in the provoking attack. Whatever other policy preferences may be expressed later in the national security decision-making process, the basic legal requirement is to ensure the balance of military advantage gained outweighs the damage done to noncombatants and their property.

The principle of chivalry permits ruses of war, but forbids perfidy. Perfidy is unlawfully deceiving an opponent that he is entitled to enjoy, or required to accord, protections of law that are, in fact, inapplicable in the situation. Attacking an enemy facility from a vehicle displaying the Red Cross or the U.N. flag would be examples of perfidiously deceiving the enemy into according protections under law (not attacking the vehicle) which are not applicable. Broadcasting a false notice of cease-fire or calling for a *parlementaire* before attacking the unwary opponent would be an example of perfidiously deceiving one's opponent that he is entitled to a protection of law, which in fact, he is not. Ruses of war, on the other hand, are all other wartime deceptions that convince an enemy to launch or withhold an attack based on other than perceived legal, but rather tactical, reasons. Fighter aircraft simulating transport aircraft radio "chatter" in the hopes of luring enemy fighters into attacking is a permissible ruse, as is erecting dummy missile sights in the hopes of decoying enemy attacks against them.

Treaty law reflects the four basic principles of discrimination, necessity, proportionality, and chivalry. Treaties and other international agreements codify and implement customary international law. They occasionally lead international law by affirmatively proposing new norms and standards, at first applicable to only those nations which sign the agreements. Such new standards may become binding on other nations over time, if those other nations comply with the new norms, and do so out of a sense of legal obligation. Treaties applicable to one age or type of warfare may nonetheless be useful in charting the course of the law of information conflict, in that analogies may be drawn from them and applied to operations in cyberspace.

Computer Network Espionage (CNE) and Computer Network Attack (CNA) are located on two places along the spectrum of conflict, and have distinctly different legal characteristics. Computer Network Espionage, like any form of pure espionage, is lawful under international law, but is usually not lawful under the domestic law of the target state. CNE usually involves little or no force, and involves only as much intrusion as is necessary to collect the required information from the adversary's systems. Computer Network Attack, on the other hand, involves some kind of destruction with consequences in the physical world. CNA should be analyzed as any other type of use of force, and, depending on the scope, duration, and intensity of the force employed, may rise to the level of an armed attack.

A portion of the peacetime regime survives hostilities between the belligerents, and informs the legal analysis of proposed information operations. The principle areas are divided geographically into the law of the sea, air law, space law, and foreign domestic (or host nation) law. The law of the sea addresses the metaphor of "innocent passage" in cyberspace, unauthorized broadcasting, and liability for damage done in wartime. Air law covers the metaphor of defining airspace (as applied to defining cyberspace), the legal restrictions on interfering with aircraft communications or navigation, as well as physical attacks on aircraft. Space law is useful in the distinction of air and space, the use of space platform for peaceful and non-peaceful purposes, and liability of those who cause damage against or through the use of space platforms. Finally, foreign domestic law is most applicable in the area of stationing arrangements and the limitations they place on the employment of weapons and forces forward deployed.

All planned state activities in cyberspace should be reviewed by assigned government counsel. The complexity of the application of international, US domestic, and foreign law during the conduct of state activities in cyberspace requires prudent decision-makers to have their attorneys present whenever possible in all phases of planning and operations. In the military, a legal review of all information operations is required. An instruction from the Chairman of the Joint Chiefs of Staff provides that information operations

> conducted throughout the range of military operations may involve complex legal issues. The complexities introduced by domestic, foreign, and international law make it critical that at the appropriate level of command legal counsel is involved in policy development as well as the development and employment of various techniques and capabilities. Military technologies, tactics, techniques, and procedures for conducting IO must undergo a legal review in accordance with

> reference to ensure that they are consistent with the domestic and international legal obligations of the United States.[819]

This complexity permits legal advisers to support the proposed activities of their clients with many meritorious arguments of lawfulness. They must also, however, advise their clients of potential allegations of unlawfulness to prevent adverse consequences that may undermine their client's positions and national goals. Finally, legal advisers must fully understand the science and technology that is behind the military applications of cyberspace—simply knowing the law is not enough.

Technology and the practice of states will shape the evolution of law that governs information operations. The international community must be guided by principles that evolve with technology and state practice. If good faith is exercised in applying the existing principles of the law of conflict management to the use of force in cyberspace, then a reasonable regime will emerge via state practice.

Although there is a considerable body of international law that governs the use of force in cyberspace, technology does not allow a state to completely prevent computer espionage or computer network attack, nor does it assure that a state can reliably determine the identity of an intruder or attacker in any timely manner. This shortcoming in technology limits a state's ability to use force in self-defense. The real exigency facing warfighters and decision-makers is whether technology will allow a state to respond by identifying an intruding or attacking state, not whether international law will permit a use of force in self-defense.

---

[819] Chairman of the Joint Chiefs of Staff Instruction 3210.01A, *Joint Information Operations Policy*, Nov. 6, 1998, Encl. A, para. 3h. Reference e, as referred to in the quote, is Department of Defense Directive 5100.77, *DoD Law of War Program*, Jul. 10, 1979.

Table 1

USE OF FORCE PRINCIPLES

Maintaining a credible ability to use force in cyberspace is lawful and a fundamentally important aspect of deterrence and international peace and security.
International and national law governs, but does not prohibit *per se*, state activities that involve coercion and the use of force in cyberspace.
States have an international obligation to act in good faith in settling their disputes by peaceful means.
Articles 2(4), 39, and 51 of the Charter of the United Nations now redefine and codify the contemporary *jus ad bellum* in its entirety.
If a state activity is a use of force within the meaning of Article 2(4), it is unlawful unless it is an exercise of that state's inherent right of self-defense or unless it is authorized by the Security Council under its coercive Chapter VII authority.
What constitutes a prohibited 'threat or use of force' in cyberspace is a question of fact that must be *subjectively* analyzed in each and every case in the context of all relevant law and circumstances.
The best way to accurately predict what may be considered an armed attack within the meaning of Article 51 is by studying state practice.
An international armed conflict exists upon the declaration of war, the occurrence of *de facto* hostilities between two or more contracting parties even if the state of war is not recognized by one of them, and in all cases of partial or total occupation even if met with no armed resistance.
De facto hostilities exist between two states when the scope, duration, and intensity of force between them reaches the level of an armed attack within the meaning of Article 51 of the Charter.
A use of force within the meaning of Article 2(4) may not always rise to the scope, duration, and intensity threshold of an armed attack that invokes a *de facto* international armed conflict and the *jus in bello*.
A state never loses its right to use force in self-defense in response to a use of force within the meaning of Article 2(4), however, the right of self-defense under customary international law may not always justify an armed response.
State activities in cyberspace that constitute a use of force within the meaning of Article 2(4) may be conducted by any state agent—not just the military.
Computer espionage is lawful under international law and is not a use of force within the meaning of Article 2(4), however, it may demonstrate the requisite hostile intent of the intruding state necessary to invoke the victim state's right of anticipatory self-defense.

Almost *any* CNA that intentionally causes almost *any* destructive effect within the sovereign territory of another state is a probably unlawful use of force within the meaning of Article 2(4) that may produce the effects of an armed attack prompting the right of self-defense.
The principle of discrimination requires that, prior to any use of force, the attacker distinguish between combatants and noncombatants, between military objectives and civilian objects.
The principle of necessity permits the use of all force required for mission accomplishment and force protection, but no superfluous force or unnecessary suffering.
The principle of proportionality requires the military commander to balance the collateral damage (against civilians and their property) of a planned attack against the concrete and direct military advantage expected to be gained.
The principle of chivalry permits ruses of war, but forbids perfidy.
Treaty law reflects the four basic principles of discrimination, necessity, proportionality, and chivalry.
Computer Network Espionage (CNE) and Computer Network Attack (CNA) are located on two places along the spectrum of conflict, and have distinctly different legal characteristics.
A portion of the peacetime regime survives hostilities between the belligerents, and informs the legal analysis of proposed information operations.
All planned state activities in cyberspace should be reviewed by government counsel.
Technology and the practice of states will shape the evolution of law that governs IO.

ABBREVIATIONS

CA	Civil affairs
COMINT	Communications intelligence
CNA	Computer network attack
CNE	Computer network espionage
ELINT	Electronics intelligence
EA	Electronic attack
EW	Electronic warfare
EP	Electronic protection
ES	Electronic warfare support
FISINT	Foreign instrumentation signals intelligence
ICJ	International Court of Justice
ICRC	International Committee of the Red Cross
IMINT	Imagery intelligence
IO	Information operations
IW	Information warfare
LOAC	Law of armed conflict
LOCM	Law of conflict management
LOIC	Law of information conflict
OPSEC	Operations security
OSINT	Open-source intelligence
PA	Public affairs
PSYOPS	Psychological operations
ROE	Rules of engagement
SIGINT	Signals intelligence
SROE	Standing Rules of Engagement
WWW	World Wide Web

GLOSSARY

Aggressive use of force. Generally, aggressive refers to beginning a dispute by being the first to either threaten or use force. *See* BLACK'S LAW DICTIONARY, *supra* note 18, at 60. For the purposes of this book, an 'aggressive use of force' is a shorthand term used to refer to any use of force within the meaning of Article 2(4) that is not justified by a state's right of self-defense or authorized by the Security Council under its coercive Chapter VII powers.

Anticipatory self-defense. Authorizes the use of force in self-defense at that point in time before an attack occurs when "the necessity of that self-defence is instant, overwhelming and leaving no choice of means, and no moment for deliberation." *See* THE CHARTER OF THE UNITED NATIONS: A COMMENTARY, *supra* note 70, at 675. See "hostile intent," this glossary.

Armed attack. A use of force that rises to a certain scope, duration, and intensity threshold within the meaning of Article 51 of the Charter of the United Nations that invokes a state's inherent right of self-defense.

Armed conflict. A state of *de facto* hostilities invoked by the use of force by one party without any formal declaration of war. DOCUMENTS ON THE LAWS OF WAR, *supra* note 4, at 1-2. See "war," this glossary.

Armed force. Generally used to refer to military force involving the use of weapons, but is interpreted very broadly under international law to include non-military physical force, indirect force, and other forms of coercion that are tantamount in effect to an armed attack.

Civil affairs (CA). The activities of a commander that establish, maintain, influence, or exploit relations between military forces and civil authorities, both governmental and non-governmental, and the civilian populace in a friendly, neutral, or hostile area of operations in order to facilitate military operations and consolidate operational objectives. Civil affairs may include performance by military forces of activities and functions normally the responsibility of local government. These activities may occur prior to, during, or subsequent to other military actions. They may also occur, if directed, in the absence of other military operations. JOINT PUB 1-02, *supra* note 292, at 91.

Common Article 2. An article common to the four Geneva Conventions of 1949 that is widely accepted as the threshold test for when an international armed conflict exists, and consequently, when the law of armed conflict applies. The Common Article 2 threshold thus determines the transition point between peace and armed conflict.

Communications intelligence (COMINT). Technical and intelligence information derived from foreign communications by other than the intended recipients. JOINT PUB 1-02, *supra* note 292, at 109.

Computer espionage. The act through the medium of cyberspace of obtaining, delivering, transmitting, communicating, or receiving information about the national defense of a state with an intent, or reason to believe, that the information may be used to the injury of that state or to the advantage of any foreign nation.

Computer network attack (CNA). Operations that disrupt, deny, degrade, or destroy information resident in computers and computer networks or the computers and networks themselves. JOINT PUB 3-13, *supra* note 45, at I-9. *See also* JOINT PUB 1-02, *supra* note 292, at 112.

Covert operation. An operation that is so planned and executed as to conceal the identity of or permit plausible denial by the sponsor. A covert operation differs from a clandestine operation in that emphasis is placed on concealment of identity of sponsor rather than on concealment of the operation. JOINT PUB 1-02, *supra* note 292, at 129.

Cyberspace. The environment created by the confluence of cooperative networks of computers, information systems, and telecommunication infrastructures commonly referred to as the internet and the World Wide Web.

Dicta. Opinions of a judge that do not embody the resolution or determination of the court. Expressions in court's opinion that go beyond the facts before court and therefore are individual views of author of opinion and not binding in subsequent cases. BLACK'S LAW DICTIONARY, *supra* note 18, at 408.

Electronics intelligence (ELINT). Technical and geolocation intelligence derived from foreign non-communications electromagnetic radiations emanating from other than nuclear detonation or radioactive sources. JOINT PUB 1-02, *supra* note 292, at 168.

Electronic warfare (EW). Any military action involving the use of electromagnetic and directed energy to control the electromagnetic spectrum or to attack the enemy. The three major subdivisions within electronic warfare are: electronic attack, electronic protection, and electronic warfare support.

a. *Electronic attack (EA).* That division of electronic warfare involving the use of electromagnetic, directed energy, or antiradiation weapons to attack personnel, facilities, or equipment with the intent of degrading, neutralizing, or destroying enemy combat capability. EA includes:

   1) actions taken to prevent or reduce an enemy's effective use of the electromagnetic spectrum, such as jamming and electromagnetic deception, and

   2) employment of weapons that use either electromagnetic or directed energy as their primary destructive mechanism (lasers, radio frequency weapons, particle beams).

b. *Electronic protection (EP).* That division of electronic warfare involving actions taken to protect personnel, facilities, and equipment from any effects of friendly or enemy employment of electronic warfare that degrade, neutralize, or destroy friendly combat capability.

c. *Electronic warfare support (ES).* That division of electronic warfare involving actions tasked by, or under direct control of, an operational commander to search for, intercept, identify, and locate sources of intentional and unintentional radiated electromagnetic energy for the purpose of immediate threat recognition. Thus, electronic warfare support provides information required for immediate decisions involving electronic warfare operations and other tactical actions such as threat avoidance, targeting, and homing. Electronic warfare support data can be used to produce signals intelligence, both communications intelligence, and electronics intelligence.

JOINT PUB 1-02, *supra* note 292, at 168.

Espionage. The act of obtaining, delivering, transmitting, communicating, or receiving information about the national defense of a state with an intent, or reason to believe, that the information may be used to the injury of that state or to the advantage of any foreign nation. *See* JOINT PUB 1-02, *supra* note 292, at 174.

Foreign instrumentation signals intelligence (FISINT). Technical and intelligence information derived from the intercept of foreign electromagnetic emissions associated with the testing and operational deployment of non-US aerospace, surface, and subsurface systems such as telemetry, beaconry, electronic interrogators, and video data links. JOINT PUB 1-02, *supra* note 292, at 195.

Hostile act. An attack or other use of force by a foreign force or terrorist unit (organization or individual) against the United States, US forces, and in certain circumstances, US citizens, their property, US commercial assets, and other designated non-US forces, foreign nationals and their property. It is also force used directly to preclude or impede the mission and/or duties of US forces, including the recovery of US personnel and vital US Government property. When a hostile act is in progress, the right exists to use proportional force, including armed force, in self-defense by all necessary means available to deter or neutralize the potential attacker or, if necessary, to destroy the threat. CJCSI 3121.01, *supra* note 240, Encl. A, para. 5e.

Hostile intent. The threat of imminent use of force by a foreign force or terrorist unit (organization or individual) against the United States, US forces, and in certain circumstances, US citizens, their property, US commercial assets, or other designated non-US forces, foreign nationals and their property. When hostile intent is present, the right exists to use proportional force, including armed force, in self-defense by all necessary means available to deter or neutralize the potential attacker or, if necessary, to destroy the threat. CJCSI 3121.01, *supra* note 240, Encl. A, para. 5f.

Imagery intelligence (IMINT). Intelligence derived from the exploitation of collection by visual photography, infrared sensors, lasers, electro-optics, and radar sensors such as synthetic aperture radar wherein images of objects are reproduced optically or electronically on film, electronic display devices, or other media. JOINT PUB 1-02, *supra* note 292, at 227.

Information operations (IO). Actions taken to affect adversary information and information systems while defending one's own information and information systems. JOINT PUB 1-02, *supra* note 292, at 234.

Information warfare (IW). Information operations conducted during time of crisis or conflict to achieve or promote specific objectives over a specific adversary or adversaries. JOINT PUB 1-02, *supra* note 292, at 234.

Infra. Below, under, beneath, underneath. This word occurring by itself in a book refers the reader to a later part of the book. BLACK'S LAW DICTIONARY, *supra* note 18, at 701.

International armed conflict. An armed conflict between two or more sovereign states.

International Court of Justice (ICJ). The judicial arm of the United Nations. It has the jurisdiction to give advisory opinions on matters of law and treaty construction when requested by the General Assembly, Security Council or any other international agency so authorized by the General Assembly. It has jurisdiction, also, to settle legal disputes between nations when voluntarily submitted to it. Its judgments may be enforced by the Security Council. BLACK'S LAW DICTIONARY, *supra* note 18, at 732.

International law. The law that regulates the intercourse of nations; the law of nations. The customary law that determines the rights and regulates the intercourse of independent nations in peace and war. BLACK'S LAW DICTIONARY, *supra* note 18, at 733.

internet. A cooperatively run global network of computers with a common addressing scheme linked by telecommunication infrastructures.

Ipso facto. By the fact itself; by the mere fact. By the mere effect of an act or a fact. BLACK'S LAW DICTIONARY, *supra* note 18, at 743.

Jus. The whole body of law. BLACK'S LAW DICTIONARY, *supra* note 18, at 770.

Jus ad bellum. Latin for the law of conflict management.

Jus in bello. Latin for the law of armed conflict.

Law of armed conflict (LOAC). A set of rules under international law that govern the actual conduct of hostilities. Also generally referred to as the law of war. See "law of war," this glossary.

Law of conflict management (LOCM). A set of rules under international law that govern the use of force between states and determine whether a conflict is lawful or unlawful in its inception.

Law of information conflict (LOIC). The composite of the peacetime regime of international law, the law of conflict management, and the law of armed conflict that regulates the conduct of all state activities in cyberspace.

Law of War. This term comprises the body of rules and principles observed by civilized nations for the regulation of matters inherent in, or incidental to, the conduct of a public war; such, for example, as the relations of neutrals and belligerents, blockades, captures, prizes, truces and armistices, capitulations, prisoners, and declarations of war and peace. BLACK'S LAW DICTIONARY, *supra* note 18, at 1419-20. Also generally referred to as the law of armed conflict.

Line of belligerency. The point on the use of force spectrum at which international armed conflicts begin and the law of armed conflict applies. This point is triggered by the declaration of war, the occurrence of *de facto* hostilities, and all cases of partial or total occupation.

Military deception. Actions executed to deliberately mislead adversary military decision-makers as to friendly military capabilities, intentions, and operations, thereby causing the adversary to take specific actions (or inactions) that will contribute to the accomplishment of the friendly mission. The five categories of military deception are strategic military deception, operational military deception, tactical military deception, Service military deception, and military deception in support of operations security. JOINT PUB 1-02, *supra* note 292, at 298.

Open-source intelligence (OSINT). Information of potential intelligence value that is available to the general public. JOINT PUB 1-02, *supra* note 292, at 340.

Operations security (OPSEC). A process of identifying critical information and subsequently analyzing friendly actions attendant to military operations and other activities to:

a. Identify those actions that can be observed by adversary intelligence systems.

b. Determine indicators hostile intelligence systems might obtain that could be interpreted or pieced together to derive critical information in time to be useful to adversaries.

c. Select and execute measures that eliminate or reduce to an acceptable level the vulnerabilities of friendly actions to adversary exploitation.

JOINT PUB 1-02, *supra* note 292, at 343.

Prima facie. At first sight; on the face of it; a fact presumed to be true unless disproved by some evidence to the contrary. BLACK'S LAW DICTIONARY, *supra* note 18, at 1071.

Psychological operation (PSYOP). Planned operations to convey selected information and indicators to foreign audiences to influence their emotions, motives, objective reasoning, and ultimately the behavior of foreign governments, organizations, groups, and individuals. The purpose of psychological operations is to induce or reinforce foreign attitudes and behavior favorable to the originator's objectives. JOINT PUB 1-02, *supra* note 292, at 375.

Public affairs (PA). Those public information, command information, and community relations activities directed toward both the external and internal publics with interest in the Department of Defense. JOINT PUB 1-02, *supra* note 292, at 375.

Reprisal. An act that is unlawful *per se,* unless it can be justified as a countermeasure triggered by an unlawful act, and is designed to induce the offending state to return to full compliance with the law. The right of reprisal does not include the right to unlawfully use force contrary to Article 2(4) of the Charter. THE CHARTER OF THE UNITED NATIONS: A COMMENTARY, *supra* note 70, at 105.

Retortion. Generally, retortion refers to when a state unlawfully subjects citizens of another state to severe and stringent regulations or harsh treatment in response to similar unlawful mistreatment of its own citizens. *See* BLACK'S LAW DICTIONARY, *supra* note 18, at 1183.

Rules of engagement (ROE). Directives issued by competent military authority which delineate the circumstances and limitations under which United States forces will initiate and/or continue combat engagement with other forces encountered. JOINT PUB 1-02, *supra* note 292, at 403.

Self-defense. The protection of person or property against some injury attempted by another. [Generally, under US domestic law,] the law of "self-defense" justifies an act done in the reasonable belief of immediate danger, and, if an injury was done by defendant in justifiable self-defense, he can never be punished criminally nor held responsible for damages in a civil action. BLACK'S LAW DICTIONARY, *supra* note 18, at 1219-20.

Signals intelligence (SIGINT). Intelligence information comprising either individually or in combination all communications intelligence, electronics intelligence, and foreign instrumentation signals intelligence, however transmitted. JOINT PUB 1-02, *supra* note 292, at 422.

Standing Rules of Engagement for US Forces (SROE). The document approved by the US Secretary of Defense that provides implementation guidance on the inherent right and obligation of self-defense and the application of force for mission accomplishment. The SROE establish fundamental policies and procedures governing the actions to be taken by US force commanders during all military operations, contingencies, or prolonged conflicts. CJCSI 3121.01, *supra* note 240, Encl. A, para. 1a.

Stare decisis. To abide by, or adhere to, decided cases. Policy of courts to stand by precedent and not to disturb a settled point of law. BLACK'S LAW DICTIONARY, *supra* note 18, at 1261.

Supra. Above; upon. This word occurring by itself in a book refers the reader to a previous part of the book. BLACK'S LAW DICTIONARY, *supra* note 18, at 1291.

Threat of force. An express or implied promise by a government of a resort to force conditional on non-acceptance of certain demands of that government. BROWNLIE, *supra* note 91, at 364.

Use of force. A state activity that threatens the territorial integrity or political independence of another state within the meaning of Article 2(4) of the Charter of the United Nations.

War. A state of *de jure* hostilities invoked by a formal declaration of one party (DOCUMENTS ON THE LAWS OF WAR, *supra* note 4, at 1-2) that creates an international armed conflict as a matter of law (FM 27-10, *supra* note 156, § 8). A contest by force between two or more nations, carried on for any purpose, or armed conflict of sovereign powers or declared and open hostilities, or the state of nations among whom there is an interruption of pacific relations, and a general contention by force, authorized by the sovereign. BLACK'S LAW DICTIONARY, *supra* note 18, at 1419. See "armed conflict," this glossary.

World Wide Web (WWW). A collection of files and databases linked by hypertext that was created in 1989 as the primary platform of the internet that translates diverse computer protocols into standard format.

APPENDIX A

SOURCES OF THE LAW OF ARMED CONFLICT[†]

The general rule of international law is that states are only bound by those international laws to which they themselves have agreed to be bound.[820] This general principle is limited only to the extent that state practice develops into customary international law, which is binding on all states.[821]

As described in Chapter 5, the sources of the law of armed conflict may be found in Article 38(1) of the Vienna Convention. These four sources are international conventions, international custom, general principles of law, and judicial decisions and the teachings of the most highly qualified publicists.

Only the first two sources of law listed in Article 38(1), however, are actually sources of international law. General principles of law recognized by civilized nations are a source of law for the International Court of Justice to settle a specific dispute between two states, but these general principles are not customary international law binding on all states. Similarly, judicial decisions and the teachings of qualified publicists are not binding customary international law themselves, but "are considered to be indicative of the interpretation and application" of international law.[822] When an international treaty or a general

---

[†] This appendix is drawn substantially from work by Prof. Walter Gary Sharp, Sr., for Aegis Research Corporation.

[820] *See* THE CHARTER OF THE UNITED NATIONS: A COMMENTARY, *supra* note 70, 83 (Bruno Simma ed., 1994).

[821] *See id.*

[822] *See id.* at 992. International and national judicial bodies have played an important role in both the clarification and development of the law of armed conflict. *See* ROBERTS & GUELFF, *supra* note 4, at 6. In addition to the International Military Tribunals of Nuremberg and Tokyo, for example, the national and military courts of the occupying states prosecuted the overwhelming majority of those accused of war crimes during World War II. *Id.*

principle of customary international law does not resolve an issue under the law of armed conflict, legal advisers should consider general principles of law recognized by civilized nations, judicial decisions, and the teachings of qualified publicists as guidelines to determine applicable rules of law that govern the conduct of belligerents.

References for Customary International Law

Customary international law is inherently defined by state practice, undertaken out of a sense of legal obligation (*opinio juris*). Accordingly, such state practice must be examined to determine the customary international law that may be applicable in any given circumstance. References that compile and interpret customary international law are not a source of international law, but they are evidence of state practice. The most useful references provide examples of state practice and a corresponding governmental interpretation of the applicable treaties and customary international law, providing evidence of the presence or absence of *opinio juris*. The Office of the Legal Adviser for the US Department of State, for example, maintains an excellent collection of legal opinions and editorial comments entitled CUMULATIVE DIGEST OF UNITED STATES PRACTICE IN INTERNATIONAL LAW.[823] This authoritative compilation contains an extensive chapter on the legal regulation on the use of force that provides the official view of the United States in the context of specific uses of force by the United States and other states.[824]

The most concise and useful references for the law of armed conflict are military manuals. The US Department of the Navy has published THE COMMANDER'S HANDBOOK ON THE LAW OF NAVAL OPERATIONS.[825] This is an excellent manual that extracts from treaty and customary international law a

---

[823] *See, e.g.*, CUMULATIVE DIGEST, *supra* note 162.

[824] *See, e.g.*, *Id.* at 3388-3659. In addition to the law of armed conflict, this chapter addresses issues that include the resort to war and armed force, civil war, neutrality and nonbelligerency, United Nations and regional peacekeeping, military sanctions, arms control and disarmament, war and emergency powers, and military sales. *Id.*

[825] *See generally* COMMANDER'S HANDBOOK, *supra* note 109.

superb distillation of the law of armed conflict.[826] Another excellent reference, albeit dated, which distills customary and treaty law into an operational manual is the US Department of the Army's field manual entitled THE LAW OF LAND WARFARE.[827] There is not an official US manual for air warfare, but the definitive work in this area that is widely used within the Department of Defense is published in the US Air Force law journal.[828]

Evidence of customary law of armed conflict can also be found in draft treaty texts that were not adopted and other documents that are universally considered to reflect customary international law. A number of these documents can be found in the treaty compilations discussed below. For example, the 1923 Hague draft Rules of Aerial Warfare[829] were never adopted in legally binding form, but they are regarded as an authoritative application of the general principles of the law of armed conflict to aerial warfare and bombardment.[830] Similarly, extracts from the 1946 Judgment of the International Military Tribunal[831] at Nuremberg which are considered to reflect customary international law[832] are often reprinted, as well as other documents such as the International

---

826 *See generally id.*

827 *See generally* FM 27-10, *supra* note 154.

828 *See* W. Hays Parks, *Air War and the Law of War*, 32 A.F. L. REV. 1, 1-225 (1990).

829 Rules of Aerial Warfare, *supra* note 534, *reprinted in* ROBERTS & GUELFF, *supra* note 4, at 123.

830 *See* ROBERTS & GUELFF, *supra* note 4, at 121-22.

831 *See, e.g.*, Trial of the Major War Criminals before the International Military Tribunal, Nuremberg, 1948, vol. XXII, IMT Secretariat, Nuremberg, excerpts *reprinted in* ROBERTS & GUELFF, *supra* note 4, at 155.

832 *See* ROBERTS & GUELFF, *supra* note 4, at 153-54.

Law Commission's summary of international law principles recognized by the Charter and Judgment of the Nuremberg Tribunal.[833]

References for Treaty Texts

Treaties are binding on their respective parties, and, to the extent they reflect customary international law, are binding on all states. DOCUMENTS ON THE LAWS OF WAR[834] is one of the most authoritative and useful treaty compilations of the law of armed conflict. The editors, Professor Adam Roberts and Professor Richard Guelff, provide a concise but rich history that provides a situational context for each of the twenty-nine documents they include in their book.[835] This collection is well-indexed, and is a superb operational resource for law of armed conflict treaties in force and a detailed list of their respective state parties as of approximately January 1989.[836] It also contains other documents, which even though they are not formal treaties, are authoritative accounts of ustomary international.[837]

The most comprehensive treaty compilation in one volume is THE LAWS OF ARMED CONFLICTS,[838] edited by Professor Dietrich Schindler and Dr. Jiri Toman. This compilation is an outstanding resource, reproducing the texts of eighty-seven treaties, draft treaties, and resolutions on the law of armed conflict dating back to the first law of armed conflict treaty in 1856.[839] Each of these

---

833 Principles of International Law Recognized in the Charter of the Nuremberg Tribunal and in the Judgment of the Tribunal [1950] 2 Y.B. Int'l L. Comm'n 374, *reprinted in* SCHINDLER & TOMAN, ARMED CONFLICTS, *supra* note 5, at 923.

834 *See generally* ROBERTS & GUELFF, *supra* note 4.

835 *See generally id.*

836 *See generally* ROBERTS & GUELFF, *supra* note 4.

837 *See generally id.*

838 SCHINDLER & TOMAN, ARMED CONFLICTS, *supra* note 5.

839 *Id.* at vii, xxiii.

documents is preceded by an excellent introductory note that provides the historical context of the document. This text is also well-indexed, and contains a detailed list of the state parties to each treaty as of approximately December 1987.

Finally, the International Committee of the Red Cross[840] has a free online database of the law of armed conflict, which it refers to as international humanitarian law.[841] This compilation is a user-friendly, searchable database that contains the texts and analyses of ninety-one treaties and other documents, as well as an up-to-date list of signatures, ratifications, and the full text of all reservations.[842] The ICRC also has its database of international humanitarian law available on CD ROM, with all the data updated as of December 31, 1998.[843] The ICRC also publishes an INDEX OF INTERNATIONAL HUMANITARIAN LAW.[844] This index is an extremely useful tool for researching

---

[840] The mission statement of the ICRC is found on its homepage:

The International Committee of the Red Cross (ICRC) is an impartial, neutral and independent organization whose exclusively humanitarian mission is to protect the lives and dignity of victims of war and internal violence and to provide them with assistance. It directs and coordinates the international relief activities conducted by the Movement in situations of conflict. It also endeavours to prevent suffering by promoting and strengthening humanitarian law and universal humanitarian principles. Established in 1863, the ICRC is at the origin of the International Red Cross and Red Crescent Movement.

The International Committee of the Red Cross Homepage (visited Mar. 26, 2000) <http://www.icrc.org>.

[841] *See id.*

[842] *See id.*

[843] *See id.*

[844] INDEX OF INTERNATIONAL HUMANITARIAN LAW (Waldemar A. Solf & J. Ashley Roach eds., 1987).

the Hague Conventions of 1907, the four Geneva Conventions of 1949, and the two Protocols of 1977 to the four Geneva Conventions of 1949.

APPENDIX B

THE REISMAN-BAKER GUIDELINES FOR COVERT OPERATIONS

1. Try, as far as possible, to accomplish overtly what subordinates propose to do covertly.[845]

2. Never say that you will never do things covertly.[846]

---

[845] W. MICHAEL REISMAN and JAMES E. BAKER, REGULATING COVERT ACTION: PRACTICES, CONTEXTS, AND POLICIES OF COVERT COERCION ABROAD IN INTERNATIONAL AND AMERICAN LAW 140 (1992). Reisman and Baker continue:

> This recommendation is subject to a broad margin of appreciation, for the property of being "covert" is slippery and can be achieved in many ways, for example, a very "general" statement of plans to an oversight agency may be claimed after the fact as full revelation despite the fact that at the time its effect was to shroud in secrecy whatever was on the drawing boards. By "overt," here, we mean overt, first, in terms of disclosure to the appropriate oversight agency; second, in terms of disclosure to the international community; third, in terms of disclosure to the target. The *specificity* of disclosure will necessarily be a function of many variables and will involve democratic considerations, efficiency considerations, and international political considerations.

Id.

[846] *Id.* Reisman and Baker explain:

> It may be very tempting to take the "moral high ground" and favor diplomacy, mutual understanding, accommodation, and economic sanctions. But there may come a time when reasonableness will have been exhausted and philosophy and poetry manifestly will not prevail over a grave evil or danger. Poverty may be at the root of all political instability. But there are moments of crisis calling for resolute action when talk of poverty as the root of evil and the need for economic and developmental assistance is escapism and may border on madness. At certain levels and for certain phases of operations, secrecy will continue to be required and will be a part of governmental operations, in the foreign policy arena. When secrecy is exposed after public promises of "never,"

3. Do not confuse covert actions or phases of actions with coercive actions or phases of actions.[847]

4. But all executives should eschew secrecy for its own sake.[848]

5. An act accomplished covertly should be overtly lawful.[849]

> the net effect will be to damage public confidence in governmental operations, in general, and the credibility of the incumbents in particular. It is best to acknowledge beforehand that this is one of the lawful tools at the disposal of the executive, that it has been authorized by legislation, that appropriate procedures have been prescribed, but that because of the special costs of its use in a democracy, it will be resorted to no more than necessary. In short, the responsible executive official should be on record that there are instances when its use may be lawful and appropriate and that not to use it then would be wrong.

Id. at 140-41.

[847] *Id.* at 141. Reisman and Baker elaborate:

> Though some may find some philosophical cogency or rhetorical mileage in arguments that concealing anything from another, under any circumstances, constitutes an act of violence against that other, we think it is more sensible to assume that concealing one's motives, preparatory actions, or implementation from someone else, although not necessarily nice or ethically optimal and hence not generally desirable, nevertheless is not the same as an act of coercion against another and that its lawfulness or morality may turn on many contextual factors.

Id.

[848] *Id.* Reisman and Baker amplify:

> The political advantages of secrecy are obvious. It enhances the power of the party using it, and where power sharing is called for, secrecy circumvents it. The costs of secrecy in a democracy should rule it out in all but those cases in which other compelling democratic values would be lost if secrecy were not used.

Id.

6. In determining what is lawful under international law, it is important to use contextual and consequential methods of inquiry rather than methods of textual and logical derivation.[850]

---

[849] *Id.* Reisman and Baker continue:

> Assuming that secrecy is not per se unlawful (in some legislative contexts, it may be), the action or phase of action that is being accomplished secretly should itself be lawful. In other words, secrecy should not be used to evade the requirements of law or to conceal unlawful actions. Covert actions and, in particular, interstate collaborative covert actions should be tested by international standards as if they were actions to enforce international obligations. The outcome of that inquiry should be a critical factor in deciding to proceed with the action. But the test should be conducted honestly and not evaded by the use euphemisms. Earlier we indicated why we think state-sponsored assassination is bad policy, though we acknowledge that tyrannicide has been a morally difficult issue through the ages. But, apart from that, any operation may lead to the deaths of people. In any instance where a politically motivated killing is a possible or probable outcome of United States action or inaction, we would argue that democratic values oblige decision makers to express policy preferences with clarity and specificity, not with such obfuscating terms as *personal defense weapons, remove,* and *neutralize,* terms that appeared during the Trujillo and Chilean episodes and that seem to recur with some regularity in secret government documents [footnote omitted]. The use of such terms subverts personal moral examination and critical political review.

Id.

[850] *Id.* Reisman and Baker expand:

> Two points are noteworthy. First, international law is structurally different from developed domestic systems and perforce uses a different method for assessing lawfulness. Hence the consequentialist mode, based on a thorough contextual examination, rather than a textual and rule-oriented approach should be employed. Second, not everything that is permitted in international law should be automatically deemed permissible in particular cases. One should not pretend that the defects of international law are a virtue. The low-level of regulation in many strategic modes is lamentable because it does not serve world order. A state actor that refrains from some theretofore licit practices may contribute, by its own abstinence, to the formation and installation of a more appropriate norm.
>
> The implementation of this recommendation requires that lawyers who have the necessary background, but who are not in the direct chain of

7. [S]tructure internally an examination process as vigorous and critical as would be the public debate were it to occur. In particular, if you think circumstances require departure from national law and procedure, carefully review your motives to be sure that irrational factors are not operating.[851]

8. When the educative or informative function of a strategy is important, the ultimate implementation of the strategy should not be covert.[852]

> command have an opportunity to submit their written views, which become part of the record.

Id. at 141-42.

[851] *Id.* at 142. Reisman and Baker preface this:

> One of the costs of secrecy is that a vigorous and open examination of reasons for, and likelihood of success of, an operation will not take place. Instead, a process characterized by what Janis has called "groupthink" may operate [footnote omitted]. Hence the need in a non-public decision-making process contemplating covert action to [follow guideline number seven].

Id.

[852] *Id.* Reisman and Baker dilate:

> This recommendation is driven, obviously, by efficiency considerations. Many particularly coercive strategies are designed to inform and deter and not simply to inflict some deprivation or punishment on the target. If adequate information about the message, the messenger, and the sender is not available, the real purpose of the strategic operation is frustrated. Equally obviously, the overt quality of the operation may make it more costly to the initiating party.
>
> In discussion with other scholars in the initial presentation of our views in the Institute of Peace, this guideline raised many questions. Secrecy and "untraceability" may be useful when the object is to destroy a target after persuasive means have failed. The elimination of a target such as the reported chemical weapons factory at Rabta is not an informative or educative project. Since its purpose is destruction, untraceability may be desirable, especially because the target government will predictably try to avenge itself, perhaps by terrorism. But when the United States, after protests about mining and targeting of shipping in the Persian Gulf, decided to punish Iran by destroying general oil installations by covert destruction that could have been interpreted as an

9. We strongly endorse the refinement of methods of congressional oversight, though we would think that the oversight, for reasons of both executive and congressional efficiency, should be retrospective and judgmental rather than prospective and executive.[853]

> accident would have been far less educational than an overt operation in which the identity of the actors, the demonstration of their capacity, and the reasons were unmistakable. In addition, overt action could incorporate warnings to personnel in the targeted facilities and minimize the loss of life.

Id. at 142-43.

[853] *Id.* at 143. Reisman and Baker explain:

> There is a tendency to view executive branches as monarchical vestiges. As a wide range of organizational phenomena indicate, specialized agencies for policy implementation are required in all but the simplest organizations. Hence the recurring development of separate institutions designed to make (including to appraise the fulfillment of policy objectives) and implement policy.
>
> Covert actions, in our view, require an executive. If there is a national decision to have a capacity for such action, it should not be frustrated by an inefficient system of implementation. But democracies also require that executive actions be subjected to oversight and appraisal. If those charged with these functions have also been involved in the actions, they are unlikely to appraise themselves very vigorously. Hence, in the interest of the national community, we recommend a clean division and specialization of function in the various successive oversight regimes we anticipate.

Id.

10. In contemplating any covert operation, whether from the standpoint of the operator or those providing the oversight, assume that it will become public knowledge much sooner than you would like and decide if you can live with the consequences before you discover you have to.[854]

[854] *Id.* Reisman and Baker conclude:

> The illusion of secrecy can provide a false and treacherous sense of security. A game-theory proposition may be a useful *vade mecum* for the political operator: always assume your adversary knows everything you are doing so that when he does find out, you will already have discounted the cost.

Id.

APPENDIX C

THE CHARTER OF THE UNITED NATIONS[855]

Table of Contents

[855] 26 June 1945, 59 Stat. 1031, T.S. No. 993, 3 Bevans 1153; as amended [arts. 23, 27, and 61] on 17 December 1963, 16 U.S.T. 1134, T.I.A.S. 5857, 557 U.N.T.S. 143; as amended [art. 109] on 20 December 1965, 19 U.S.T. 5450, T.I.A.S. 652; as amended [art. 61] on 20 December 1971, 24 U.S.T. 2225, T.I.A.S. 7739. The author gratefully acknowledges the availability of the electronic version of this document located at www.tufts.edu/fletcher/multilaterals.html which was provided by the Multilaterals Project of Tufts University's Fletcher School of Law and Diplomacy. Chapters included in this annex appear in bold face.

PREAMBLE

WE THE PEOPLES
OF THE UNITED NATIONS
DETERMINED

to save succeeding generations from the scourge of war, which twice in our lifetime has brought untold sorrow to mankind, and

to reaffirm faith in fundamental human rights, in the dignity and worth of the human person, in the equal rights of men and women and of nations large and small, and

to establish conditions under which justice and respect for the obligations arising from treaties and other sources of international law can be maintained, and

to promote social progress and better standards of life in larger freedom,

AND FOR THESE ENDS

to practice tolerance and live together in peace with one another as good neighbours, and

to unite our strength to maintain international peace and security, and

to ensure, by the acceptance of principles and the institution of methods, that armed force shall not be used, save in the common interest, and

to employ international machinery for the promotion of the economic and social advancement of all peoples,

HAVE RESOLVED TO COMBINE OUR EFFORTS TO ACCOMPLISH THESE AIMS

Accordingly, our respective Governments, through representatives assembled in the city of San Francisco, who have exhibited their full powers found to be in good and due form, have agreed to the present Charter of the United Nations and do hereby establish an international organization to be known as the United Nations.

CHAPTER I

PURPOSES AND PRINCIPLES

Article I

The Purposes of the United Nations are:

1. To maintain international peace and security, and to that end: to take effective collective measures for the prevention and removal of threats to the peace, and for the suppression of acts of aggression or other breaches of the peace, and to bring about by peaceful means, and in conformity with the principles of justice and international law, adjustment or settlement of international disputes or situations which might lead to a breach of the peace;

2. To develop friendly relations among nations based on respect for the principle of equal rights and self-determination of peoples, and to take other appropriate measures to strengthen universal peace;

3. To achieve international co-operation in solving international problems of an economic, social, cultural, or humanitarian character, and in promoting and encouraging respect for human rights and for fundamental freedoms for all without distinction as to race, sex, language, or religion; and

4. To be a centre for harmonizing the actions of nations in the attainment of these common ends.

Article 2

The Organization and its Members, in pursuit of the Purposes stated in Article I, shall act in accordance with the following Principles:

1. The Organization is based on the principle of the sovereign equality of all its Members.

2. All Members, in order to ensure to all of them the rights and benefits resulting from membership, shall fulfil in good faith the obligations assumed by them in accordance with the present Charter.

3. All Members shall settle their international disputes by peaceful means in such a manner that international peace and security, and justice, are not endangered.

4. All Members shall refrain in their international relations from the threat or use of force against the territorial integrity or political independence of any state, or in any other manner inconsistent with the Purposes of the United Nations.

5. All Members shall give the United Nations every assistance in any action it takes in accordance with the present Charter, and shall refrain from giving assistance to any state against which the United Nations is taking preventive or enforcement action.

6. The Organization shall ensure that states which are not Members of the United Nations act in accordance with these Principles so far as may be necessary for the maintenance of international peace and security.

7. Nothing contained in the present Charter shall authorize the United Nations to intervene in matters which are essentially within the domestic jurisdiction of any state or shall require the Members to submit such matters to settlement under the present Charter; but this principle shall not prejudice the application of enforcement measures under Chapter VII.

CHAPTER II

MEMBERSHIP

Article 3

The original Members of the United Nations shall be the states which, having participated in the United Nations Conference on International Organization at San Francisco, or having previously signed the Declaration by United Nations of 1 January 1942, sign the present Charter and ratify it in accordance with Article 110.

Article 4

1. Membership in the United Nations is open to all other peace-loving states which accept the obligations contained in the present Charter and, in the judgment of the Organization, are able and willing to carry out these obligations.

2. The admission of any such state to membership in the United Nations will be effected by a decision of the General Assembly upon the recommendation of the Security Council.

Article 5

A Member of the United Nations against which preventive or enforcement action has been taken by the Security Council may be suspended from the exercise of the rights and privileges of membership by the General Assembly upon the recommendation of the Security Council. The exercise of these rights and privileges may be restored by the Security Council.

Article 6

A Member of the United Nations which has persistently violated the Principles contained in the present Charter may be expelled from the Organization by the General Assembly upon the recommendation of the Security Council.

CHAPTER III

ORGANS

Article 7

1. There are established as the principal organs of the United Nations: a General Assembly, a Security Council, an Economic and Social Council, a Trusteeship Council, an International Court of Justice, and a Secretariat.

2. Such subsidiary organs as may be found necessary may be established in accordance with the present Charter.

Article 8

The United Nations shall place no restrictions on the eligibility of men and women to participate in any capacity and under conditions of equality in its principal and subsidiary organs.

CHAPTER IV

THE GENERAL ASSEMBLY

COMPOSITION

Article 9

1. The General Assembly shall consist of all the Members of the United Nations.

2. Each Member shall have not more than five representatives in the General Assembly.

FUNCTIONS AND POWERS

Article 10

The General Assembly may discuss any questions or any matters within the scope of the present Charter or relating to the powers and functions of any organs provided for in the present Charter, and, except as provided in Article 12, may make recommendations to the Members of the United Nations or to the Security Council or to both on any such questions or matters.

Article 11

1. The General Assembly may consider the general principles of co-operation in the maintenance of international peace and security, including the principles governing disarmament and the regulation of armaments, and may make recommendations with regard to such principles to the Members or to the Security Council or to both.

2. The General Assembly may discuss any questions relating to the maintenance of international peace and security brought before it by any Member of the United Nations, or by the Security Council, or by a state which is not a Member of the United Nations in accordance with Article 35, paragraph 2, and, except as provided in Article 12, may make recommendations with regard to any such questions to the state or states concerned or to the Security Council or to both. Any such question on which action is necessary shall be referred to the Security Council by the General Assembly either before or after discussion.

3. The General Assembly may call the attention of the Security Council to situations which are likely to endanger international peace and security.

4. The powers of the General Assembly set forth in this Article shall not limit the general scope of Article 10.

Article 12

1. While the Security Council is exercising in respect of any dispute or situation the functions assigned to it in the present Charter, the General Assembly shall not make any recommendation with regard to that dispute or situation unless the Security Council so requests.

2. The Secretary-General, with the consent of the Security Council, shall notify the General Assembly at each session of any matters relative to the maintenance of international peace and security which are being dealt with by the Security Council and shall similarly notify the General Assembly, or the Members of the United Nations if the General Assembly is not in session, immediately the Security Council ceases to deal with such matters.

Article 13

1. The General Assembly shall initiate studies and make recommendations for the purpose of:

a. promoting international co-operation in the political field and encouraging the progressive development of international law and its codification;

b. promoting international co-operation in the economic, social, cultural, educational, and health fields, and assisting in the realization of human rights and fundamental freedoms for all without distinction as to race, sex, language, or religion.

2. The further responsibilities, functions and powers of the General Assembly with respect to matters mentioned in paragraph 1 (b) above are set forth in Chapters IX and X.

Article 14

Subject to the provisions of Article 12, the General Assembly may recommend measures for the peaceful adjustment of any situation, regardless of origin, which it deems likely to impair the general welfare or friendly relations among nations, including situations resulting from a violation of the provisions of the present Charter setting forth the Purposes and Principles of the United Nations.

Article 15

1. The General Assembly shall receive and consider annual and special reports from the Security Council; these reports shall include an account of the measures that the Security Council has decided upon or taken to maintain international peace and security.

2. The General Assembly shall receive and consider reports from the other organs of the United Nations.

Article 16

The General Assembly shall perform such functions with respect to the international trusteeship system as are assigned to it under Chapters XII and XIII, including the approval of the trusteeship agreements for areas not designated as strategic.

Article 17

1. The General Assembly shall consider and approve the budget of the Organization.

2. The expenses of the Organization shall be borne by the Members as apportioned by the General Assembly.

3. The General Assembly shall consider and approve any financial and budgetary arrangements with specialized agencies referred to in Article 57 and shall examine the administrative budgets of such specialized agencies with a view to making recommendations to the agencies concerned.

VOTING

Article 18

1. Each member of the General Assembly shall have one vote.

2. Decisions of the General Assembly on important questions shall be made by a two-thirds majority of the members present and voting. These questions shall include: recommendations with respect to the maintenance of international peace and security, the election of the non-permanent members of the Security Council, the election of the members of the Economic and Social Council, the election of members of the Trusteeship Council in accordance with paragraph 1 (c) of Article 86, the admission of new Members to the United Nations, the

suspension of the rights and privileges of membership, the expulsion of Members, questions relating to the operation of the trusteeship system, and budgetary questions.

3. Decisions on other questions, including the determination of additional categories of questions to be decided by a two-thirds majority, shall be made by a majority of the members present and voting.

Article 19

A Member of the United Nations which is in arrears in the payment of its financial contributions to the Organization shall have no vote in the General Assembly if the amount of its arrears equals or exceeds the amount of the contributions due from it for the preceding two full years. The General Assembly may, nevertheless, permit such a Member to vote if it is satisfied that the failure to pay is due to conditions beyond the control of the Member.

PROCEDURE

Article 20

The General Assembly shall meet in regular annual sessions and in such special sessions as occasion may require. Special sessions shall be convoked by the Secretary-General at the request of the Security Council or of a majority of the Members of the United Nations.

Article 21

The General Assembly shall adopt its own rules of procedure. It shall elect its President for each session.

Article 22

The General Assembly may establish such subsidiary organs as it deems necessary for the performance of its functions.

CHAPTER V

THE SECURITY COUNCIL

COMPOSITION

Article 23

1. The Security Council shall consist of fifteen Members of the United Nations. The Republic of China, France, the Union of Soviet Socialist Republics, the United Kingdom of Great Britain and Northern Ireland, and the United States of America shall be permanent members of the Security Council. The General Assembly shall elect ten other Members of the United Nations to be non-permanent members of the Security Council, due regard being specially paid, in the first instance to the contribution of Members of the United Nations to the maintenance of international peace and security and to the other purposes of the Organization, and also to equitable geographical distribution.

2. The non-permanent members of the Security Council shall be elected for a term of two years. In the first election of the non-permanent members after the increase of the membership of the Security Council from eleven to fifteen, two of the four additional members shall be chosen for a term of one year. A retiring member shall not be eligible for immediate re-election.

3. Each member of the Security Council shall have one representative.

FUNCTIONS AND POWERS

Article 24

1. In order to ensure prompt and effective action by the United Nations, its Members confer on the Security Council primary responsibility for the maintenance of international peace and security, and agree that in carrying out its duties under this responsibility the Security Council acts on their behalf.

2. In discharging these duties the Security Council shall act in accordance with the Purposes and Principles of the United Nations. The specific powers granted to the Security Council for the discharge of these duties are laid down in Chapters VI, VII, VIII, and XII.

3. The Security Council shall submit annual and, when necessary, special reports to the General Assembly for its consideration.

Article 25

The Members of the United Nations agree to accept and carry out the decisions of the Security Council in accordance with the present Charter.

Article 26

In order to promote the establishment and maintenance of international peace and security with the least diversion for armaments of the world's human and economic resources, the Security Council shall be responsible for formulating, with the assistance of the Military Staff Committee referred to in Article 47, plans to be submitted to the Members of the United Nations for the establishment of a system for the regulation of armaments.

VOTING

Article 27

1. Each member of the Security Council shall have one vote.

2. Decisions of the Security Council on procedural matters shall be made by an affirmative vote of nine members.

3. Decisions of the Security Council on all other matters shall be made by an affirmative vote of nine members including the concurring votes of the permanent members; provided that, in decisions under Chapter VI, and under paragraph 3 of Article 52, a party to a dispute shall abstain from voting.

PROCEDURE

Article 28

1. The Security Council shall be so organized as to be able to function continuously. Each member of the Security Council shall for this purpose be represented at all times at the seat of the Organization.

2. The Security Council shall hold periodic meetings at which each of its members may, if it so desires, be represented by a member of the government or by some other specially designated representative.

3. The Security Council may hold meetings at such places other than the seat of the Organization as in its judgment will best facilitate its work.

Article 29

The Security Council may establish such subsidiary organs as it deems necessary for the performance of its functions.

Article 30

The Security Council shall adopt its own rules of procedure, including the method of selecting its President.

Article 31

Any Member of the United Nations which is not a member of the Security Council may participate, without vote, in the discussion of any question brought before the Security Council whenever the latter considers that the interests of that Member are specially affected.

Article 32

Any Member of the United Nations which is not a member of the Security Council or any state which is not a Member of the United Nations, if it is a party to a dispute under consideration by the Security Council, shall be invited to participate, without vote, in the discussion relating to the dispute. The Security Council shall lay down such conditions as it deems just for the participation of a state which is not a Member of the United Nations.

CHAPTER VI

PACIFIC SETTLEMENT OF DISPUTES

Article 33

1. The parties to any dispute, the continuance of which is likely to endanger the maintenance of international peace and security, shall, first of all, seek a solution by negotiation, enquiry, mediation, conciliation, arbitration, judicial settlement, resort to regional agencies or arrangements, or other peaceful means of their own choice.

2. The Security Council shall, when it deems necessary, call upon the parties to settle their dispute by such means.

Article 34

The Security Council may investigate any dispute, or any situation which might lead to international friction or give rise to a dispute, in order to determine whether the continuance of the dispute or situation is likely to endanger the maintenance of international peace and security.

Article 35

1. Any Member of the United Nations may bring any dispute, or any situation of the nature referred to in Article 34, to the attention of the Security Council or of the General Assembly.

2. A state which is not a Member of the United Nations may bring to the attention of the Security Council or of the General Assembly any dispute to which it is a party if it accepts in advance, for the purposes of the dispute, the obligations of pacific settlement provided in the present Charter.

3. The proceedings of the General Assembly in respect of matters brought to its attention under this Article will be subject to the provisions of Articles 11 and 12.

Article 36

1. The Security Council may, at any stage of a dispute of the nature referred to in Article 33 or of a situation of like nature, recommend appropriate procedures or methods of adjustment.

2. The Security Council should take into consideration any procedures for the settlement of the dispute which have already been adopted by the parties.

3. In making recommendations under this Article the Security Council should also take into consideration that legal disputes should as a general rule be referred by the parties to the International Court of Justice in accordance with the provisions of the Statute of the Court.

Article 37

1. Should the parties to a dispute of the nature referred to in Article 33 fail to settle it by the means indicated in that Article, they shall refer it to the Security Council.

2. If the Security Council deems that the continuance of the dispute is in fact likely to endanger the maintenance of international peace and security, it shall

decide whether to take action under Article 36 or to recommend such terms of settlement as it may consider appropriate.

Article 38

Without prejudice to the provisions of Articles 33 to 37, the Security Council may, if all the parties to any dispute so request, make recommendations to the parties with a view to a pacific settlement of the dispute.

CHAPTER VII

ACTION WITH RESPECT TO THREATS TO THE PEACE, BREACHES OF THE PEACE, AND ACTS OF AGGRESSION

Article 39

The Security Council shall determine the existence of any threat to the peace, breach of the peace, or act of aggression and shall make recommendations, or decide what measures shall be taken in accordance with Articles 41 and 42, to maintain or restore international peace and security.

Article 40

In order to prevent an aggravation of the situation, the Security Council may, before making the recommendations or deciding upon the measures provided for in Article 39, call upon the parties concerned to comply with such provisional measures as it deems necessary or desirable. Such provisional measures shall be without prejudice to the rights, claims, or position of the parties concerned. The Security Council shall duly take account of failure to comply with such provisional measures.

Article 41

The Security Council may decide what measures not involving the use of armed force are to be employed to give effect to its decisions, and it may call upon the Members of the United Nations to apply such measures. These may include complete or partial interruption of economic relations and of rail, sea, air, postal, telegraphic, radio, and other means of communication, and the severance of diplomatic relations.

Article 42

Should the Security Council consider that measures provided for in Article 41 would be inadequate or have proved to be inadequate, it may take such action by air, sea, or land forces as may be necessary to maintain or restore international peace and security. Such action may include demonstrations, blockade, and other operations by air, sea, or land forces of Members of the United Nations.

Article 43

1. All Members of the United Nations, in order to contribute to the maintenance of international peace and security, undertake to make available to the Security Council, on its call and in accordance with a special agreement or agreements, armed forces, assistance, and facilities, including rights of passage, necessary for the purpose of maintaining international peace and security.

2. Such agreement or agreements shall govern the numbers and types of forces, their degree of readiness and general location, and the nature of the facilities and assistance to be provided.

3. The agreement or agreements shall be negotiated as soon as possible on the initiative of the Security Council. They shall be concluded between the Security Council and Members or between the Security Council and groups of Members and shall be subject to ratification by the signatory states in accordance with their respective constitutional processes.

Article 44

When the Security Council has decided to use force it shall, before calling upon a Member not represented on it to provide armed forces in fulfilment of the obligations assumed under Article 43, invite that Member, if the Member so desires, to participate in the decisions of the Security Council concerning the employment of contingents of that Member's armed forces.

Article 45

In order to enable the United Nations to take urgent military measures, Members shall hold immediately available national air-force contingents for combined international enforcement action. The strength and degree of readiness of these contingents and plans for their combined action shall be determined within the limits laid down in the special agreement or agreements referred to in Article 43, by the Security Council with the assistance of the Military Staff Committee.

Article 46

Plans for the application of armed force shall be made by the Security Council with the assistance of the Military Staff Committee.

Article 47

1. There shall be established a Military Staff Committee to advise and assist the Security Council on all questions relating to the Security Council's military requirements for the maintenance of international peace and security, the employment and command of forces placed at its disposal, the regulation of armaments, and possible disarmament.

2. The Military Staff Committee shall consist of the Chiefs of Staff of the permanent members of the Security Council or their representatives. Any Member of the United Nations not permanently represented on the Committee shall be invited by the Committee to be associated with it when the efficient discharge of the Committee's responsibilities requires the participation of that Member in its work.

3. The Military Staff Committee shall be responsible under the Security Council for the strategic direction of any armed forces placed at the disposal of the Security Council. Questions relating to the command of such forces shall be worked out subsequently.

4. The Military Staff Committee, with the authorization of the Security Council and after consultation with appropriate regional agencies, may establish regional subcommittees.

Article 48

1. The action required to carry out the decisions of the Security Council for the maintenance of international peace and security shall be taken by all the Members of the United Nations or by some of them, as the Security Council may determine.

2. Such decisions shall be carried out by the Members of the United Nations directly and through their action in the appropriate international agencies of which they are members.

Article 49

The Members of the United Nations shall join in affording mutual assistance in carrying out the measures decided upon by the Security Council.

Article 50

If preventive or enforcement measures against any state are taken by the Security Council, any other state, whether a Member of the United Nations or not, which finds itself confronted with special economic problems arising from the carrying out of those measures shall have the right to consult the Security Council with regard to a solution of those problems.

Article 51

Nothing in the present Charter shall impair the inherent right of individual or collective self-defence if an armed attack occurs against a Member of the United Nations, until the Security Council has taken measures necessary to maintain international peace and security. Measures taken by Members in the exercise of this right of self-defence shall be immediately reported to the Security Council and shall not in any way affect the authority and responsibility of the Security Council under the present Charter to take at any time such action as it deems necessary in order to maintain or restore international peace and security.

CHAPTER VIII

REGIONAL ARRANGEMENTS

Article 52

1. Nothing in the present Charter precludes the existence of regional arrangements or agencies for dealing with such matters relating to the maintenance of international peace and security as are appropriate for regional action provided that such arrangements or agencies and their activities are consistent with the Purposes and Principles of the United Nations.

2. The Members of the United Nations entering into such arrangements or constituting such agencies shall make every effort to achieve pacific settlement of local disputes through such regional arrangements or by such regional agencies before referring them to the Security Council.

3. The Security Council shall encourage the development of pacific settlement of local disputes through such regional arrangements or by such regional agencies either on the initiative of the states concerned or by reference from the Security Council.

4. This Article in no way impairs the application of Articles 34 and 35.

Article 53

1. The Security Council shall, where appropriate, utilize such regional arrangements or agencies for enforcement action under its authority. But no enforcement action shall be taken under regional arrangements or by regional agencies without the authorization of the Security Council, with the exception of measures against any enemy state, as defined in paragraph 2 of this Article, provided for pursuant to Article 107 or in regional arrangements directed against renewal of aggressive policy on the part of any such state, until such time as the Organization may, on request of the Governments concerned, be charged with the responsibility for preventing further aggression by such a state.

2. The term enemy state as used in paragraph 1 of this Article applies to any state which during the Second World War has been an enemy of any signatory of the present Charter.

Article 54

The Security Council shall at all times be kept fully informed of activities undertaken or in contemplation under regional arrangements or by regional agencies for the maintenance of international peace and security.

CHAPTER XVI

MISCELLANEOUS PROVISIONS

Article 102

1. Every treaty and every international agreement entered into by any Member of the United Nations after the present Charter comes into force shall as soon as possible be registered with the Secretariat and published by it.

2. No party to any such treaty or international agreement which has not been registered in accordance with the provisions of paragraph 1 of this Article may invoke that treaty or agreement before any organ of the United Nations.

Article 103

In the event of a conflict between the obligations of the Members of the United Nations under the present Charter and their obligations under any other international agreement, their obligations under the present Charter shall prevail.

Article 104

The Organization shall enjoy in the territory of each of its Members such legal capacity as may be necessary for the exercise of its functions and the fulfilment of its purposes.

Article 105

1. The Organization shall enjoy in the territory of each of its Members such privileges and immunities as are necessary for the fulfilment of its purposes.

2. Representatives of the Members of the United Nations and officials of the Organization shall similarly enjoy such privileges and immunities as are necessary for the independent exercise of their functions in connexion with the Organization.

3. The General Assembly may make recommendations with a view to determining the details of the application of paragraphs 1 and 2 of this Article or may propose conventions to the Members of the United Nations for this purpose.

APPENDIX D

GENERAL ASSEMBLY RESOLUTION 3314 (1974) DEFINITION OF AGGRESSION[856]

The General Assembly,

Having considered the report of the Special committee on the Question of Defining Aggression, established pursuant to its resolution 2330 (XXII) of 18 December 1967, covering the work of its seventh session held from 11 March to 12 April 1974, including the draft Definition of Aggression adopted by the Special Committee by consensus and recommended for adoption by the General Assembly.

Deeply convinced that the adoption of the Definition of Aggression would contribute to the strengthening of international peace and security,

1. *Approves* the Definition of Aggression, the text of which is annexed to the present resolution;

2. *Expresses its appreciation* to the Special Committee on the Question of Defining Aggression for its work which resulted in the elaboration of the Definition of Aggression;

3. *Calls upon* all States to refrain from all acts of aggression and other uses of force contrary to the Charter of the United Nations and the Declaration on Principles of International Law concerning Friendly Relations and Co-operation among States in accordance with the Charter of the United Nations;

4. *Calls the attention* of the Security Council to the Definition of Aggression, as set out below, and recommends that it should, as appropriate, take account of that Definition as guidance in determining, in accordance with the Charter, the existence of an act of aggression.

---

[856] G.A. Res. 3314, U.N. GAOR, 29th Sess., U.N. Doc. A/RES/3314 (1974).

2319th plenary meeting
14 December 1974

ANNEX
Definition of Aggression

The General Assembly,

Basing itself on the fact that one of the fundamental purposes of the United Nations is to maintain international peace and security and to take effective collective measures for the prevention and removal of threats to the peace, and for the suppression of acts of aggression or other breaches of the peace,

Recalling that the Security Council, in accordance with Article 39 of the Charter of the United Nations, shall determine the existence of any threat to the peace, breach of the peace or act of aggression and shall make recommendations, or decide what measures shall be taken in accordance with Articles 41 and 42, to maintain or restore international peace and security,

Recalling also the duty of States under the Charter to settle their international disputes by peaceful means in order not to endanger international peace, security and justice,

Bearing in mind that nothing in this Definition shall be interpreted as in any way affecting the scope of the provisions of the Charter with respect to the functions and powers of the organs of the United Nations,

Considering also that, since aggression is the most serious and dangerous form of the illegal use of force, being fraught, in the conditions created by the existence of all types of weapons of mass destruction, with the possible threat of a world conflict and all its catastrophic consequences, aggression should be defined at the present stage,

Reaffirming the duty of States not to use armed force to deprive peoples of their right to self-determination, freedom and independence, or to disrupt territorial integrity,

Reaffirming also that the territory of a State shall not be violated by being the object, even temporarily, of military occupation or of other measures of force taken by another State in contravention of the Charter, and that it shall not be the object of acquisition by another State resulting from such measures or the threat thereof,

Reaffirming also the provisions of the Declaration on Principles of International Law concerning Friendly Relations and Co-operation among States in accordance with the Charter of the United Nations,

Convinced that the adoption of a definition of aggression ought to have the effect of deterring a potential aggressor, would simplify the determination of acts of aggression and the implementation of measures to suppress them and would also facilitate the protection of the rights and lawful interests of, and the rendering of assistance to, the victim,

Believing that, although the question whether an act of aggression has been committed must be considered in the light of all the circumstances of each particular case, it is nevertheless desirable to formulate basic principles as guidance for such determination,

Adopts the following Definition of Aggression:

Article 1

Aggression is the use of armed force by a State against the sovereignty, territorial integrity or political independence of another State, or in any other manner inconsistent with the Charter of the United Nations, as set out in this Definition.

Explanatory note: In this Definition the term "State":

(a) Is used without prejudice to questions of recognition or to whether a State is a member of the United Nations;

(b) Includes the concept of a "group of States" where appropriate.

Article 2

The first use of armed force by a State in contravention of the Charter shall constitute *prima facie* evidence of an act of aggression although the Security Council may, in conformity with the Charter, conclude that a determination that an act of aggression has been committed would not be justified in the light of other relevant circumstances, including the fact that the acts concerned or their consequences are not of sufficient gravity.

Article 3

Any of the following acts, regardless of a declaration of war, shall, subject to and in accordance with the provisions of article 2, qualify as an act of aggression:

(a) The invasion or attack by the armed forces of a State of the territory of another State, or any military occupation, however temporary, resulting from such invasion or attack, or any annexation by the use of force of the territory of another State or part thereof;

(b) Bombardment by the armed forces of a State against the territory of another State or the use of any weapons by a State against the territory of another State;

(c) The blockade of the ports or coasts of a State by the armed forces of another State;

(d) An attack by the armed forces of a State on the land, sea or air forces, or marine and air fleets of another State;

(e) The use of armed forces of one State which are within the territory of another State with the agreement of the receiving State, in contravention of the conditions provided for in the agreement or any extension of their presence in such territory beyond the termination of the agreement;

(f) The action of a State in allowing its territory, which it has placed at the disposal of another State, to be used by that other State for perpetrating an act of aggression against a third State;

(g) The sending by or on behalf of a State of armed bands, groups, irregulars or mercenaries, which carry out acts of armed force against another State of such gravity as to amount to the acts listed above, or its substantial involvement therein.

Article 4

The acts enumerated above are not exhaustive and the Security Council may determine that other acts constitute aggression under the provisions of the Charter.

Article 5

1. No consideration of whatever nature, whether political, economic, military or otherwise, may serve as a justification for aggression.

2. A war of aggression is a crime against international peace. Aggression gives rise to international responsibility.

3. No territorial acquisition or special advantage resulting from aggression is or shall be recognized as lawful.

Article 6

Nothing in this Definition shall be construed as in any way enlarging or diminishing the scope of the Charter, including its provisions concerning cases in which the use of force is lawful.

Article 7

Nothing in this Definition, and in particular article 3, could in any way prejudice the right to self-determination, freedom and independence, as derived from the Charter, of peoples forcibly deprived of that right and referred to in the Declaration on Principles of International Law concerning Friendly Relations and Co-operations among States in accordance with the Charter of the United Nations, particularly peoples under colonial and racist regimes or other forms of alien domination; nor the right of these peoples to struggle to that end and to seek and receive support, in accordance with the principles of the Charter and in conformity with the above-mentioned Declaration.

Article 8

In their interpretation and application the above provisions are interrelated and each provision should be construed in the context of the other provisions.

APPENDIX E

CHAIRMAN OF THE JOINT CHIEFS OF STAFF INSTRUCTION 3121.01 STANDING RULES OF ENGAGEMENT FOR US FORCES[857]

ENCLOSURE A STANDING RULES OF ENGAGEMENT FOR US FORCES

1. Purpose and Scope

a. The purpose of these SROE is to provide implementation guidance on the inherent right and obligation of self-defense and the application of force for mission accomplishment. The SROE establish fundamental policies and procedures governing the actions to be taken by US force commanders during all military operations, contingencies, or prolonged conflicts. In order to provide uniform training and planning capabilities, this document is authorized for distribution to commanders at all levels to be used as fundamental guidance for training and directing their forces.

b. Except as augmented by supplemental rules of engagement for specific operations, missions, or projects, the policies and procedures established herein remain in effect until rescinded.

c. US forces operating with multinational forces:

(1) US forces assigned to the operational control (OPCON) of a multinational force will follow the ROE of the multinational force unless otherwise directed by the National Command Authorities

---

[857] Chairman of the Joint Chiefs of Staff Instruction 3121.01, *Standing Rules of Engagement for US Forces,* Oct. 1, 1994, as amended Dec. 22, 1994, *reprinted in* OPERATIONAL LAW HANDBOOK, Chap. 8, at pp. 8-8 through 8-14, US Army Judge Advocate General's School (1997). The author gratefully acknowledges the availability of the electronic version of this document provided by the United States Army Judge Advocate General's School.

(NCA). US forces will be assigned and remain OPCON to a multinational force only if the combatant commander and higher authority determine that the ROE for that multinational force are consistent with the policy guidance on unit self-defense and with the rules for individual self-defense contained in this document.

(2) When US forces, under US OPC0N, operate in conjunction with a multinational force, reasonable efforts will be made to effect common ROE. If such ROE cannot be established, US forces will exercise the right and obligation of self-defense contained in this document while seeking guidance from the appropriate combatant command. To avoid mutual interference, the multinational forces will be informed prior to US participation in the operation of the US forces' intentions to operate under these SROE and to exercise unit self-defense. For additional guidance concerning peace operations, see Appendix A to Enclosure A.

(3) Participation in multinational operations may be complicated by varying national obligations derived from international agreements; i.e., other members in a coalition may not be signatories to treaties that bind the United States, or they may be bound by treaties to which the United States is not a party. US forces still remain bound by US treaty obligations even if the other members in a coalition are not signatories to a treaty and need not adhere to its terms.

d. Commanders of US forces subject to international agreements governing their presence in foreign countries (e.g., Status of Forces Agreements) are not relieved of the inherent authority and obligation to use all necessary means available and to take all appropriate action for unit self-defense.

e. US forces in support of operations not under operational or tactical control of a combatant commander or performing missions under direct control of the NCA, Military Departments, or other US government departments/agencies (i.e., marine security guards, certain special security forces) will operate under use-of-force policies or ROE promulgated by those departments or agencies.

f. US Coast Guard (USCG) units and units under USCG OPCON conducting law enforcement operations, and USCG personnel using their law enforcement authority, will follow the use-of-force policy issued by the

Commandant, USCG. Nothing in the USCG use-of-force policy negates a commander's inherent authority and obligation to use all necessary means available and to take all appropriate action for unit self-defense in accordance with these SROE.

g. The guidance in this document does not cover US forces deployed to assist federal and local authorities during times of civil disturbance within the territorial jurisdiction of any state, the District of Columbia, Commonwealths of Puerto Rico and the Northern Marianas, US possessions, and US territories. Forces in these situations will follow use-of-force policy found in DoD Civil Disturbance Plan, "Garden Plot" (Appendix I to Annex C of Garden Plot).

h. US forces deployed to assist foreign, federal, and local authorities in disaster assistance missions, such as earthquakes and hurricanes, will follow use-of-force guidelines as set forth in the mission's execute order and subsequent orders.

i. US forces will always comply with the Law of Armed Conflict. However, not all situations involving the use of force are armed conflicts under international law. Those approving operational rules of engagement must determine if the internationally recognized Law of Armed Conflict applies. In those circumstances when armed conflict, under international law, does not exist, Law of Armed Conflict principles may nevertheless be applied as a matter of national policy. If armed conflict occurs, the actions of US forces will be governed by both the Law of Armed Conflict and rules of engagement.

2. Policy

a. THESE RULES DO NOT LIMIT A COMMANDER'S INHERENT AUTHORITY AND OBLIGATION TO USE ALL NECESSARY MEANS AVAILABLE AND TO TAKE ALL APPROPRIATE ACTION IN SELF-DEFENSE OF THE COMMANDER'S UNIT AND OTHER US FORCES IN THE VICINITY.

b. US national security policy serves to protect the United States, US forces, and, in certain circumstances, US citizens and their property, US commercial assets, and other designated non-US forces, foreign nationals, and their property from hostile attack. US national security policy is guided, in part, by the need to maintain a stable international environment compatible with US national security interests. In addition, US national security interests guide our global objectives of deterring armed attack against the United States across the range of military operations, defeating an attack should deterrence fail, and preventing or neutralizing hostile efforts to intimidate or coerce the United States by the threat or use of armed force or terrorist actions. Deterrence requires clear and evident capability and resolve to fight at any level of conflict and, if necessary, to increase deterrent force capabilities and posture deliberately so that any potential aggressor will assess its own risks as unacceptable. US policy, should deterrence fail, provides flexibility to respond to crises with options that:

(1) Are proportional to the provocation.

(2) Are designed to limit the scope and intensity of the conflict.

(3) Will discourage escalation.

(4) Will achieve political and military objectives

3. Intent. These SROE are intended to:

a. Provide general guidelines on self-defense and are applicable worldwide to all echelons of command.

b. Provide guidance governing the use of force consistent with mission accomplishment.

c. Be used in operations other than war, during transition from peacetime to armed conflict or war, and during armed conflict in the absence of superseding guidance.

4. Combatant Commanders' SROE

a. Combatant commanders may augment these SROE as necessary to reflect changing political and military policies, threats, and missions specific to their AOR. When specific standing rules governing the use of force in a combatant commander's AOR are required that are different from these

SROE, they will be submitted to the Chairman of the Joint Chiefs of Staff for NCA approval as necessary and promulgated by the Joint Staff as an Annex to Enclosure C of these SROE.

b. Combatant commanders will distribute these SROE to subordinate commanders and units for compliance. The mechanism for disseminating ROE supplemental measures is set forth in Enclosure B.

5. Definitions

a. **Inherent Right of Self-Defense.** A commander has the authority and obligation to use all necessary means available and to take all appropriate action to defend that commander's unit and other US forces in the vicinity from a hostile act or demonstrated hostile intent. Neither these rules, nor the supplemental measures activated to augment these rules, limit this inherent right and obligation. At all times, however, the requirements of necessity and proportionality as amplified in these SROE will be the basis for the judgment of the commander as to what constitutes an appropriate response to a particular hostile act or demonstration of hostile intent.

b. **National Self-Defense.** National self-defense is the act of defending the United States, US forces, and in certain circumstances, US citizens and their property, US commercial assets, and other designated non-US forces, foreign nationals and their property, from a hostile act or hostile intent. Once a force or terrorist unit is declared hostile by appropriate authority exercising the right and obligation of national self-defense (see paragraph 2 of Appendix A to Enclosure A), individual US units do not need to observe a hostile act or determine hostile intent before engaging that force.

NOTE: **Collective Self-Defense.** Collective self-defense, as a subset of national self-defense, is the act of defending other designated non-US forces, personnel and their property from a hostile act or demonstration of hostile intent. Only the NCA may authorize US forces to exercise collective self-defense.

c. **Unit Self-Defense.** Unit self-defense is the act of defending a particular unit of US forces, including elements or personnel thereof, and other US forces in the vicinity, against a hostile act or hostile intent. The need to exercise unit self-defense may arise in many situations such as localized low-level conflicts, humanitarian efforts, peace enforcement actions, terrorist response, or prolonged engagements. Individual self-defense is a subset of unit self-defense: see the glossary for a definition of individual self-defense.

d. **Elements of Self-Defense**. The application of armed force in self-defense requires the following two elements:

(1) **Necessity**. A hostile act occurs or a force or terrorist unit exhibits hostile intent.

(2) **Proportionality**. The force used must be reasonable in intensity, duration, and magnitude, based on all facts known to the commander at the time, to decisively counter the hostile act or hostile intent and to ensure the continued safety of US forces.

e. **Hostile Act**. A hostile act is an attack or other use of force by a foreign force or terrorist unit (organization or individual) against the United States, US forces, and in certain circumstances, US citizens, their property, US commercial assets, and other designated non-US forces, foreign nationals and their property. It is also force used directly to preclude or impede the mission and/or duties of US forces, including the recovery of US personnel and vital US Government property. When a hostile act is in progress, the right exists to use proportional force, including armed force, in self-defense by all necessary means available to deter or neutralize the potential attacker or, if necessary, to destroy the threat. (See definitions in the Glossary for amplification.)

f. **Hostile Intent**. Hostile intent is the threat of imminent use of force by a foreign force or terrorist unit (organization or individual) against the United States, US forces, and in certain circumstances, US citizens, their property, US commercial assets, or other designated non-US forces, foreign nationals and their property. When hostile intent is present, the right exists to use proportional force, including armed force, in self-defense by all necessary means available to deter or neutralize the potential attacker or, if necessary, to destroy the threat. (See definitions in the Glossary for amplification.)

g. **Hostile Force**. Any force or terrorist unit (civilian, paramilitary, or military), with or without national designation, that has committed a hostile act, demonstrated hostile intent, or has been declared hostile.

6. **Declaring Forces Hostile.** Once a force is declared hostile by appropriate authority, US units need not observe a hostile act or a demonstration of hostile intent before engaging that force. The responsibility for exercising the right and obligation of national self-defense and declaring a force hostile is a matter of the utmost importance demanding considerable judgment of command. All available intelligence, the status of international relationships, the requirements of international law, the possible need for a political decision, and the potential consequences for the United States must be carefully weighed. Exercising the right and obligation of national self-defense by competent authority is in addition to and does not supplant the right and obligation to exercise unit self-defense. The authority to declare a force hostile is limited as amplified in Appendix A to Enclosure A.

7. **Authority to Exercise Self-Defense**

a. **National Self-Defense.** The authority to exercise national self-defense is outlined in Appendix A to Enclosure A.

b. **Collective Self-Defense.** Only the NCA may authorize the exercise of collective self-defense.

c. **Unit Self-Defense.** A unit commander has the authority and obligation to use all necessary means available and to take all appropriate action to defend the unit, including elements and personnel thereof, or other US forces in the vicinity, against a hostile act or hostile intent. In defending against a hostile act or hostile intent under these SROE, unit commanders should use only that degree of force necessary to decisively counter the hostile act or hostile intent and to ensure the continued safety of US forces.

8. **Action in Self-Defense**

a. **Means of Self-Defense.** All necessary means available and all appropriate actions may be used in self-defense. The following guidelines apply for unit or national self-defense:

(1) **Attempt to Control Without the Use of Force.** The use of force is normally a measure of last resort. When time and circumstances permit, the potentially hostile force should be warned and given the opportunity to withdraw or cease threatening actions. (See Appendix A to Enclosure A for amplification.)

(2) **Use Proportional Force To Control the Situation.** When the use of force in self-defense is necessary, the nature, duration, and scope of the engagement should not exceed that which is required to decisively counter the hostile act or hostile intent and to ensure the continued safety of US forces or other protected personnel or property.

(3) **Attack To Disable or Destroy.** An attack to disable or destroy a hostile force is authorized when such action is the only prudent means by which a hostile act or hostile intent can be prevented or terminated. When such conditions exist, engagement is authorized only until the hostile force no longer poses an imminent threat.

b. **Immediate Pursuit of Hostile Foreign Forces.** In self-defense, US forces may pursue and engage a hostile force that has committed a hostile act or demonstrated hostile intent and that remains an imminent threat. (See Appendix A to Enclosure A for amplification.)

c. **Defending US Citizens, Property, and Designated Foreign Nationals**

(1) **Within a Foreign Nation's US Recognized Territory or Territorial Airspace.** A foreign nation has the principal responsibility for defending US citizens and property within these areas. (See Appendix A to Enclosure A for amplification.)

(2) **At Sea.** Detailed guidance is contained in Annex A to Appendix B of this enclosure.

(3) **In International Airspace.** Protecting civil aircraft in international airspace is principally the responsibility of the nation of registry. Guidance for certain cases of actual or suspected hijacking of airborne US or foreign civil aircraft is contained in MCM-102-92, 24 July 1992, "Hijacking of Civil Aircraft."

(4) **Terrorism.** Terrorist attacks are usually undertaken by civilian or paramilitary organizations, or by individuals under circumstances in which a determination of hostile intent may be difficult. The definitions of hostile act and hostile intent set forth above will be used in situations where terrorist attacks are likely. The term "hostile force" includes terrorist units when used in this document. When

circumstances and intelligence dictate, supplemental ROE will be used to meet this special threat.

(5) **Piracy**. Piracy is defined as an illegal act of violence, depredation (i.e., plundering, robbing, or pillaging), or detention in or over international waters committed for private ends by the crew or passengers of a private ship or aircraft against another ship or aircraft or against persons or property on board such ship or aircraft. US warships and aircraft have an obligation to repress piracy on or over international waters directed against any vessel, or aircraft, whether US or foreign flagged. If a pirate vessel or aircraft fleeing from pursuit proceeds into the territorial sea, archipelagic waters, or superjacent airspace of another country every effort should be made to obtain the consent of nation sovereignty [*sic*] to continue pursuit. Where circumstances permit, commanders will seek guidance from higher authority before using armed force to repress an act of piracy.

d. Operations Within or in the Vicinity of Hostile Fire or Combat Zones Not Involving the United States

(1) US forces should not enter, or remain in, a zone in which hostilities (not involving the United States) are imminent or occurring between foreign forces unless directed by proper authority.

(2) If a force commits a hostile act or demonstrates hostile intent against US forces in a hostile fire or combat zone, the commander is obligated to act in unit self-defense in accordance with SROE guidelines.

e. Right of Assistance Entry

(1) Ships, or under certain circumstances aircraft, have the right to enter a foreign territorial sea or archipelagic waters and corresponding airspace without the permission of the coastal or island state to engage in legitimate efforts to render emergency assistance to those in danger or distress from perils of the sea.

(2) Right of assistance extends only to rescues where the location of those in danger is reasonably well known. It does not extend to

entering the territorial sea, archipelagic waters, or national airspace to conduct a search.

(3) For ships and aircraft rendering assistance on scene, the right and obligation of self-defense extends to and includes persons, vessels, or aircraft being assisted. The right of self-defense in such circumstances does not include interference with legitimate law enforcement actions of a coastal nation. However, once received on board the assisting ship or aircraft, persons assisted will not be surrendered to foreign authority unless directed by the NCA.

(4) Further guidance for the exercise of the right of assistance entry is contained in CJCS Instruction 2410.01, 20 July 1993, "Guidance for the Exercise of Right of Assistance Entry."

APPENDIX F

AN ASSESSMENT OF INTERNATIONAL LEGAL ISSUES IN INFORMATION OPERATIONS

MAY 1999

Department of Defense
Office of General Counsel

TABLE OF CONTENTS

[Page numbers have been altered to correspond to pages in this book.]

I. INTRODUCTION

A. Sources and Application of International Law.

International law consists of binding legal obligations among sovereign states. Two of the basic principles of the international legal system are that sovereign states are legally equal and independent actors in the world community, and that they generally assume legal obligations only by affirmatively agreeing to do so. The most effective instruments in creating international law are international agreements, which may be either bilateral or multilateral. Some of these agreements, such as the United Nations Charter, establish international institutions that the parties agree to invest with certain authority. It is also generally accepted that there is a body of customary international law, which consists of practices that have been so widely followed by the community of nations, with the understanding that compliance is mandatory, that they are considered to be legally obligatory.

International institutions have legislative authority to create legal obligations for nations only when their member nations have agreed to give them that authority. The most prominent example is the power of the UN Security Council to pass resolutions requiring individual nations to perform or refrain from certain actions in order to protect or restore international peace and security in the context of a particular situation. The decisions of the International Court of Justice are binding upon nations that have accepted the jurisdiction of the Court and are parties to litigation before it. Other international institutions can also be given the power to impose binding obligations upon nations that agree to submit to their authority. In addition, certain actions of some international institutions, such as the International Court of Justice and the UN General Assembly, are considered to be persuasive evidence of the existence of principles of customary international law.

As with domestic law, the primary mechanism that makes international law effective is voluntary compliance. Also as with domestic law, the threat of sanctions is often required as well. The international legal system provides institutional enforcement mechanisms such as international litigation before the International Court of Justice and other judicial and arbitral tribunals, as well as the right to petition the United Nations Security Council to authorize coercive measures to protect or restore international peace and security. The international legal system also provides self-help enforcement mechanisms such as the right to use force in individual and collective self-defense and the right in some circumstances to repudiate treaty obligations which have been violated by another party. An aggrieved nation may always withdraw from voluntary relationships involving diplomatic representation and most kinds of commerce. Even the right to publicly complain about another nation's illegal behavior may provide an effective enforcement mechanism if such complaints generate diplomatic costs for the offending nation.

Chief Justice Oliver Wendell Holmes once wrote, "The life of the law has not been logic; it has been experience." It seldom happens that a legislature foresees a problem before it arises and puts into place a legislative solution before it is needed. More typically, legislators react to a problem that has already manifested itself. The international legal system operates in the same manner. The international community ordinarily does not negotiate treaties to deal with problems until their consequences have begun to be felt. This is not all bad, since the solution can be tailored to the actual problems that have occurred, rather than to a range of hypothetical possibilities. One consequence, however, is that the resulting law, whether domestic or international, may be sharply influenced by the nature of the events that precipitate legal developments, together with all their attendant policy and political considerations.

The development of international law concerning artificial earth satellites provides a good example. If the nations had sat down with perfect foresight and asked themselves, "Should we permit those nations among us that have access to advanced technology to launch satellites into orbit that will pass over the territory of the rest of us and take high-resolution imagery, listen in our telecommunications, record weather information, and broadcast information directly to telephones and computers within our borders?", a very restrictive regime of space law might have resulted. Instead, what happened was that the first satellites launched by the Soviet Union and the United States were seen as entirely benign devices engaged in scientific research, and it was also perfectly clear that no nation had the capability to interfere with them as they passed over its territory. In these circumstances, it quickly became accepted customary international law, soon enshrined in the Outer Space Treaty, that objects in orbit were beyond the territorial claims of any nation, and that outer space is available for exploitation by all.

The history of space law contrasts sharply with that of air law. Much of the early development of heavier-than-air aviation coincided with the First World War, during which the military power of aircraft for intelligence gathering, attacking ground forces, and bombing enemy cities was clearly demonstrated. The result was a highly restricted [sic] regime of air law in which any entry into a nation's airspace without its permission was to be regarded as a serious violation of its sovereignty and territorial integrity.

Similarly, we can make some educated guesses as to how the international legal system will respond to information operations, but the direction that response actually ends up taking may depend a great deal on the nature of the events that draw the nations' attention to the issue. If information operations techniques are seen as just another new technology that does not greatly threaten the nations' interests, no dramatic legal developments may occur. If they are seen as a revolutionary threat to the security of nations and the welfare of their citizens, it will be much more likely that efforts will be made to restrict or prohibit information operations by legal means. These are considerations that national leaders should understand in making decisions on using information operations techniques in the current formative period, but it should also be understood that the course of future events is often beyond the control of statesmen.

The actors in the international legal system are sovereign states. International legal obligations and international enforcement mechanisms generally do not apply to individual persons except where a nation enforces certain principles of international law through its domestic criminal law, or in a very limited class of serious offenses (war crimes, genocide, crimes against humanity, and crimes against peace) that the nations have agreed may be tried and punished by international criminal tribunals.

B. Essentials of Treaty Law.

In domestic US law there are important distinctions between treaties and executive agreements. This distinction primarily involves issues of Constitutional authority within the US government, but it is of little importance internationally. Treaties and executive agreements are equally binding between the United States and the other party or parties to an international agreement. We will use the term "treaty" in this paper as a shorthand way of referring to all forms of legally binding state-to-state international agreements.

Treaty obligations are binding on their parties, but international law recognizes certain circumstances in which a nation can regard a treaty obligation as being suspended, modified, or terminated. The parties can always modify or terminate a treaty by mutual consent. Some international agreements expire by their own terms after a fixed period of time. Generally, unless the terms of the agreement establish a right of unilateral withdrawal, a nation may not unilaterally repudiate or withdraw from a treaty unless it has a basis for doing so that is recognized under international law. Treaty obligations are reciprocal in nature. If one of the parties commits a material breach of its obligations under the treaty, the other may be entitled to suspend its own compliance, or to withdraw from the agreement entirely. Also, a fundamental change in circumstances may justify a decision by one of the parties to regard its treaty obligations as suspended or terminated.

One of these fundamental changes of circumstance is the initiation of armed hostilities between the parties. Some international agreements specifically provide that they will remain in effect during armed conflict between the parties, such as law of war treaties and the United Nations Charter. Most treaties, however, are silent on whether or not they will continue to apply during hostilities between the parties. Many peacetime agreements facilitate tourism, transportation, commerce, and other relationships the continuation of which would be fundamentally inconsistent with a state of armed conflict between the parties. Agreements on other subjects, such as boundary settlements and reciprocal rights of inheritance of private property, may be unrelated to the existence of hostilities, and may ultimately be determined to remain in full force. The issues involved may be particularly complicated when the treaty concerned is multilateral, rather than bilateral. When two parties to a multilateral treaty are engaged in armed conflict, the result may well be that the effect of the treaty is suspended between the belligerents, but remains in effect among each belligerent and the other parties. We will see later in this paper that the United States is a party to a variety of bilateral and multilateral agreements containing obligations that may affect information operations. One of our tasks will be to determine as best we can which of these agreements are likely to remain in effect during hostilities. The tests we will apply are (1) whether there is specific language in the treaty addressing its effect during hostilities between the parties, and (2) if there is no such language, whether the object and purpose of the treaty is or is not compatible with a state of armed hostilities between the parties.

C. New Legal Challenges Presented by Information Operations.

Many traditional military activities are included in current concepts of "information operations" and "information warfare," including physical attacks on information systems by traditional military means, psychological operations, military deception, and "electronic warfare" operations such as jamming radar and radio signals. The application of international law to these traditional kinds of operations is reasonably well settled. Similarly, electro-magnetic pulse (EMP) weapons and directed-energy weapons such as lasers, micro-wave devices, and high energy radio frequency (HERE) guns will probably operate in a manner similar enough to that of traditional weapons that one could apply existing legal principles to them without much difficulty. It will not be as easy to apply existing international law principles to **information attack,** a term used to describe the use of electronic means to gain access to or change information in a targeted information system without necessarily damaging its physical components. One of the principal forms of information attack is likely to be **computer network attack,** or in today's vernacular, the "hacking" of another nation's computer systems.

The proliferation of global electronic communications systems and the increased interoperability of computer equipment and operating systems have greatly improved the utility of all kinds of information systems. At the same time, these developments have made information systems that are connected to any kind of network, whether it be the internet or some other radio or hard-wired communications system, vulnerable to computer network attacks. Moreover, global communications are almost seamlessly interconnected and virtually instantaneous, as a result of which distance and geographical boundaries have become essentially irrelevant to the conduct of computer network attacks. The result is that many information systems are subject to computer network attack anywhere and anytime. The attacker may be a foreign state, an agent of a foreign state, an agent of a non-governmental entity or group, or an individual acting for purely private purposes. The equipment necessary to launch a computer network attack is readily available and inexpensive, and access to many computer systems can be obtained through the internet or another network to which access is easily obtained.

One major implication is that it may be very difficult to attribute a particular computer network attack to a foreign state, and to characterize its intent and motive. For the purposes of analysis we will initially assume away issues of attribution and characterization, returning to them near the end of the analysis. Another major implication is that an attacker may not be physically present at the place where the effects of the attack are felt. The means of attack may not be tangibly present either, except in the form of anonymous and invisible radio waves or electrons. This will complicate the application of traditional rules of international law that developed in response to territorial invasions and attacks by troops, aircraft, vehicles, vessels, and kinetic weapons that the victim could see and touch, and whose sponsor was usually readily apparent.

II. THE LAW OF WAR

A. Essentials of the Law of War.

The terms "law of war" and "law of armed conflict" are synonymous. The latter term has the virtue that it more clearly applies to all international armed conflicts, whether or not they are formally declared wars. "Law of war" is shorter and more familiar, and we will use it in this paper. The application of the law of war does not generally depend on which of the parties was at fault in starting the conflict. The law of war applies whenever there is a state of international armed conflict, and it applies in the same manner to all the parties to the conflict. There is a small subset of the law of war that applies to noninternational armed conflicts such as civil wars, but those sorts of conflict are not immediately relevant to this paper and will not be discussed. As with other branches of international law, the law of war is composed of treaties and customary international law. The United States is a party to eighteen law of war treaties, along with their various annexes and protocols, and several more law of war agreements are pending before the Senate. The United States also recognizes the existence of a considerable body of customary law of war.

The general principles of the law of war have been expressed in various ways, but their essence can be said to be as follows:

- Distinction of combatants from noncombatants: With very limited exceptions, only members of a nation's regular armed forces are entitled to use force against the enemy. They must distinguish themselves from noncombatants, and they must not use noncombatants or civilian property to shield themselves from attack. If lawful combatants are captured by the enemy they may not be punished for their combatant acts, so long as they complied with the law of war. They are required to be treated humanely in accordance with agreed standards for the treatment of prisoners of war, and they must be released promptly at the cessation of hostilities. Persons who commit combatant acts without authorization are subject to criminal prosecution.

- Military necessity: Enemy military forces are declared hostile. They may be attacked at will, along with their equipment and stores. Civilians and civilian property that make a direct contribution to the war effort may also be attacked, along with objects whose damage or destruction would produce a military advantage because of their nature, location, purpose, or use. A corollary of this principle is that noncombatants and civilian objects making no direct contribution to the war effort, and whose destruction would provide no significant military advantage to the attacker, are immune from deliberate attack.

- Proportionality: When an attack is made against a lawful military target, collateral injury and damage to noncombatants and civilian property may be unavoidable. Attacks may be carried out against lawful military targets even if some amount of collateral damage is foreseeable, unless the foreseeable collateral damage is disproportionate to the military advantage likely to be attained. The military advantage to be gained from an attack refers to an attack considered as a whole rather than only from isolated or particular parts of an attack. Generally, "military advantage" is not restricted to tactical gains, but is linked to the full context of war strategy. The commander ordering the attack is responsible for making the proportionality judgment. The calculus may be affected somewhat if the enemy has failed to carry out his duty to separate his troops and equipment from noncombatants and civilian property, since in such circumstances the defender must shoulder much of the blame for any collateral damage that results. A corollary of the principle of proportionality is that the attacker has a responsibility to take reasonable steps to find out what collateral damage a contemplated attack may cause.

- Superfluous injury: The nations have agreed to ban certain weapons because they cause superfluous injury. Among these are "dum-dum" bullets, projectiles filled with glass or other nondetectable fragments, poisoned weapons, and laser weapons specifically designed to cause permanent blindness to unenhanced vision.

- Indiscriminate weapons: The nations have agreed to ban certain other weapons because they cannot be directed with any precision against combatants. Among these are bacteriological weapons and poison gas.

- Perfidy: The law of war provides certain visual and electronic symbols to identify persons and property that are protected from attack. Among these are prisoners of war and prisoner of war camps, the wounded and sick, and medical personnel, vehicles, aircraft, and vessels. Any misuse of these protected symbols to immunize a lawful military target from attack constitutes the war crime of perfidy. Suppression of such acts is necessary to preserve the effectiveness of such symbols, since known misuse may lead the combatants to disregard them. For similar reasons, it is unlawful to feign surrender, illness, or death to gain an advantage in combat, as well as to broadcast a false report of a cease-fire or armistice.

- Neutrality: Nations not engaged in a conflict may declare themselves to be neutral. A neutral nation is entitled to immunity from attack by the belligerents, so long as the neutral nation satisfies its obligation not to assist either side. If a neutral nation is unable or unwilling to halt the use of its territory by one of the belligerents in a manner that gives it a military advantage, the other belligerent may have a right to attack its enemy in the neutral's territory. There is considerable support for the argument that the concept of neutrality has no application during a conflict in which one of the belligerents is a nation or coalition of nations authorized by the UN Security Council to use armed force to protect or restore international peace and security. This conclusion is based upon Article 49 of the Charter, which provides, "The Members of the United Nations shall join in affording mutual assistance in carrying out the measures decided upon by the Security Council." In other situations, however, as when a nation uses armed force in individual or collective self-defense without the benefit of a Security Council mandate, it would appear that nations not involved in the conflict retain the option of declaring themselves to be neutral.

B. Application to Information Operations.

It is by no means clear what information operations techniques will end up being considered to be "weapons," or what kinds of information operations will be considered to constitute armed conflict. On the other hand, those issues may not end up being particularly important to the analysis of law of war issues. If the deliberate actions of one belligerent cause injury, death, damage, and destruction to the military forces, citizens, and property of the other belligerent, those actions are likely to be judged by applying traditional law of war principles.

- Distinction of combatants from noncombatants: This rule grew up when combatants could see each other and make a judgment of whether or not to open fire based in part on whether or not the individual in the sights wore an enemy uniform. When the unit of combat came to be a vessel, tank, truck, or aircraft, it became more important that such vehicles be properly marked than that their occupants wear a distinctive uniform. If a computer network attack is launched from a location far from its target, it may be of no practical significance whether the "combatant" is wearing a uniform. Nevertheless, the law of war requires that lawful combatants be trained in the law of war, that they serve under effective discipline, and that they be under the command of officers responsible for their conduct. This consideration argues for retaining the requirement that combatant information operations during international armed conflicts be conducted only by members of the armed forces. If combatant acts are conducted by unauthorized persons, their government may be in violation of the law of war, depending on the circumstances, and the individuals concerned are at least theoretically subject to criminal prosecution either by the enemy or by an international war crimes tribunal. The long-distance and anonymous nature of computer network attacks may make detection and prosecution unlikely, but it is the firmly established policy of the United States that US forces will fight in full compliance with the law of war.

- Military necessity: In developed nations both military and civilian infrastructures are vulnerable to computer network attacks. During an armed conflict virtually all military infrastructures will be lawful targets, but purely civilian infrastructures must not be attacked unless the attacking force can demonstrate that a definite military advantage is expected from the attack. Stock exchanges, banking systems, universities, and similar civilian infrastructures may not be attacked simply because a belligerent has the ability to do so. In a long and protracted conflict, damage to the enemy's economy and research and development capabilities may well undermine its war effort, but in a short and limited conflict it may be hard to articulate any expected military advantage from attacking economic targets. Targeting analysis must be conducted for computer network attacks just as it traditionally has been conducted for attacks using traditional weapons.

- Proportionality: During Desert Storm, one of the earliest targets of the coalition bombing campaign was the electrical power system in Baghdad. Considering the important military uses being made of electricity from that system, it was clearly a lawful military target. The Iraqi government then made a public pronouncement that the coalition's attack on the city's electrical power system constituted an act of attempted genocide. The logic of this position was that the city's sewage system depended on electric pumping stations, so when the electricity went out the sewage system backed up and created a threat of epidemic disease. No one took this claim very seriously, but this incident highlights the fact that when an attack is made on an infrastructure that is being used for both military and civilian purposes the commander will not be in a proper position to weigh the proportionality of the expected military advantage against the foreseeable collateral damage unless the commander has made a reasonable effort to discover whether the system is being used for civilian purposes that are essential to public health and safety. This principle operates in exactly the same way whether the attack is carried out using traditional weapons or in the form of a computer network attack.

As stated above, the law of war places much of the responsibility for collateral damage on a defending force that has failed to properly separate military targets from noncombatants and civilian property. When military officials decide to use civilian infrastructure for military purposes (or vice-versa), they ought to consider the fact that such action may make that infrastructure a lawful military target. There may be no choice, as when military traffic has to move on civilian highways and railroads. There may be little alternative to military use of civilian communications systems, since it is impractical to put into place dedicated military communications systems that have sufficient capacity to carry all military communications. Where there is a choice, however, military systems should be kept separate from infrastructures used for essential civilian purposes.

Military command and control systems have long been recognized as lawful military targets. Civilian media generally are not considered to be lawful military targets, but circumstances may make them so. In both Rwanda and Somalia, for example, civilian radio broadcasts urged the civilian population to commit acts of violence against members of other tribes, in the case of Rwanda, or against UN-authorized forces providing humanitarian assistance, in the case of Somalia. When it is determined that civilian media broadcasts are directly interfering with the accomplishment of a military force's mission, there is no law of war objection to using the minimum necessary force to shut them down. The extent to which force can be used for purely psychological operations purposes, such as shutting down a civilian radio station for the sole purpose of undermining the morale of the civilian population, is an issue that has yet to be addressed authoritatively by the international community.

- Superfluous injury: We are not aware that any weapon or device yet conceived specifically for use in information operations has any potential for causing superfluous injury, but new systems should always be reviewed with an eye to their potential for causing catastrophic and untreatable injuries to human beings to an extent not required by military necessity.

- Indiscriminate weapons: The prohibition on indiscriminate weapons may apply to information operations techniques such as malicious logic, as when malicious logic launched against a military information system spreads to other information systems being used to provide essential services to noncombatants. It might also apply if malicious logic spreads to information systems belonging to neutral or friendly nations. Finally, it might be applied indirectly if the consequence of a computer network attack is to release dangerous forces, such as opening the floodgates of a dam, causing an oil refinery in a populated area to explode in flames, or causing the release of radioactivity.

- Perfidy: It may seem attractive for a combatant vessel or aircraft to avoid being attacked by broadcasting the agreed identification signals for a medical vessel or aircraft, but such actions would be a war crime. Similarly, it might be possible to use computer "morphing" techniques to create an image of the enemy's chief of state informing his troops that an armistice or cease-fire agreement had been signed. If false, this would also be a war crime.

- Neutrality: If a neutral nation permits its information systems to be used by the military forces of one of the belligerents, the other belligerent generally has a right to demand that it stop doing so. If the neutral refuses, or if for some reason it is unable to prevent such use by an belligerent, the other belligerent may have a limited right of self-defense to prevent such use by its enemy. It is quite foreseeable, for example, that a belligerent might demand that a neutral nation not provide satellite imagery of the belligerent's forces to its enemy, or that the neutral cease providing real-time weather information or precision navigation services.

There appears, however, to be a limited exception to this principle for communications relay systems. The primary international agreement concerning neutrality, the 1907 *Hague Convention Respecting the Rights and Duties of Neutral Powers and Persons in Case of War on Land,* to which the United States is a party, provides in Articles 8 and 9 that "A neutral Power is not called upon to forbid or restrict the use on behalf of the belligerents of telegraph or telephone cables or of wireless telegraph apparatus belonging to it or to Companies or private individuals," so long as such facilities are provided impartially to both belligerents. The plain language of this agreement would appear to apply to communication satellites as well as to ground-based facilities.

There is nothing in this agreement, however, that would suggest that it applies to systems that generate information, rather than merely relay communications. These would include the satellite imagery, weather, and navigation systems mentioned above, as well as other kinds of intelligence-producing systems such as signals intelligence and hydrophonic systems. For example, if a belligerent nation demanded that the US government deny GPS navigation services to its enemy, and if the US were unable or unwilling to comply, the belligerent may have the right to take necessary and proportional acts in self-defense, such as jamming the GPS signal in the combat area.

International consortia present special problems. Information systems built around space-based components require such huge investments and access to such advanced technology that even developed nations prefer to share the costs with other nations. Where an international communications system is developed by a military alliance such as NATO, few neutrality issues are likely to arise. Other international consortia, however, provide satellite communications and weather data that are used for both civilian and military purposes, and they have a breath [sic] of membership that virtually guarantees that not all members of the consortium will be allies in future conflicts. Some current examples are INTELSAT, INMARSAT, ARABSAT, EUTELSAT, and EUMETSAT.

The members of some these consortia have attempted to deal with the possibility that one or more of the member nations will be involved in an armed conflict by limiting the use that may be made of the system during armed conflict. The INMARSAT agreement, for example, provides that the mobile communications service provided by the system may be used "exclusively for peaceful purposes." This provision provides less than a perfect solution, however, since the member nations and the INMARSAT staff have concluded that this language permits use of INMARSAT by UN peacekeeping or peacemaking forces acting under the auspices of the UN Security Council, even if they are engaged in armed conflict to accomplish their missions.

C. Assessment.

There are novel features of information operations that will require expansion and interpretation of the established principles of the law of war. Nevertheless, the outcome of this process of extrapolation appears to be reasonably predictable. The law of war is probably the single area of international law in which current legal obligations can be applied with the greatest confidence to information operations.

III. INTERNATIONAL LEGAL REGULATION OF THE USE OF FORCE IN "PEACETIME"

A. International Law Concerning the Use of Force Among Nations.

As discussed above, the law of war authorizes a nation engaged in an international armed conflict to employ armed force to attack lawful military targets belonging to the enemy. Resolutions of the United Nations Security Council (UNSC) may also authorize the use of armed force as provided in the UN Charter. The focus of this section, however, is on the application of international law principles in circumstances where there is neither a state of armed conflict nor a UNSC mandate—*i.e.* in peacetime, including the conduct of military operations other than war.

An exploration of the manner in which international law on the use of force among nations is likely to apply to peacetime computer intrusions will serve three distinct purposes: (1) it will enable a government that is resolved to conduct itself in scrupulous compliance with international law to avoid activities that are likely to be regarded by the target nation and the world community as violations of international law; (2) it will enable a government contemplating activities that might be considered to violate international law to weigh the risks of such actions; and (3) it will enable a government that is the victim of an information attack to identify the remedies afforded to it by international law, including appeals to the Security Council, the use of force in self-defense, and other self-help remedies not involving the use of force.

The frequently heard question, "Is a computer network attack an **act of war**?" invokes an obsolete concept not mentioned in the UN Charter and seldom heard in modern diplomatic discourse. An act of war is a violation of another nation's rights under international law that is so egregious that the victim would be justified in declaring war. Declarations of war have fallen into disuse, and the act of war concept plays no role in the modern international legal system. In any event, significant sanctions may follow from much less serious violations of another nation's rights that would not be regarded as acts of war.

The members of the United Nations have agreed in Article 2 (4) of the UN Charter to "refrain in their international relations from the threat **or use of force** against the territorial integrity or political independence of any state, or in any other manner inconsistent with the Purposes of the United Nations."

This obligation is elaborated in the *Declaration on Principles of International Law Concerning Friendly Relations and Cooperation among States in Accordance with the Charter of the United Nations*, General Assembly Resolution 2625 (1970), which provides in part:

"**A war of aggression** constitutes a crime against the peace for which there is responsibility under international law."

"States have a duty to refrain from **acts of reprisal involving the use of force.**"

"Every State has the duty to refrain from organizing, instigating, assisting **or participating in acts of civil strife or terrorist** acts in another State or acquiescing in organized activities within its territory directed towards the commission of such acts, **when the acts referred to in the present paragraph involve a threat or use of force.**"

"Nothing in the foregoing paragraphs shall be construed as enlarging or diminishing in any way the scope of the provisions of the Charter concerning cases in which the use of force is lawful."

NOTE: The United States has often expressed the view that most General Assembly resolutions are only recommendations, but that in exceptional cases particular General Assembly resolutions that are meant to be declaratory of international law, are adopted with the support of all members, and are observed by the practice of states, are persuasive evidence of customary international law on a particular subject. Representatives of the United States have on several occasions publicly endorsed the Declaration on Friendly Relations as one of the few General Assembly resolutions that the United States regards as an authoritative restatement of customary international law, at least until the practice of states fails to demonstrate that they consider its principles to be legally binding.

In its 1974 "Definition of Aggression" Resolution, the General Assembly further provided:

- Article 1. Aggression is the use of **armed force** by a State against the sovereignty, territorial integrity or political independence of another State, or in any other manner inconsistent with. the Charter of the United Nations, as set out in this Definition.

- Article 2. The first use of armed force by a State in contravention of the Charter shall constitute prima facie evidence of an act of aggression although the Security Council may, in conformity with the Charter, conclude that a determination that an act of aggression has been committed would not be justified in the light of other relevant circumstances, including the fact that the acts concerned or their consequences are not of sufficient gravity.

- Article 3. Any of the following acts, regardless of a declaration of war, shall, subject to and in accordance with the provisions of Article 2, qualify as an act of aggression:

  (a) The invasion or attack by the armed forces of a State of the territory of another State, or any military occupation, however temporary, resulting from such invasion or attack, or any **annexation by the use of force of the territory** of another State or part thereof;

  (b) **Bombardment by the armed forces** of a State against the territory of another State or the use of any weapons by a State **against the territory** of another State;

  (c) The **blockade of the ports** or coasts of a State **by the armed forces** of another State;

  (d) An **attack by the armed forces** of a State **on the land, sea or air forces, or marine and air fleets** of another State;

  (e) The use of armed forces of one State which are within the territory of another State with the agreement of the receiving State, in contravention of the conditions provided for in the agreement or any extension of their presence in such territory beyond the termination of the agreement;

  (f) The action of a State in allowing its territory, which it has placed at the disposal of another State, to be used by that other State for perpetrating an act of aggression against a third State;

  (g) The sending by or on behalf of a State **of armed bands, groups, irregulars or mercenaries, which carry out acts of armed force** against another State of such gravity as to amount to the acts listed above, or its substantial involvement therein.

NOTE: The United States delegation noted that the text of this resolution reflected hard bargaining among the 35 states that were members of the Special Committee on the Question of Defining Aggression. After the resolution was adopted by the General Assembly without a vote, the US delegation stated the view that the resolution did not establish rights and obligations of states, but that it was "likely to provide useful guidance" to the Security Council. Translated, this statement appears to indicate that the United States does not regard the language of this resolution as a completely authoritative restatement of customary international law, but that its essential concepts are correct. In any event, the question of what constitutes an "act of aggression" is unlikely to be as useful for our purposes as is the question, what kinds of information attacks are likely to be considered by the world community to be "armed attacks" and "uses of force."

Turning to the question of when force may lawfully be used by nations, the United Nations Charter provides that in some circumstances the Security Council may authorize the use of coercive measures, including military force:

Article 39. The Security Council shall determine the existence of any threat to the peace, breach of the peace, or act of aggression and shall make recommendations, or decide what measures shall be taken in accordance with Articles 41 and 42, to maintain or restore international peace and security.

Article 41. The Security Council may decide what measures not involving the use of armed force are to be employed to give effect to its decisions, and it may call upon the Members of the United Nations to apply such measures. These may include complete or partial interruption of economic relations and of rail, sea, air, postal, telegraphic, radio, and other means of communication, and the severance of diplomatic relations.

Article 42. Should the Security Council consider that measures provided for in Article 41 would be inadequate or have proved to be inadequate, it may take such action by air, sea, or land forces as may be necessary to maintain or restore international peace and security. Such action may include demonstrations, blockade, and other operations by air, sea, or land forces of Members of the United Nations.

Perhaps most significantly, the Charter also provides in Article 51, "Nothing in the present Charter shall impair the inherent right of individual or collective self-defense if an **armed attack** occurs against a Member of the United Nations, until the Security Council has taken measures necessary to maintain international peace and security."

Read together, these provisions of the Charter and the related General Assembly resolutions provide a myriad of terms and concepts concerning prohibited uses of force among nations, including the threat or use of force, acts of aggression, wars of aggression, the use of armed force, acts of armed force, invasion, attack, bombardment, and blockade. These acts may be directed at the victim nation's territorial integrity or political independence, or against its military forces or marine or air fleets. They all have in common the presence of troops and the use of traditional military weapons. The question before us is how they are likely to apply to computer network attacks.

Further, when one looks for provisions describing a sanction or remedy, only two provisions present themselves: the authority of the Security Council to authorize various sanctions, including the use of the members' armed forces, when it finds there is a "threat to the peace, breach of the peace, or act of aggression;" and Article 51 's recognition of the inherent right of self defense "if an armed attack occurs.

There is no requirement that a "threat to the peace" take the form of an armed attack, a use of force, or any other condition specified in the charter. The Security Council has the plenary authority to conclude that virtually any kind of conduct or situation constitutes a "threat to the peace" in response to which it can authorize remedial action of a coercive nature. Nothing would prevent the Security Council from finding that a computer network attack was a "threat to the peace" if it determined that the situation warranted such action. It seems unlikely that the Security Council would take action based on an isolated case of state-sponsored computer intrusion producing little or no damage, but a computer network attack that caused widespread damage, economic disruption, and loss of life could well precipitate action by the Security Council. The debate in such a case would more likely center on the offender's intent and the consequences of the offending action than on the mechanism by which the damage was done.

The language of Article 51, on the other hand, requires an "armed attack." A close parsing of the language would tend to limit its effect to attacks and invasions using traditional weapons and forces. On the other hand, there is a well-established view that Article 51 did not create the right of self-defense, but that it only recognized a pre-existing and inherent right that is in some respects broader than the language of Article 51.

History has also seen the emergence of such derivative doctrines as "anticipatory self-defense" and "self-defense in neutral territory," both of which have been relied upon by the United States in certain circumstances. "Anticipatory self-defense" permits a nation to strike the first blow if it has good reason to conclude that it is about to be attacked. The JCS Standing Rules of Engagement implement this doctrine in their authorization of the use of force in response to a demonstration of "hostile intent" by an adversary. "Self-defense in neutral territory" is the right to use force to neutralize a continuing threat located in the territory of a neutral state, but not acting on its behalf, when the neutral state is unable or unwilling to execute its responsibility to prevent the use of its territory as a base or sanctuary for attacks on another nation. This doctrine has venerable roots in US foreign and defense policy, dating at least to the *Caroline* incident. In December 1837, Canada, which was still a British colony, was fighting an insurrection. More than 1,000 insurgents were encamped on both the Canadian and US sides of the Niagara River. A small steamer, the *Caroline,* was used by the insurgents to travel across and along the river. On the night of December 19, 1837, a party of British troops crossed the Niagara and attacked the *Caroline* in the port of Schlosser, New York, setting the vessel on fire and casting it adrift over the Niagara Falls. One US citizen was killed on the dock, another was missing, and several others were wounded. The United States demanded reparations. The British Government responded that it had acted in self-defense. Secretary of State Daniel Webster agreed that the doctrine of self-defense in neutral territory was a valid principle of international law, but asserted that it did not apply in the circumstances of this case. Britain continued to maintain that its action was legal, but nonetheless apologized for the invasion of US territory. No reparations were paid.

In 1986 the United States bombed Libya as a response to Libya's continuing support for terrorism against US military forces and other US interests. In June 1993 US forces attacked the Iraqi military intelligence headquarters because the government of Iraq had conspired to assassinate former President Bush. In August 1998 US cruise missiles struck a terrorist training camp in Afghanistan and a chemical plant in Sudan in which chemical weapons had been manufactured. The rationale articulated for each of these actions was self-defense. Acts of self-defense must satisfy the tests of necessity and proportionality, but there is no requirement that an act of self-defense use the same means as the provocation, that the object of the attack be either a similar type of target or the means used in the offending attacks, or that the action taken be contemporaneous with the provocation, particularly if the attacker is responding to a continuing course of conduct.

B. Acts Not Amounting to the Use of Force.

In its 1949 decision in the Corfu Channel Case, the ICJ ruled that the intrusion of British warships into Albanian territorial waters, which it found to have been without justification under any principle of international law, constituted a violation of Albania's territorial sovereignty. The result seems to be recognition of a general international law of trespass, although the remedy may be limited to a declaratory judgment that the victim's rights have been violated.

The ICJ's predecessor, the Permanent Court of International Justice, in its 1928 Chorzow Factory Decision, declared that reparations were due to any nation whose rights under international law were violated by another nation. This concept is often referred to as the doctrine of state responsibility.

There is also a general recognition of the right of a nation whose rights under international law have been violated to take countermeasures against the offending state, in circumstances where neither the provocation nor the response involves the use of armed force. For example, an arbitral tribunal in 1978 ruled that the United States was entitled to suspend French commercial air flights into Los Angeles after the French had suspended US commercial air flights into Paris. Discussions of the doctrine of countermeasures generally distinguish between countermeasures that would otherwise be violations of treaty obligations or of general principles of international law (in effect, reprisals not involving the use of armed force) and retorsions—actions that may be unfriendly or even damaging, but which do not violate any international legal obligation. The use of countermeasures is subject to the same requirements of necessity and proportionality as apply to self-defense. Some examples of countermeasures that have been generally accepted as lawful are the suspension of diplomatic relations, trade and communications embargoes, cutting off foreign aid, blocking assets belonging to the other nation, and prohibiting travel to or from the other nation.

The international law doctrines of self-defense, reprisal, and countermeasures all require that a nation invoking them do so with the intent of protecting itself against further harm, either by directly blocking further hostile acts against itself or by persuading its tormentor to cease and desist. The motive must be protection of the nation or its citizens or other national interests from further harm—the satisfaction of extracting revenge, by itself, is not acceptable. These doctrines also demand that a state do only what is necessary and proportional in the circumstances.

In summary, it appears that one trend in international law is to provide some kind of remedy for every violation of a nation's rights under international law. Some of these remedies are in the nature of self-help, such as armed self-defense, the interruption of commercial or diplomatic relations, or public protest. Other remedies may be sought from international institutions, such as an imposition of coercive measures by the Security Council, or a declaratory judgment or an order to make reparations from an international tribunal. The issue for the victim is to choose the most effective available sanction. The issue for a nation contemplating an action that may be considered to violate the rights of another nation under international law is to accurately predict what sanctions such action may provoke.

C. Application to Computer Network Attacks.

There is no way to be certain how these principles of international law will be applied by the international community to computer network attacks. As with other developments in international law, much will depend on how the nations and international institutions react to the particular circumstances in which these issues are raised for the first time. If we were to limit ourselves to the language of Article 51, the obvious question would be, "Is a computer network attack an 'armed attack' that justifies the use of force in self-defense?" If we focused on the means used, we might conclude that electronic signals imperceptible to human senses don't closely resemble bombs, bullets, or troops. On the other hand, it seems likely that the international community will be more interested in the consequences of a computer network attack than in its mechanism. It might be hard to sell the notion that an unauthorized intrusion into an unclassified information system, without more, constitutes an armed attack. On the other hand, if a coordinated computer network attack shuts down a nation's air traffic control system along with its banking and financial systems and public utilities, and opens the floodgates of several dams resulting in general flooding that causes widespread civilian deaths and property damage, it may well be that no one would challenge the victim nation if it concluded that it was a victim of an armed attack, or of an act equivalent to an armed attack. Even if the systems attacked were unclassified military logistics systems, an attack on such systems might seriously threaten a nation's security. For example, corrupting the data in a nation's computerized systems for managing its military fuel, spare parts, transportation, troop mobilization, or medical supplies may seriously interfere with its ability to conduct military operations. In short, the consequences are likely to be more important than the means used.

If the international community were persuaded that a particular computer network attack or a pattern of such attacks should be considered to be an "armed attack," or equivalent to an armed attack, it would seem to follow that the victim nation would be entitled to respond in self-defense either by computer network attack or by traditional military means in order to disable the equipment and personnel that were used to mount the offending attack. In some circumstances it may be impossible or inappropriate to attack the specific means used in an attack (e.g., because the specific equipment and personnel used cannot be reliably identified or located, or an attack on the specific means used would not be effective, or an effective attack on the specific means used might result in disproportionate collateral damage). Where the specific means cannot be effectively attacked, any legitimate military target could be attacked, including intelligence and military leadership targets, as long as the purpose of the attack is to dissuade the enemy from further attacks or to degrade the enemy's ability to undertake them.

There has been some support for the proposition that a nation has an inherent right to use force in self-defense against acts that do not constitute a classic armed attack. This view is supported by the inclusion in the General Assembly's definition of aggression of acts that do not entail armed attacks by a nation's armed forces, such as the unlawful extension of the presence of visiting forces, or allowing a nation's territory to be used by another state "for perpetrating an act of aggression against a third State." US practice also support this position, as demonstrated in the 1986 bombing of Libyan command and leadership targets to persuade Libya to stop sponsoring terrorist attacks against US interests, and in the 1998 attack on the Iraqi military intelligence headquarters to persuade Iraq to desist from assassination plots against former President Bush. A contrary view was expressed in the International Court of Justice's 1986 ruling in Nicaragua v. US that the provision of arms by Nicaragua to the leftist rebels in El Salvador did not constitute an armed attack on El Salvador, so it could not form the basis of a collective self-defense argument that would justify armed attacks in response, such as laying of mines in Nicaraguan waters or certain attacks on Nicaraguan ports, oil installations and a naval base—acts that were "imputable" to the United States. The Court also said it had insufficient evidence to determine whether certain cross-border incursions by Nicaraguan military forces into the territory of Honduras and Costa Rico [sic] constituted armed attacks. The extent to which Nicaragua's conduct would justify El Salvador and its ally the United States in responding in ways that did not themselves constitute an armed attack was not before the Court. The opinion of the court nevertheless provides some support for the proposition that the provocation must constitute an armed attack before it will justify an armed attack in self-defense. It seems safe to say that the issue of whether traditional armed force may be used in self-defense in response to provocations that are not technically regarded as armed attacks is far from settled, and that the positions taken by states may be sharply influenced by the nature of the events concerned, together with all attendant policy and political considerations.

By logical implication, to the extent that a nation chooses to respond to a computer network attack by mounting a similar computer network attack of its own, the issue of whether the initial provocation constituted an armed attack may become a tautology. If the provocation is considered to be an armed attack, the victim may be justified in launching its own armed attack in self-defense. If the provocation is not considered to be an armed attack, a similar response will also presumably not be considered to be an armed attack. Accordingly, the question of the availability of the inherent right of self-defense in response to computer network attacks comes into sharpest focus when the victim of a computer network attack considers acting in self-defense using traditional military means. The issue may also arise if the response causes disproportionately serious effects (e.g., if a state responded to a computer network attack that caused only minor inconvenience with its own computer network attack that caused multiple deaths and injuries). As in all cases when a nation considers acting in self-defense, the nation considering such action will have to make its best judgment on how world opinion, or perhaps a body such as the International Court of Justice (ICJ) or the UNSC, is likely to apply the doctrine of self defense to electronic attacks. As with many novel legal issues, we are likely to discover the answer only from experience.

It seems beyond doubt that any unauthorized intrusion into a nation's computer systems would justify that nation at least in taking self-help actions to expel the intruder and to secure the system against reentry. An unauthorized electronic intrusion into another nation's computer systems may very well end up being regarded as a violation of the victim's sovereignty. It may even be regarded as equivalent to a physical trespass into a nation's territory, but such issues have yet to be addressed in the international community. Furthermore, the act of obtaining unauthorized access to a nation's computer system creates a vulnerability, since the intruder will have had access to the information in the system and he may have been able to corrupt data or degrade the operating system. Accordingly, the discovery that an intrusion has occurred may call into question the reliability of the data and the operating system and thus reduce its utility. If an unauthorized computer intrusion can be reliably characterized as intentional and it can be attributed to the agents of another nation, the victim nation will at least have the right to protest, probably with some confidence of obtaining a sympathetic hearing in the world community.

D. An "Active Defense" Against Computer Network Attacks.

A persistent foreign intruder who gains repeated unauthorized entry into a nation's computer systems by defeating a variety of security measures or who gains entry into a number of computer systems may demand a different response. Such behavior may indicate both that there is a continuing danger and that coercive measures are necessary to stop the intruder's pattern of conduct. Similarly, there may be a right to use force in self defense against a single foreign electronic attack in circumstances where significant damage is being done to the attacked system or the data stored in it, when the system is critical to national security or to essential national infrastructures, or when the intruder's conduct or the context of the activity clearly manifests a malicious intent.

If it is capable of doing so, in such circumstances the victim nation may be justified in launching a computer attack in response, intended to disable the equipment being used by the intruder. Disabling one computer may or may not defeat a state-sponsored operation. It may, however, serve as a "shot across the bow" warning of more serious consequences if the offending behavior continues. It is also an action unlikely to come to public attention unless one of the two governments announces it, making it a potentially useful measure for conflict avoidance. Conducting a responsive computer network attack as a measure of self-defense against foreign computer network attacks would have the major advantage that it would minimize issues of proportionality, which would be more likely to arise if traditional military force were used, such as firing a cruise missile at the building from which a computer network attack is being conducted. Either response would likely be analyzed on the basis of the traditional criteria of necessity and proportionality.

If it is impractical to focus an attack on the equipment used in the provocation, any legitimate military target may be attacked. The primary value of being able to demonstrate a nexus between the provocation and the response is to be able to argue the likely therapeutic effect of the force used in self-defense. As a practical matter, the next most attractive target after the equipment used in the provocation may be the offending nation's communications systems, or its military or intelligence chain of command. The consequences of a large-scale campaign of computer network attacks might well justify a large-scale traditional military response. A Russian academic took this argument to its extreme in a published statement to the effect that Russia reserves the right to respond to an information warfare attack with nuclear weapons.

As stated above, the discussion up to this point has assumed we know who an intruder is, and that we are confident in characterizing his intent. In practice, this is seldom the case, at least in the early stages of responding to computer intrusions. The above legal analysis may change if the identity and location of an intruder is uncertain, or if his intent is unclear.

Identification of the originator of an attack has often been a difficult problem, especially when the intruder has used a number of intermediate relay points, when he has used an anonymous bulletin board" whose function is to strip away all information about the origin of messages it relays, or when he has used a device that generates false origin information. Progress has been made, however, in solving the technical problem of identifying the originator of computer messages, and reliable identification of the computer that originated a message may soon be routinely available. Attribution may also be provided by intelligence from other sources, or it might be reliably inferred from the relationship of the attack to other events.

Locating the computer used by the intruder does not entirely solve the attribution problem, however, since it may have been used by an unauthorized person, or by an authorized user for an unauthorized purpose. A parent may not know that the family computer is being used for unlawful attacks on government computer systems. Universities, businesses, and other government agencies may be similarly unaware that their computer systems are being misused. The owner of a computer system may have some responsibility to make sure it is not being used for malicious purposes, but the extent of such responsibility, and the consequences of failing to meet it, have apparently not been addressed in any US or foreign statute or court decision. These considerations should make us cautious in implementing any "active defense" system for government computer systems. Nevertheless, circumstances may arise in which the urgency of protecting critical information systems from serious damage may warrant adoption of a properly designed "active defense."

Similarly, characterization of an intruder's intentions may be difficult. Nevertheless, such factors as persistence, sophistication of methods used, targeting of especially sensitive systems, and actual damage done may persuasively indicate both the intruder's intentions and the dangers to the system in a manner that would justify use of an "active defense." As with attribution, there may be useful intelligence on this issue from other sources, or it may be possible to reliably infer the intent of the intruder from the relationship of the attack to other events.

A determination that an intrusion comes from a foreign country is only a partial solution to the attribution problem, since the attack may or may not be state-sponsored. State-sponsored attacks may well generate the right of self-defense. State sponsorship might be persuasively established by such factors as signals or human intelligence, the location of the offending computer within a state-controlled facility, or public statements by officials. In other circumstances, state sponsorship may be convincingly inferred from such factors as the state of relationships between the two countries, the prior involvement of the suspect state in computer network attacks, the nature of the systems attacked, the nature and sophistication of the methods and equipment used, the effects of past attacks, and the damage which seems likely from future attacks.

Attacks that cannot be shown to be state-sponsored generally do not justify acts of self defense in another nation's territory. States jealously guard their sovereign prerogatives, and they are intolerant of the exercise of military, law-enforcement, and other "core sovereign powers" by other states within their territory without their consent. When individuals carry out malicious acts for private purposes against the interests of one state from within the territory of a second state, the aggrieved state does not generally have the right to use force in self-defense against either the second state itself or the offending individual. Even if it were possible to conduct a precise computer network attack on the equipment used by such individual actors, the state in which the effects of such an attack were felt, if it became aware of it, could well take the position that its sovereignty and territorial integrity had been violated. The general expectation is that a nation whose interests are damaged by the private conduct of an individual who acts within the territory of another nation will notify the government of that nation and request its cooperation in putting a stop to such conduct.

Only if the requested nation is unwilling or unable to prevent recurrence does the doctrine of self-defense permit the injured nation to act in self-defense inside the territory of another nation. The US cruise missile strikes against terrorists camps in Afghanistan on 20 August 1998 provides a close analogy in which the United States attacked camps belonging to a terrorist group located in the territory of a state which had clearly stated its intention to continue to provide a refuge for the terrorists. At some point, providing safe refuge for those who conduct attacks against another nation becomes complicity in those attacks. At a minimum, the offended nation is authorized to attack its tormenters, the terrorists. As complicity shades into the kinds of active support and direction that are commonly called "state sponsorship," military and leadership targets of the host state may themselves become lawful targets for acts of self-defense.

Attacks on insurgents or on terrorists and other criminals using a neutral nation's territory as a refuge may also be justified when the neutral state is unable to satisfy its obligations. During the Vietnam war, the United States attacked North Vietnamese military supply lines and base camps in Cambodia after the Cambodian government took the position that it was unable to prevent North Vietnam from making such use of its territory. This principle might justify using active defense measures against a computer intruder located in a neutral nation if the government of the neutral nation declared it had no way to locate the intruder and make him stop, or if its behavior made it clear that it could not or would not act, or even if the circumstances did not allow time for diplomatic representations to be effective. As an analogy, it seems unlikely that a nation would complain very loudly if its neighbor nation returned fire against a terrorist sniper firing from its territory.

In summary, the international law of self-defense would not generally justify acts of "active defense" across international boundaries unless the provocation could be attributed to an agent of the nation concerned, or until the sanctuary nation has been put on notice and given the opportunity to put a stop to such private conduct in its territory and has failed to do so, or the circumstances demonstrate that such a request would be futile. Nevertheless, in some circumstances the National Command Authorities (NCA) might decide to defend US information systems by attacking a computer system overseas, and take the risk of having to make an apology or pay compensation to the offended government. Among the factors the NCA would probably consider would be the danger presented to US national security from continuing attacks, whether immediate action is necessary, how much the sanctuary nation would be likely to object, and how the rest of the world community would be likely to respond.

There need be less concern for the reaction of nations through whose territory or communications systems a destructive message may be routed. If only the nation's public communications systems are involved, the transited nation will normally not be aware of the routing such a message has taken. Even if it becomes aware of the transit of such a message and attributes it to the United States, there would be no established principle of international law to which it could point as being violated. As discussed above, even during an international armed conflict international law does not require a neutral nation to restrict the use of its public communications networks by belligerents. Nations generally consent to the free use of their communications networks on a commercial or reciprocal basis. Accordingly, use of a nation's communications networks as a conduit for an electronic attack would not be a violation of its sovereignty in the same way that would be a flight through its airspace by a military aircraft.

A transited state would have somewhat more right to complain if the attacking state obtained unauthorized entry into its computer systems as part of the communications path to the target computer. It would be even more offended if malicious logic directed against a target computer had some harmful effect against the transited state's own equipment, operating systems, or data. The possibility of such collateral damage would have to be carefully considered by the state launching any such attack. If there were a high potential for such collateral damage to transited systems, the weapon might even be considered to be an "indiscriminate" weapon incapable of being reliably directed against a legitimate target.

There are at least two ways in which the availability of improved technology may affect the active-defense equation. First, it might be argued that as a government acquires the ability to build better firewalls and other security systems it will be harder to argue that an active defense is "necessary." This argument might be raised even if the target government has failed to install all possible technological security measures on the system that is under attack. This demanding approach to "necessity" finds little support in the practice of nations. The focus of self-defense analysis is on events as they unfold, and not as they might have been if different budgeting and acquisition decisions had been made sometime in the past. If such systems are in place, however, their apparent effectiveness should be taken into account in deciding whether active defense measures are necessary. This does not mean that a nation has no right of self-defense where a first attempted intrusion fails, or even when a series of intrusions fall. If an attacker is permitted to continue mounting a campaign of such attacks it may learn by trial and error, it may employ other capabilities, or it may stumble onto a point of vulnerability. Just as an infantry unit exercising the right of self-defense may pursue a force that breaks off an attack and attempts to retreat until the attacker ceases to be a threat, decisions on taking measures of self-defense against computer network attacks must take into account the extent to which an attacker continues to present a threat of continuing attacks.

Another possible implication of a defender's technological prowess may arise when a nation has the capacity for graduated self-defense measures. Some may argue that a nation having such capabilities must select a response that will do minimal damage. This is a variant of the argument that a nation possessing precision-guided munitions must always use them whenever there is a potential for collateral damage. That position has garnered little support among nations and has been strongly rejected by the United States. There is broad recognition that the risk of collateral damage is only one of many military considerations that must be balanced by military authorities planning an attack. One obvious consideration is that a military force that goes into a protracted conflict with a policy of always using precision-guided munitions whenever there is any potential for collateral damage will soon exhaust its supply of such munitions. Similarly, military authorities must be able to weigh all relevant military considerations in choosing a response in self-defense against computer network attacks. These considerations will include the probable effectiveness of the means at their disposal, the ability to assess their effects, and the "fragility" of electronic means of attack (i.e., once they are used, an adversary may be able to devise defenses that will render them ineffective in the future). In the process of reasoning by analogy to the law applicable to traditional weapons, it must always be kept in mind that computer network attacks are likely to present implications that are quite different from the implications presented by attacks with traditional weapons. These different implications may well yield different conclusions.

It may be possible to specify certain information systems that are vital to national security—both government systems and key civilian infrastructure systems. This process should serve both to give such systems high priority for security measures and also to identify a class of systems any attack on which would immediately raise the issue of whether an active defense should be employed. This should not, of course, eliminate consideration of using an active defense against attacks on systems not on such a "vital systems" list where the circumstances justify such action. For example, a vigorous attack that threatens to overwhelm an information system not on the "vital systems" list but that performs an important national security function could be a more valid occasion to use active defense measures than would be a trivial and easily defeated attack on a designated "vital system." A list of "vital systems" would serve primarily as a alert mechanism that would bring about a prompt high-level evaluation of all the circumstances.

In addition, it would be useful to create a process for determining when the response to a computer intrusion should shift from the customary law enforcement and counter-intelligence modes to a national defense mode. Such a process should include (1) a statement of general criteria to be applied; (2) identification of officials or agencies that will be involved in making the decision; and (3) procedures to be followed.

There are of course a variety of treaty obligations that will have to be considered before adopting an "active defense" against foreign computer network attacks, and these will be discussed below. There are also a variety of domestic legal concerns that will have to be addressed, and these will be discussed in the companion assessment of domestic law issues in information operations.

E. Assessment.

It is far from clear the extent to which the world community will regard computer network attacks as "armed attacks" or "uses of force," and how the doctrines of self-defense and countermeasures will be applied to computer network attacks. The outcome will probably depend more on the consequences of such attacks than on their mechanisms. The most likely result is an acceptance that a nation subjected to a state-sponsored computer network attack can lawfully respond in kind, and that in some circumstances it may be justified in using traditional military means in self-defense. Unless the nations decide to negotiate a treaty addressing computer network attacks, which seems unlikely anytime in the near future, international law in this area will develop through the actions of nations and through the positions the nations adopt publicly as events unfold. US officials must be aware of the implications of their own actions and statements in this formative period.

IV. SPACE LAW

A. Introduction.

International law regulating activities in outer space is important to the information operator because space segments are critical to so many important information systems. These systems perform such functions as communications relay, imagery collection, missile warning, navigation, weather forecasting, and signals intelligence. In fact, it can be said that at the current stage of space activity, the exclusive functions of both military and civilian satellites are to gather and relay information. In the conduct of information operations, there will be strong imperatives to interfere with the space-based information systems belonging to an adversary, and to defend one's own.

One approach to attacking space systems is by targeting their ground stations. Another approach is to jam or "spoof" their communications links. Such actions are subject to the normal international law principles governing other terrestrial activity. Sometimes, however, it may be more effective to attack the satellite or satellites that form the space segment of the system. As we will see, activities in space are subject both to general principles of international law and to a number of treaty obligations that apply specifically to space activities.

B. Space Law Treaties.

There is probably no other field of human endeavor that produced so much international law in such a short period. Within twenty years after the first Sputnik launch in 1957, international diplomatic conferences produced four major widely accepted multilateral space law treaties. Taken together, these treaties provide the foundations of existing space law.

The Treaty on Principles Governing the Activities of States in the Exploration and Use of Outer Space, including the Moon and Other Celestial Bodies (the Outer Space Treaty, 1967)

The Agreement on the Rescue of Astronauts, Return of Astronauts, and the Return of Objects Launched into Outer Space (the Rescue and Return Agreement, 1968)

The Convention on International Liability for Damages Caused by Space Objects (the Liability Convention, 1972)

The Convention on the Registration of Objects Launched into Outer Space (the Registration Convention, 1975)

NOTE: There is another treaty called the Moon Agreement of 1979 that the United States has never signed and which has attracted only 9 parties, among whom only France is active in space operations. In addition, several provisions of the 1980 Environmental Modification Convention apply to space activity. These agreements are not directly relevant to information operations, however, and they will not be discussed further here.

The four major space treaties together establish the following principles that are directly relevant to information operations. These principles have been so widely accepted that they are generally regarded as constituting binding customary international law, even for non-parties to these agreements.

- Space is free for exploration and use by all nations. It is not subject to national appropriation by claim of sovereignty, use, occupation, or any other means.

- Activities in space shall be conducted with due regard for the interests of other states.

- States that launch space objects are liable for any damage they may do in space, in the air, or on the surface of the Earth. Different standards of liability are established for damage done to other items in space, for which a "fault" standard applies, and damage done on the surface of the Earth and to aircraft in flight, for which absolute liability applies.

- Space activities are subject to general principles of international law, including the UN Charter.

Several conclusions are apparent from these general principles. The first is that the rules on the use of force discussed in Section III of this paper apply fully to activities in outer space. Among these are that nations are obliged not to use force in their relations with each other unless they are acting in self-defense or when authorized to do so by the UN Security Council. Once again, however, as with other forms of information operations, one has to consider what actions by or against objects in space will be considered to be uses of force. The world community would probably not hesitate to regard as a use of force the destruction of a satellite by a missile or a laser. It would probably react similarly if it could be proven that one nation took over control of another nation's satellite by electronic means and caused it to fire its retro rockets and fall out of orbit. In such a case, the consequences will probably matter more than the mechanism used. The reaction of the world community to lesser kinds of interference is hard to predict. For example, if one nation were able by electronic means to suspend the operations of another nation's satellite for a brief period, after which it returned to service undamaged, it seems likely that the world community would consider such action as a breach of the launching nation's sovereign rights, but not as a use of armed force.

One could argue, however, that this argument is unimportant because the space treaties create a specific obligation not to interfere with the space activities of other nations, and to pay reparations for any damages resulting from such interference. This argument appears to have considerable force, at least in peacetime. During an international armed conflict between the two nations concerned, however, the law of armed conflict would apply unless it was trumped by the principle of noninterference with space systems. Resolution of this issue depends largely on whether the four space treaties will be considered to apply during an armed conflict. None of them has any specific provision that indicates whether the parties intended that the agreement apply in wartime.

There appears to be a strong argument that the principle of noninterference established by these agreements is inconsistent with a state of hostilities, at least where the systems concerned are of such high military value that there is a strong military imperative for the adversary to be free to interfere with them, even to the extent of destroying the satellites in the system. As indicated in the discussion of treaty law in the introduction to this paper, the outcome of this debate may depend on the circumstances in which it first arises in practice. Nevertheless, it seems most likely that these agreements will be considered to be suspended between the belligerents for the duration of any armed conflict, as least to the extent necessary for the conduct of the conflict.

If the principle of noninterference is regarded as suspended for the period of the conflict, it also seems likely that the liability provisions in these agreements would also be suspended, at least between the parties. This would not, however, excuse the belligerents from liability to neutral nations if their actions caused damage to their citizens or property.

C. Specific Prohibitions of Military Activities in Space

There is a popular notion that military activities in space are prohibited—that space is a place a little closer to heaven into which the nations have agreed not to introduce weapons and human conflict. There is a germ of truth in this notion, supported by high flights of rhetoric in international fora, but the existing treaty restrictions on military operations in space are in fact very limited. These restrictions are included in both the space treaties listed above and in various arms control agreements.

The Outer Space Treaty provides that the parties will not "place in orbit around the Earth any objects carrying nuclear weapons or any other kinds of weapons of mass destruction, install such weapons on celestial bodies [i.e., the moon, planets, and asteroids], or station such weapons in outer space in any other manner." The treaty permits placing in orbit weapons other than nuclear weapons and other weapons of mass destruction. Also, the treaty contains no prohibition against nuclear weapons transiting outer space, as long as they do not enter into an earth orbit and they do not explode in outer space.

The Outer Space Treaty also prohibits the establishment of military bases, the testing of weapons, and the conduct of military maneuvers on the moon or other celestial bodies. It permits these activities in orbit around the Earth, and in other places in outer space. Similarly, there is no prohibition against establishing military space stations or operating other satellites with offensive or defensive capabilities.

The *Treaty Banning Nuclear Weapons Tests in the Atmosphere, in Outer Space and Under Water* (the Limited Test Ban Treaty, 1963) prohibits all nuclear explosions in outer space. Accordingly, a party to this agreement may not lawfully explode a nuclear device in outer space in order to disable an adversary's satellites by means of the electromagnetic pulse generated by a nuclear explosion, or by its other effects. A nation operating its own satellite systems is unlikely to take such an action in any event, since its own satellites would be subject to the same effects as those belonging to its adversary.

The *Treaty on the Limitation of Anti-Ballistic Missile Systems* (the ABM Treaty, 1972) provides that no party may "develop, test or deploy space-based ABM systems or components."

Under a 1997 theater missile defense (TMD) agreement not yet ratified by the Senate, the United States and Russia have agreed not to place in space theater missile defense interceptor missiles "or space-based components based on other physical principles, whether or not part of a system, that are capable of substituting for such interceptor missiles."

A number of arms control agreements provide that no party will interfere with the others' "national technical means of verification." Translated, this means no interference with the orbiting imaging systems used to monitor the strategic arms of another party.

Read together, these agreements permit the development, testing, and deployment of anti-satellite and satellite-defense systems unless they involve either the stationing or testing of nuclear devices in outer space or the orbiting of systems that also have ABM or ATM capabilities. Their use is subject only to (1) the general principles of international law relating to the use of force; (2) the principle of noninterference with the space systems of other nations in peacetime, subject to the right to use force in self-defense and when authorized by the UN Security Council; (3) the law of war during international armed conflicts; and (4) obligations under relevant arms-control agreements not to interfere with other parties' national technical means of verification. This leaves a very broad range of permissible "space-control" systems and operations.

In a non-nuclear conflict, the parties might very well determine that the treaty prohibitions against placing nuclear weapons in orbit, against exploding nuclear devices in outer space, and against placing ABM components and ATM interceptors in orbit remain consistent with a state of limited armed conflict. Those obligations may well serve to avoid escalation of the conflict to the nuclear level. The parties' conclusions as to the obligation not to interfere with other parties' national technical means of verification will probably depend to a great extent on the circumstances of the conflict.

D. Domestic Law and Policy.

A federal statute, 18 USC 1367, makes it a felony to intentionally or maliciously interfere with a communications or weather satellite, or to obstruct or hinder any satellite transmission. The application of this statute to national security information operations is discussed in the companion assessment of domestic legal issues.

US domestic policy on developing space control capabilities has been inconsistent at best. By the early 1980s the US Air Force had developed an anti-satellite missile with an explosive warhead that was carried aloft by an F- 15 fighter and launched at high altitude. A test of this system was conducted in 1985 against a US satellite whose useful life had expired. Congress soon thereafter decreed that no appropriated funds were to be used to test any weapon against an object in orbit. In 1987 the USAF program was terminated. At the time, it appeared that members of Congress voting for the ban had done so for a variety of reasons, among which were: (1) support for the broad principle that space should be free from human conflict; (2) dismay that the first test had generated 285 pieces of trackable space debris; (3) concern that further testing of an anti-satellite capability might interfere with continuing strategic arms control negotiations; and (4) concern that the United States should not press ahead with testing an anti-satellite system when the nation had yet to decide where its own long-term interests lie. Concerning this last point, it was obvious that there is a military interest in being able to defend your own space systems and having the ability to interfere with your adversary's, but there was also a contrary consideration that the long-term interests of the United States—as the nation that depends most heavily on space systems—may be better served by promoting the development of a regime of international law that prohibits any interference by one nation with the space systems of another, and inhibits the acquisition of the capability to do so. That fundamental debate has yet to be pursued to a definitive conclusion.

Later, when public attention was drawn to the possible use of lasers as anti-satellite weapons, Congress prohibited the use of appropriated funds to illuminate any object in orbit with a laser. This restriction was removed in 1995. In October 1997 the US Army conducted a test in which it illuminated an Air Force satellite nearing the end of its useful life with the MIRACL laser, located at White Sands, New Mexico. Despite public announcements that the purpose of the experiment was purely defensive in nature—to observe the effects of the laser on the satellite's optical sensors in order to better protect US satellites from deliberate or accidental laser illumination—a public furor ensued. Shortly thereafter President Clinton exercised his short-lived item veto authority to delete funds from the FY 98 DoD Authorization Act for development of an Army Kinetic Energy Anti-Satellite Missile and two other projects that he considered to be related to space control. Congress approved additional funds for space control projects in the FY 1999 DoD Authorization Act and urged expenditure of the FY 98 funds that were restored after the Supreme Court ruled that the item veto was unconstitutional.

At this point, it seems fair to say that the United States has not arrived at a consensus on the fundamental policy issues concerning space control. It seems likely for the near future that the development of such systems will continue, with renewed controversy to be expected as soon as a decision is imminent on the deployment, or even advanced testing, of an operational system.

E. International Efforts to Control "Weaponization of Space."

Over the last decade there has been strong support in the UN General Assembly for negotiation in the Conference on Disarmament (CD) of a draft treaty banning weapons in space. The most recent action by the General Assembly was its adoption on 4 December 1998 by a vote of *165-0-4* of a resolution entitled "Prevention of an arms race in outer space." This resolution calls for reestablishment by the CD of an Ad Hoc Committee on the Prevention of an Arms Race in Outer Space that existed in prior years. Canada and Egypt are actively promoting consideration of a "no weapons in space" treaty in the CD, but so far they have garnered little active support among the other CD members. Both Russia and China have also announced their support for negotiations to ban "weaponization of space," but neither has advanced a specific proposal with much vigor. In summary, there appears to be widespread lukewarm support for the general idea of a treaty banning an "arms race in space," but the subject enjoys a low priority at the moment and no draft treaty has garnered significant support. This may all change if and when a nation or nations are known to have deployed operational space control systems, or are on the verge of doing so.

Chinese and Russian support for a ban on "weaponization of space" is seen in some quarters as ironic, since China is reported to be developing a ground-based anti-satellite laser system and Russia is the only nation known to have once had an operational anti-satellite missile. There have been a number of reports that the Soviet Union developed a "co-orbital ASAT" that was launched into orbit, where it maneuvered close enough to a target satellite to destroy the target by exploding. Reportedly, the Soviet system was tested against objects in space 20 times and became operational in 1978. Russia consistently denied that it had tested or deployed such a system until September 1997, when press reports indicate that President Yeltsin said in a letter to President Clinton that Russia at one time possessed an anti-satellite capability, but that it had since "renounced" it.

F. Assessment.

There is no legal prohibition against developing and using space control weapons, whether they would be employed in orbit, from an aircraft in flight, or from the Earth's surface. The primary prohibition is against weapons that entail the placing of nuclear weapons in orbit or that would employ a nuclear explosion in outer space. The use of space control systems in peacetime would be subject to both the general principles of international law and to treaty obligations not to interfere with other nations' space systems and national technical means of verification. These obligations would probably be suspended during an international armed conflict, during which the parties' conduct would be governed primarily by the law of war. US domestic policy on space control, however, is at best unsettled.

V. COMMUNICATIONS LAW

A. International Communications Law.

International communications law consists primarily of a number of bilateral and multilateral communications treaties. The most significant of these treaties is the *International Telecommunications Convention of 1982* (ITC), which has over 140 parties and which became effective for the United States in 1986. This agreement, often referred to as the Nairobi Convention, is the latest in a series of widely adhered to multilateral telecommunications conventions signed in this century, which were preceded by multilateral agreements in the late 1800s providing protection for submarine cables. The current series of agreements establishes the International Telecommunication Union (ITU), which has the status of a specialized agency of the United Nations, and they invest the ITU with the authority to formulate telegraph and telephone regulations which become binding legal obligations upon formal acceptance by ITU member nations. These agreements also establish mutual legal obligations among the parties, several of which are directly relevant to information operations.

Perhaps the most significant of these obligations is in Article 35, which provides that all radio "stations, whatever their purpose, must be established and operated in such a manner as not to cause harmful interference to the radio services or communications of other Members or of recognized private operating agencies, which carry on radio service, and which operate in accordance with the provisions of the Radio Regulations." "Harmful interference" is defined in

Annex 2 to the Convention as "interference which endangers the functioning of a radio navigation service or of other safety services or seriously degrades, obstructs or repeatedly interrupts a radio communication service operating in accordance with the Radio Regulations." One of the clearest violations of this provision would be the jamming or "spoofing" of a radio navigation service. Without speculating on all the possible permutations of the application of this provision to the broad range of information operations, suffice it to say that this provision on its face would appear to restrict many such operations that involve the use of radio broadcasting.

On the other hand, Article 38 of the ITC provides a specific exemption for military transmissions: "Members retain their entire freedom with regard to military radio installations of their army, naval and air forces." In July 1994, when the United States was considering broadcasting messages to the Haitian people from US military aircraft in international airspace urging them not to set out to sea in hazardous vessels, the Office of Legal Counsel in the Department of Justice relied on the military exemption in Article 38 as one of several bases for determining that the ITC does not prohibit such activity. Article 38 goes on to say, "Nevertheless, these installations must, so far as possible, observe.., the measures to be taken to prevent harmful interference, and the provisions of the Administrative Regulations concerning the types of emission and the frequencies to be used, according to the nature of the service performed by such installations." While this provision indicates that military installations do not have carte blanche to interfere with civilian communications, the phrase "so far as possible," read together with the specific exemption for military radio installations, provides considerable room for maneuver for information operations conducted by military forces.

The ITC also provides specific authority for its member nations to interfere with international telecommunications in certain circumstances:

Article 19 allows members to "stop the transmission of any private telegram which may appear dangerous to the security of the State or contrary to their laws, to public order or to decency, provided that they immediately notify the office of origin of the stoppage of any such telegram or part thereof, except when such notification may appear dangerous to the security of the State."

Article 19 also permits members to "cut off any other private telecommunications which may appear dangerous to the security of the State or contrary to its laws, to public order or to decency."

Article 20 reserves the right of members "to suspend the international telecommunication service for an indefinite time, either generally or only for certain relations and/or certain kinds of correspondence, outgoing, incoming or in transit, provided that it immediately notifies such action to each of the other Members through the medium of the Secretary-General."

Finally, it seems clear that the ITC' s provisions apply primarily in peacetime. The treaty does not specifically state how—if at all—it will apply during an armed conflict. Nevertheless, there is ample precedent in which nations have demonstrated conclusively that they regard the provisions of international communications conventions as being suspended between belligerents engaged in armed conflicts. Prior to the First World War, for example, all the major European nations were parties to the 1884 *Convention for Protection of Submarine Cables.* The first day of the war, the Royal Navy pulled up and cut the five major submarine cables serving Germany. Throughout all the wars of this century, communications facilities of all sorts have been regarded as priority military targets. Since some of the parties to the ITC and other multilateral communications conventions are likely to be neutrals in armed conflicts between other nations, the result may become somewhat complicated. Most ITC obligations will be considered to be suspended among the belligerents, but they will remain in effect between each belligerent and the neutral parties to the agreement, as well as among the neutral parties.

NOTE: The issue of the extent to which a neutral nation or an international communications consortium may continue to provide communications services to a belligerent is discussed in the law of war section of this paper.

The United States has not entered into bilateral communications agreements with very many nations, primarily because the ITC and the ITU provide a framework for handling most international communications issues. As one might expect, the need for bilateral communications agreements has arisen for the United States primarily with Canada and Mexico, because of the potential for interference in broadcast communications across our common borders. A number of bilateral communications agreements have also been negotiated between the United States and nations where US military forces are stationed. There is a potential for such bilateral agreements to either restrict or facilitate information operations by US military forces. The agreements concerned should be consulted when such an issue arises.

B. Domestic Communications Law.

The ITC and its predecessors obligate each Member nation to suppress acts by individuals or groups within its territory that interfere with the communications of other members. In partial satisfaction of this obligation, in 1934 Congress enacted 47 USC 502, which provides, "Any person who willfully and knowingly violates any rule, regulation, restriction, or condition made or imposed by any international radio or wire communications treaty or convention, or regulations annexed thereto, to which the United States is or may hereafter become a party, shall, in addition to any other penalties provided by law, be punished, upon conviction thereof, by a fine of not more than $500 for each and every day during which such offense occurs." In October 1993, when the United States was considering broadcasting radio messages to the people of Haiti supporting the return of democracy in that nation, the Office of Legal Counsel of the Department of Justice issued a written opinion to the effect that 47 USC 502 does not apply to the actions of US military members executing the instructions of the President acting within his constitutional powers to conduct foreign policy and to serve as Commander-in-Chief of US military forces. Further discussion of this statute can be found in the companion assessment of domestic legal issues in information operations.

C. Assessment.

International communications law contains no direct and specific prohibition against the conduct of information operations by military forces, even in peacetime. The established practice of nations provides persuasive evidence that telecommunications treaties are regarded as suspended among belligerents during international armed conflicts. Domestic communications laws do not prohibit properly authorized military information operations. Accordingly, neither international nor domestic communications law presents a significant barrier to information operations by US military forces.

VI. IMPLICATIONS OF OTHER TREATIES

The State Department's most recent published list of international agreements to which the United States is a party, *Treaties in Force,* January 1, 1997, is 485 pages long. The United States is a party to literally thousands of multilateral and bilateral international agreements. From their sheer numbers, one would think it inescapable that lurking somewhere in those agreements are provisions that will affect particular information operations activities. This section attempts only to highlight certain kinds of "typical" agreements that are likely to contain obligations relevant to the conduct of information operations.

A. Mutual Legal Assistance Agreements.

Mutual legal assistance agreements (sometimes called judicial assistance agreements) obligate each party to gather and provide evidence located in its territory concerning litigation or criminal prosecutions that occur within the jurisdiction of another party requesting such assistance. The United States is a party to several dozen mutual legal assistance agreements. Some of these agreements apply only to the management of particular litigation or to certain types of offenses such as drug trafficking and money laundering. Only a few mutual legal assistance agreements apply broadly to all law enforcement investigations and prosecutions. Such an agreement may supply the only domestic legal authority for the assisting party to investigate offenses that did not occur within its jurisdiction, and it also establishes procedures that expedite the requested assistance. To be effective in helping to suppress computer crimes and other high-tech offenses, mutual legal assistance agreements must either expressly cover such offenses or they must apply broadly to all crimes.

B. Extradition Agreements.

Extradition agreements obligate the parties in certain circumstances to deliver persons accused of crime to the other party for criminal prosecution. The United States is a party to more than a hundred bilateral extradition treaties, as well as to a 1933 Convention on Extradition to which thirteen nations in the Americas are parties. If no extradition treaty is in effect, a national government often will have neither an international obligation nor the domestic authority to deliver custody of an individual to another nation for the purpose of prosecution. It is important that the list of offenses covered by such agreements include computer intrusions and other high-tech crimes. In addition, the effectiveness of extradition treaties is often frustrated by provisions providing that the requested nation will not extradite its own citizens, or that it will not extradite persons who commit crimes for political reasons.

NOTE: The Department of Justice has undertaken a major initiative with the "G8" countries (the other seven being the United Kingdom, Germany, Japan, Italy, Canada, France, and Russia) to modernize the domestic criminal law of each nation to adequately provide for the investigation and prosecution of computer intrusions and other high-tech crimes, and to put into place any needed improvements to international agreements providing for mutual legal assistance and extradition. In December 1997 the Attorney General hosted a meeting of the G8 Justice and Interior Ministers to discuss these issues, and a number of follow-up working group meetings have been held since that time. The United States has also participated in a project undertaken by the Council of Europe to draft an international convention on "cyber-crime." Recently the United States undertook similar efforts in the Organization of American States and at the United Nations.

C. The United Nations Convention on the Law of the Sea (UNCLOS).

Many provisions of this treaty, which is before the Senate for advice and consent, are considered to express customary international law. Some of the provisions discussed here are among them, and are therefore considered to be binding on all nations whether or not they are parties to the Convention. Others constitute new obligations. One principle widely accepted as existing customary international law is the obligation in Article 19 for a vessel exercising the right of innocent passage through a nation's territorial sea not to engage in activities "prejudicial to the peace, good order, or security of the coastal State." The prejudicial activities listed in Article 19 include:

- any threat or use of force against the sovereignty, territorial integrity or political independence of the coastal State, or in any other manner in violation of the principles of international law embodied in the Charter of the United Nations

- any act aimed at collecting information to the prejudice of the defence or security of the coastal State

- any act of propaganda aimed at affecting the defence or security of the coastal State

- any act aimed at interfering with any systems of communication or any other facilities or installations of the coastal State"

Once UNCLOS is in general effect, these restrictions on activities aboard vessels in a coastal State's territorial sea will be of relatively minor importance because UNCLOS limits the width of the territorial sea a nation can claim to twelve nautical miles. At present, a number of nations claim territorial seas as wide as 200 miles. The twelve-mile limitation on the width of the territorial sea, together with other important guarantees UNCLOS establishes for the free operation of military aircraft and vessels, have led DoD to strongly support ratification of UNCLOS.

Article 109 of UNCLOS provides that all "States shall co-operate in the suppression of unauthorized broadcasting from the high seas" and defines unauthorized broadcasting, for the purposes of the Convention, as "the transmission of sound radio or television broadcasts from a ship or installation on the high seas intended for reception by the general public contrary to international regulations." The international regulations referred to consist primarily of the provisions of the Nairobi Convention and the ITU's Radio Regulations discussed in section V of this paper. This provision, which is generally regarded as establishing new law, was designed to deal with "pirate radio" broadcasting from vessels and platforms on the high seas, which became a significant problem for a number of countries in the 1960s. These broadcasts were primarily commercial in nature; by operating from the high seas they escaped the coastal state's regulation and taxation. Article 109 confers jurisdiction to prosecute persons engaged in pirate radio broadcasts upon the state whose flag the ship flies, the state where a broadcasting installation is registered, the state of which the broadcasting person is a citizen, any state where the transmissions can be received, and any state where authorized radio communication is suffering interference. Article 109 also provides that any state having jurisdiction to prosecute may "arrest any person or ship engaged in unauthorized broadcasting and seize the broadcasting apparatus."

Article 113 requires parties to adopt domestic criminal legislation punishing willful or culpably negligent damage to submarine cables belonging to other parties by ships or persons under their jurisdiction.

These UNCLOS provisions have the potential to affect only a narrow category of information operations, but they will have to be considered when decisions are made concerning those operations to which they do apply, at least in peacetime. UNCLOS does not expressly address how it will apply during an international armed conflict. In accordance with the general principles discussed in the introduction to this paper, provisions determined to be incompatible with a state of armed conflict will be regarded as suspended among the belligerents. The established practice of nations leaves no doubt that Article 19's regime governing innocent passage through the territorial sea will be suspended between belligerents. The same can be said with a high degree of confidence concerning

Article 113's protections for submarine cables. Article 109's provisions for the suppression of unauthorized radio broadcasting from the high seas are relatively new, with little established practice. Analytically, there would seem to be little reason to suspend its application to commercial broadcasters during an armed conflict, but it would almost certainly not apply to broadcasts from the high seas conducted by a belligerent for military or diplomatic purposes.

D. Treaties on Civil Aviation.

The United States is a party to a number of treaties concerning civil aviation, the most significant of which is the 1944 *Convention on International Civil Aviation.* This treaty, which has more than 180 parties, is often referred to as the Chicago Convention. It establishes the International Civil Aviation Organization (ICAO) and provides the basic legal framework for international civil aviation. The Convention does not directly apply to state aircraft, except for the obligation stated in Article 3(d): "The contracting States undertake, when issuing regulations for their state aircraft, that they will have due regard for the safety of navigation of civil aircraft." This concern for safe navigation by civil aircraft is also reflected in Article 28, which provides that each party will provide navigation and communications services as agreed upon through ICAO procedures, and in Article 37, which provides that the parties will comply with "international standards and recommended practices and procedures" on a variety of subjects including communications systems and air navigation aids. Over the years the ICAO Council has developed and adopted 18 technical Annexes to the Chicago Convention. Annex 10, Aeronautical Telecommunications, contains agreed provisions on aeronautical communications, navigation and surveillance. While military aircraft are not directly bound by these provisions, their obligation of "due regard" for the safety of civil aircraft generally includes an obligation not to interfere with these systems.

The United States is currently engaged in negotiations in ICAO concerning the role to be played by the Global Positioning System in future navigation systems for international civil aviation. In particular, an accommodation must be reached between ICAO' s interest in ensuring that navigation services essential to the safety of international civil aviation are not interrupted during an armed conflict, and the military imperative for the United States to be able to deny the use of GPS to a military adversary. Similar issues are certain to arise in the future in which information operations activities may create implications for the safety of international civil aviation.

The Chicago Convention is rare among multilateral treaties in that it has a specific provision concerning its application during armed conflict. Article 89 provides, "In case of war, the provisions of this Convention shall not affect the freedom of action of any of the contracting States affected, whether as belligerents or as neutrals. The same principle shall apply in the case of any contracting State which declares a state of national emergency and notifies the fact to the Council." Upon reflection, however, this provision is unlikely to be applied as broadly as its language indicates. It seems clear that many provisions of the Convention are inconsistent with a state of armed conflict. The most obvious is the principle that aircraft not engaged in scheduled airline service have the right to free passage into or through the airspace of other parties. Other provisions do not appear to be incompatible with a state of armed conflict among some of the parties. For example, the existence of a state of armed conflict among certain parties should not be regarded as suspending the belligerents' obligation to carry out their combatant activities with due regard for the safety of civil aviation. Accordingly, Article 89 does not provide much help in deciding what provisions of the Convention will remain applicable during an armed conflict, and resort will still be required to the general principle that only those obligations that are incompatible with a state of armed conflict will be suspended, and only among the belligerents.

E. Treaties on Diplomatic Relations.

The United States is a party to the 1961 *Vienna Convention on Diplomatic Relations,* a widely adhered to treaty establishing obligations among its parties concerning the treatment of diplomatic personnel and premises. Among the protections afforded a party's diplomatic mission in the territory of another state are the right to inviolability of the premises of the mission (Article 2); its "archives and documents" (Article 24); the private residences, papers, correspondence, and property of diplomatic agents (Article 30); and diplomatic communications (Article 27). The treaty further provides that the mission may communicate with its government and other missions and consulates of its government by "all appropriate means, including diplomatic couriers and messages in code or cipher. However, the mission may install and use a wireless transmitter only with the consent of the receiving State." Conversely, the treaty imposes certain duties on diplomatic missions. Article 41 provides that personnel of the mission must respect the laws and regulations of the receiving state, that they must not interfere in the receiving state's internal affairs, and that the "premises of the mission must not be used in any manner incompatible with the functions of the mission as laid down in the present Convention or by other rules of general international law or by any special agreements in force between the sending and the receiving State."

Article 45 provides that the duties of the receiving state continue in force even in the case of armed conflict between the parties, or if diplomatic relations are broken off between them, even though the staff of the mission is recalled. Planning for any information operations activity that involves diplomatic premises, persons, archives, documents, or communications, either as an instrument or as a target of the operation, must take into account these international legal obligations.

F. Treaties of Friendship. Commerce. and Navigation.

The United States is a party to a large number of bilateral agreements with other nations providing reciprocal arrangements for expedited tourism, trade, and transportation between the parties. These agreements have various titles, and their provisions differ somewhat. Most such agreements do not contain specific provisions on telecommunications, and they constitute perhaps the archetype of agreements that are likely to be regarded as suspended during an armed conflict because their provisions expediting free travel and trade between the parties are incompatible with hostilities between them. Nevertheless, planning for information operations, especially in peacetime, should include a review of all significant international agreements between the United States and any other nation that may be affected.

G. Status of Forces and Stationing Agreements.

When the military forces of one nation are present in the territory of another nation with its consent, it is customary for the nations involved to execute written agreements establishing the rights and obligations of the parties concerning the visiting forces. "Stationing agreements" establish the consent of the host nation to the presence of foreign troops; set agreed limits on their numbers, equipment, and activities; and identify facilities for their use. These topics may also be dealt with in a "defense cooperation agreement" or some other agreement providing for the overall defense relationship between the parties. It is also common for the parties to execute a "status of forces" agreement (SOFA) that addresses the allocation of various kinds of legal jurisdiction over the visiting forces. The best known of these agreements is the 1951 *Agreement Between the Parties to the North Atlantic Treaty Regarding the Status of Their Forces* (NATO SOFA). As of the end of 1998 the United States was a party to

103 SOFAs, most of which follow the general pattern of the NATO SOFA. SOFAs are necessary because of an overlap of legal jurisdiction exercised by the sending and receiving states. The receiving state has jurisdiction over persons and activities in its territory, while the sending state has both the right and the duty to exercise control over its armed forces, which is clearly a core sovereign function.

Since the full concurrent exercise of the normal jurisdiction of the sending and receiving states is impractical, status of forces agreements allocate criminal and civil court jurisdiction between the sending and receiving states, and also exempt the visiting force and its members from certain taxes, customs fees and procedures, immigration formalities, and most host nation licensing and inspection requirements. Typically, an administrative claims procedure is established for personal injuries and property damage caused by the visiting force. Another common provision requires that the visiting force and its members "respect" the host nation's laws. (This requirement will be discussed in detail in the next section of this paper). The NATO SOFA is implemented in most NATO countries by separate, more detailed, bilateral supplementary agreements, and by numerous other bilateral agreements on specific subjects including communications.

These agreements contain provisions that must be taken into account if US military forces intend to engage in information operations activities while present in the territory of the receiving state.

- For example, many such agreements require that the United States notify the host nation of any significant change in the capabilities or uses of installations made available for the use of US military forces. If US authorities intend to conduct information operations activities from such installations, a determination must be made as to whether the relevant agreements require notifying the host nation, and perhaps even requesting its consent.

- Stationing agreements often provide that the visiting US forces may install and use various communications equipment, but they often provide as well that such equipment must not interfere with host nation communications systems and that it must be used in accordance with host nation laws and regulations. If this equipment is to be used for information operations activities, it must be determined whether the contemplated activities are consistent with these obligations.

- Many stationing agreements authorize or even obligate the visiting force to use the receiving state's military and civilian communications systems. Commonly, there are obligations that any US use of host nation communications systems must not cause interference and that such use must be in accordance with host nation laws and regulations. The potential for information operations to cause interference with the host nation's communications system and the possible application of host nation laws and regulations must be carefully considered, along with the fact that the conduct of offensive information operations through host nation communications systems may subject them to possible countermeasures and acts of self-defense in peacetime, and may make them legitimate military targets during an armed conflict.

Finally, if a host nation discovers that its territory and facilities have been used without its knowledge as a base for US information operations of a nature that may tend to involve it against its will in a conflict or dispute, US diplomatic and military relationships with the host nation are likely to suffer. The host nation could well take the view that in principle there is little difference between using an ally's territory to launch air strikes and using it to launch computer network attacks or other information operations activities. As a practical matter, computer network attacks are much more difficult to identify, trace, and attribute. However, it will not always be impossible to do so, particularly when information on such attacks is available from intelligence sources. Accordingly, decisions concerning whether to conduct information operations from the territory of an ally, and especially whether to do so without the host nation's knowledge and consent, must be made at senior policy levels.

VII. FOREIGN DOMESTIC LAWS

A. Introduction.

Laws enacted by other nations may have important implications for information operations activities conducted by US military forces. As discussed in the companion assessment of domestic legal issues in information operations, US criminal statutes addressing computer-related offenses, space activities, communications, and the protection of classified information all raise important issues for information operations. These US statutes are part of the federal criminal code. Similarly, foreign laws affecting US information operations activities will most likely also consist of criminal statutes.

The sophistication of foreign domestic law on high-tech activities varies enormously, and it will continue to do so for the foreseeable future. The more technologically advanced countries tend to be better aware of the dangers created by computer hackers and other high-tech criminals, so they typically take the lead in putting legislation into place to criminalize such behavior. It is no accident that the Justice Department's international program to promote appropriate changes to mutual legal assistance treaties and other nations' domestic laws, which was discussed in Section VI of this paper, concentrated first on the G8 countries and the Council of Europe. There are other important variables at work besides technological advancement, however, including each nation's public opinion and policy positions concerning high-tech offenses, especially computer hacking. There are persons in every country, including the United States, who regard hackers as essentially harmless pranksters. There is a well-established minority view that the internet and all the computer systems connected to it should be free game, and that defeating attempts to gain unrestricted access to these resources or imposing regulations on personal conduct on the internet are repressive violations of the hackers' civil liberties. The argument is even advanced that hackers provide valuable assistance to the operators of the computer systems they attack, by revealing vulnerabilities that otherwise might have been exploited by sinister persons with malicious motives. On the international scene, there is the additional factor that many individuals love to see one of their fellow citizens succeed in pulling the tail of richer and more powerful nations, especially the United States.

As a result, the state of domestic laws dealing with high-tech misconduct varies enormously from country to country. This has important implications for US information operations for two basic reasons: (1) The state of a nation's domestic criminal law directly impacts the assistance that the nation's public officials can provide in suppressing certain behavior by persons operating in its territory; and (2) The state of the nation's domestic criminal law may have a significant effect on US information operations conducted in the nation's territory or involving communications routed through the nation's communications systems.

B. <u>Cooperation in Investigations and Prosecutions</u>.

It should be readily apparent that law enforcement officials cannot prosecute an individual for conduct that is not defined as a crime in the applicable criminal law. It may be less obvious, but equally important, that in most constitutional governments law enforcement officials may not use their authority to conduct criminal investigations unless the alleged conduct constitutes a crime. If a hacker in Country X uses the internet to gain access to a DoD computer in the Pentagon, copies sensitive data, deletes or corrupts data, and installs malicious logic, the law enforcement officials of Country X may be able to assist in investigating that conduct and may be able to extradite the offender to the United States only if one or more of the hacker's actions constitute a crime under that nation's law. Even where such legislation exists, the legal system may still not be able to provide either extradition or meaningful criminal punishment, as occurred in the case of a young Israeli hacker given a suspended sentence by an Israeli court after he participated in a series of unlawful intrusions into DoD computer systems in early 1998.

The domestic laws of some nations may also permit the use of devices specifically designed to frustrate attempts to trace internet communications to their source. Since geography is essentially irrelevant to communications on the internet, devices such as anonymous remailers, which strip off all information about the originator of a message, make it possible for a hacker located anywhere to avoid identification by routing his or her message through the anonymous remailer. In this way, weaknesses in the domestic law of one state may provide impunity to hackers everywhere. The weakest link therefore threatens law enforcement even in countries with robust and sophisticated laws. Accordingly, the imperative to bring domestic laws in every nation up to a reasonable standard should be readily apparent.

C. Effect of Foreign Domestic Law on Actions of US Information Operators.

If a CINC or a JTF commander decides to order execution of a certain information operations activity by forces under his or her command who are deployed in a foreign country, the commander may have to consider whether or not such activity is prohibited under local law. The answer may be important at two different levels of analysis: (1) The individuals who issue or execute such an order might be subject to prosecution in a host nation criminal court; and (2) the commander might feel obligated on a policy basis to refrain from issuing such an order.

If a US military member issued an order or performed an act in the course of his or her official duties overseas that was a crime under host nation law, the member could very well be subject to prosecution in a host nation criminal court. Under many SOFAs, an act done in the course of a military member's official duties falls within the primary right to exercise jurisdiction of the sending state, but that rule applies only when the conduct constitutes an offense under the law of both nations, or only under US law. Where the conduct alleged constitutes an offense only under the law of the host nation, the host nation has exclusive jurisdiction to prosecute. The United States has consistently taken the position that it would be intolerable for a US military member to be criminally prosecuted for performing an act that is legal under applicable US law, such as the Uniform Code of Military Justice (UCMJ), and which he or she was instructed to perform in the execution of an official duty. A similar issue arose recently in connection with the adoption by several NATO member nations of domestic laws making it a crime to possess anti-personnel land mines (APLs). There is no similar crime under the UCMJ.

In several cases, the nations concerned have agreed to permit the US forces to retain their APL stockpiles in the host nation's territory for at least some period of time. In these cases, either specific exemptions from the host nation law or agreed screening procedures for prosecutions have had to be devised to prevent prosecutions of US military members for performing their official duties.

A similar problem would arise in information operations if a host nation criminal law applied which had no counterpart under the UCMJ. Such a situation may be much more likely to occur than it would seem at first glance. If the host country has a reasonably sophisticated computer crimes law, it will probably cover a number of acts that may be performed as information operations activities, including using a computer to obtain unauthorized access to another computer, electronic communications, or data in storage; or to transmit malicious logic; or to interfere with a satellite or with the licensed radio communications of another party to the Nairobi Convention. All these acts would appear to violate US domestic statutes, but a US military member will not be instructed to perform them unless a specific statutory exemption exists or the statute in question has been authoritatively interpreted as not applying to his or her actions. (See the discussion of issues of statutory interpretation in the companion assessment of US domestic legal issues in information operations.) Accordingly, the host nation would have exclusive jurisdiction to prosecute. The lesson we need to extract from this discussion is that as we busily encourage other national governments to enact effective legislation on high-tech crime, we need to ensure that such legislation provides an effective exemption for US military forces who may be stationed in that country and who may be engaged in information operations.

In practice, such prosecutions are most unlikely because if US military authorities become aware that performance of certain information operations within the territory of a specific host nation, or that produce harmful effects within its territory, will subject military personnel to possible host nation criminal prosecution, those US military authorities are most unlikely to order that such operations be conducted. The result will be that US forces are unable to conduct certain activities they would otherwise conduct, or perhaps that they will have to use forces elsewhere to conduct the operation. The issue thus becomes not so much one of the prospect of criminal prosecution of individual service members but rather of a limitation on the conduct of US information operations.

This consideration may be not only a policy issue—it may involve binding legal obligations under a status of forces or similar agreement. For example, Article II of the NATO SOFA provides, "It is the duty of a force and its civilian component and the members thereof as well as their dependents to respect the law of the receiving State." Similar language appears in most other SOFAs to which the United States is a party. Considerable practice has accumulated concerning the application of this obligation to "respect" the law of the receiving state. It has often been argued that the drafters could have said the visiting force must "comply" with host nation law but instead chose the less definite term "respect." The product of almost fifty years of US practice in implementing SOFAs worldwide appears to be that US visiting forces will generally observe the content of host nation law, but are exempt from the law's procedural requirements such as licensing, inspection, and reporting. If US visiting forces seek to avoid the application of the substance of a foreign law, they generally request the host nation to grant them a specific exemption or at least to reach an understanding that a particular host nation law will not be enforced against the visiting forces.

If a contemplated information operation activity appears to conflict with host nation law, the commander concerned might choose to consult with host nation officials in an effort to resolve the issue. If time or other circumstances do not permit such consultations, the commander should carefully consider whether the activities in question should be conducted by forces outside the territory of the host nation concerned, and in a manner that would not make use of or affect that nation's communications systems. US military and diplomatic authorities should be able to manage host nation legal issues if we identify them early on and carefully consider the available courses of action.

VIII. IMPLICATIONS OF ESPIONAGE LAW

A brief review of the treatment of espionage under international law may be instructive in predicting how the international community will react to information operations, especially in those mission areas in which the same technical capabilities may be used for both espionage and information operations, and also in other areas where reasonably persuasive analogies present themselves.

A. Espionage Under International Law.

For our present purposes, espionage may be defined as the covert collection of intelligence about other nations. Espionage is a much narrower topic than "intelligence," much of which is collected via open source information, voluntary exchanges of information among nations, and technical means such as satellite imagery and signals intelligence that are generally accepted as legal by the international community. Roughly stated, covert methods of collecting intelligence are in most cases designed to go undetected by their target, and if detected they are designed to be unattributable to the sponsoring state. Nevertheless, discovery, attribution, and public disclosure occur fairly often.

B. Espionage During Armed Conflict.

The treatment of spies during armed conflict is well established in the law of war. A "spy" is defined in the law of war as any person who, when acting clandestinely or under false pretenses, obtains or endeavors to obtain information in the area controlled by a belligerent, with the intention of communicating it to a hostile party. A spy may be a military member or a civilian, and his or her citizenship is irrelevant. Military personnel wearing their own uniforms are not considered to be spies, even if they engage in collecting intelligence behind enemy lines. Only a person gathering intelligence while relying on protected civilian status or while wearing an enemy uniform is considered to be a spy under the law of war. Accordingly, information operations during an armed conflict will not raise any issue of spying under the law of war unless they involve the presence of individuals inside enemy-controlled territory who (1) are engaged in collecting information with the intent of communicating it to a hostile party, and (2) are wearing civilian clothing or enemy uniforms. It seems highly unlikely that the notions of "electronic presence" or "virtual presence" will ever find their way into the law of war concept of spying, for two reasons: (1) If an individual is not physically behind enemy lines he or she is not subject to capture during the mission; and (2) There will be no issue of acting under false pretenses by abusing protected civilian status or by wearing the enemy's uniform. This will exclude most information operations activities from being considered espionage in wartime. Nevertheless, behind-the-lines missions to collect information, or to install devices that enable the collection of information, may well raise wartime spying issues.

If caught in enemy territory, a spy can be punished, after an appropriate trial, under the domestic law of the captor. The punishment can include the death penalty. The nation on whose behalf the spy was acting, however, will not be considered to have violated any international legal obligation. In addition, if individuals who may have engaged in espionage but successfully complete their missions (that is, they have returned to friendly lines) and subsequently are captured while not engaged in acts of spying, they may not be punished for their previous acts of espionage.

C. Espionage in Peacetime.

Unlike the relatively well developed treatment of espionage under the law of war, there is very little authority on the treatment of espionage under international law in peacetime. There have of course been many domestic criminal trials of peacetime spies in many countries, including the United States. By contrast, there has been almost no activity concerning peacetime espionage within the international legal system except for public complaints and the expulsion of implicated diplomats. This may be because the primary harm done to the victim nation consists of the fact that certain secret information has been compromised, which is a more abstract and indirect type of injury than dead or injured citizens, property damage, or invasions of territory. The lack of strong international legal sanctions for peacetime espionage may also constitute an implicit application of the international law doctrine called *"tu quoque"* (roughly, a nation has no standing to complain about a practice in which it itself engages). Whatever the reasons, the international legal system generally imposes no sanctions upon nations for acts of espionage except for the political costs of public denunciation, which don't seem very onerous.

The consequences for individuals caught spying, however, can be very serious. Such individuals can be tried for whatever crimes their conduct may constitute under the victim nation's domestic law, whether charged as espionage, as unlawful entry into its territory, or as a common crime such as burglary, murder, theft, bribery, obtaining unauthorized access to state secrets, or unauthorized computer intrusions. This fact accounts to some extent for the widespread practice of assigning intelligence operatives to embassy staff positions in which they enjoy diplomatic immunity from prosecution. The only remedy for an offended host nation is to declare such persons to *be persona non grata,* which obligates the sending nation to remove them from the country.

The treatment of espionage under international law may help us make an educated guess as to how the international community will react to information operations activities. As discussed in Section III of this paper on the use of force, international reaction is likely to depend on the practical consequences of the activity. If lives are lost and property is destroyed as a direct consequence, the activity may very well be treated as a use of force. If the activity results only in a breach of the perceived reliability of an information system, it seems unlikely that the world community will be much exercised. In short, information operations activities are likely to be regarded much as is espionage—not a major issue unless significant practical consequences can be demonstrated.

That leaves the issue of the possible criminal liability of an information operator who may later come into the custody of a nation that has been the victim of an operation in which he or she has engaged. As with a spy, there is no evident theoretical reason why such an individual could not be prosecuted for violation of the victim nation's criminal laws. As a practical matter, however, the problems of detection and attribution of information operations activities at the national level are daunting; the likelihood of being able to prove in court that an individual engaged in a certain information operations activity—while not impossible—seems unlikely. Perhaps the best policy would be to advise information operators not to vacation in locales where the effects of their activities have been felt, at least until a decent interval has passed.

Finally, it deserves mention that there is an established division of labor within the US government between the intelligence community and the uniformed military forces concerning "covert action." Generally speaking, the intelligence community conducts covert action operations in peacetime that do not consist of traditional military activities. It remains to be seen how information operations activities will fall within this division of labor, especially when they are associated with military operations other than war.

D. Assessment.

Information operations activities are unlikely to fall within the definition of spying in wartime, although a limited category of activities related to information operations may so qualify. Information operations activities are more likely to fall within the category of peacetime espionage. Perhaps more importantly, the reaction of the world community to information operations that do not generate widespread dramatic consequences is likely to be very similar to its reaction to espionage, which has traditionally been tepid.

IX. INTERNATIONAL EFFORTS TO RESTRICT "INFORMATION WARFARE"

As soon as the concept of "information warfare" began to receive broad press coverage, discussion began of negotiating a treaty that would prohibit or restrict it. A draft treaty text that circulated on the internet in 1995 said simply, "The Parties to this Convention agree not to engage in information warfare against each other." The first public governmental initiative was a resolution tabled by Russia in the UN's First Committee in October 1998 that apparently reflected a serious effort to get the UN to focus on the subject. The Russian resolution included a call for states to report their views regarding the "advisability of elaborating international legal regimes to ban the development, production and use of particularly dangerous information weapons." The United States has taken the position that it is premature at this point to discuss negotiating an international agreement on information warfare, and that the energies of the international community would be better spent on topics of immediate concern such as helping each other to secure information systems against criminals and terrorists. So far there has been little support expressed for the Russian initiative.

There are both similarities and differences between the concept of a treaty to ban or restrict information warfare and similar efforts to prohibit "weaponization of space." One similarity is the political reality that nations lacking a significant new military capability that they perceive will be dominated by a few wealthy and powerful states have a strong incentive to agree to ban or restrict that capability. There may be an even greater incentive to prevent interference with information systems, which all nations possess to some degree, than with space systems, in which only 30 nations are currently active and which are dominated by the United States, Russia, and the European Space Agency. On the other hand, the number of nations that have any reasonable expectation of developing their own space control systems anytime soon can be counted on the fingers of one hand, while anyone with a desk-top computer and an internet connection thereby has access both to hacker tools and to a wide variety of important information targets worldwide. Accordingly, as nations appraise where their long-term national interests lie, the calculus is quite different as between international legal restriction of the "weaponization of space" and similar control of information warfare. With space systems, most states do not expect to be either an attacker or a defender in the near future. With information systems, all states can reasonably expect to be both.

As with space control, the United States has not yet addressed fundamental policy decisions about where its long-term interests lie in connection with the possible international legal restriction of information operations. On the one hand, there is an obvious military interest in being able to interfere with an adversary's information systems, and in being able to protect one's own. Used as a tool of military power, information operations capabilities have the significant advantage that they minimize both collateral damage and friendly losses of personnel and equipment. Their use may avoid unwanted escalation of a dispute or conflict. They are relatively cheap and require much less in the way of forward basing, deployment, and logistical support than do traditional weapons and their delivery platforms.

On the other hand, as the nation that relies most heavily on advanced information systems, the United States has the greatest vulnerability to attack. This concern would seem to drive US policymakers to consider the merits of international restrictions on information operations. If we could negotiate an effective international ban on certain types of information operations activities, might signing such a treaty best serve our long-term national interests?

The subject of information operations is of course much more complex than that of space control, since there are so many more information systems subject to attack, so many more ways of attacking them, so many more potential players, plus constant rapid changes in the relevant systems and technologies. As we have learned in our internal US policy deliberations, there are great difficulties in even agreeing on definitions of what ought to be included in discussions of "information warfare" and "information operations." In these circumstances, it seems unlikely that there will be much enthusiasm anytime soon for negotiating an international agreement that would significantly restrict information operations.

X. OBSERVATIONS

There seems to be little likelihood that the international legal system will soon generate a coherent body of "information operations" law. The most useful approach to the international legal issues raised by information operations activities will continue to be to break out the separate elements and circumstances of particular planned activities and then to make an informed judgment as to how existing international legal principles are likely to apply to them. In some areas, such as the law of war, existing legal principles can be applied with considerable confidence. In other areas, such the application of use of force principles to adopting an "active defense," it is much less clear where the international community will come out, and the result will probably depend much more on the perceived equities of the situations in which the issues first arise in practice. The growth of international law in these areas will be greatly influenced by what decision-makers say and do at those critical moments.

There seems to be no particularly good reason for the United States to support negotiations for new treaty obligations in most of the areas of international law that are directly relevant to information operations. The principal exception is international criminal cooperation, where current US efforts to improve mutual legal assistance and extradition agreements should continue to receive strong emphasis. Another idea that might prove fruitful is to negotiate a treaty to suppress "information terrorism," but there seems to be little concept at present how such an agreement would operate or how it would reliably contribute value to information assurance and critical infrastructure protection.

There are no "show-stoppers" in international law for information operations as now contemplated in the Department of Defense. There are, however, many areas where legal uncertainties create significant risks, most of which can be considerably reduced by prudent planning. Since so many of these potential issues are relatively novel, and since the actions taken and public positions announced by nations will strongly influence the development of international law in this area, the involvement of high-level policy officials in planning and executing information operations is much more important at present than is the case with more traditional military activities.

INDEX